P9-CFJ-332

Science for Children

A BOOK FOR TEACHERS

Willard J. Jacobson

Professor of Natural Sciences
Teachers College
Columbia University

Abby Barry Bergman

Head of the Lower School
Riverdale Country School
New York City

Prentice-Hall, Inc., Englewood Cliffs, New Jersey 07632

Library of Congress Cataloging in Publication Data

Jacobson, Willard J
Science for children.

Includes bibliographies and index.
1. Science—Study and teaching (Elementary)—Handbook, manuals, etc. I. Bergman, Abby Barry, joint author. II. Title.
LB1585.J32 372.3'5'044 79-20328
ISBN 0-13-794784-4

Printed in the United States of America

10 9 8 7 6 5 4 3 2 1

Editorial/production supervision by Marina Harrison
Interior design by Suzanne Behnke
Cover design by Suzanne Behnke
Cover art from the Collection of C. V. S. Roosevelt
Manufacturing buyer: John Hall

Picture credits:

Figure 9–2 reprinted with permission, Biological Sciences Curriculum Study. Figures 13–2, 13–3, 13–4, 13–5, 13–6, 13–9, 13–10, 13–11, 13–13, 13–15, 14–4, 14–5, 14–7, and 14–9 reprinted with permission from Charles Tanzer, *Biology and Human Progress,* 5/e, Prentice-Hall, Inc. Figures 13–7, 13–8, 13–12, 17–1, 17–2, 17–6, 17–8, 17–11, and 18–7 reprinted with permission from Ames, Baker, Leahy, *Science For Your Needs,* 2/e, Prentice-Hall, Inc. Figure 13–14 adapted from a photo by John Coulter, Figures 15–1, 15–3, 15–4, 15–5, 15–7, 15–8, 16–1, 18–1, 18–3, 18–4, 18–5, 18–6, 18–8, 18–10, 18–11, 18–12, 18–13, 18–14, 18–15, 18–18, 18–19, 18–20, 19–1, 19–2, 19–4, 19–8, and 19–12 reprinted with permission from Milton O. Pella, *Physical Science For Progress,* 3/e, Prentice-Hall, Inc. Figures 16–2 and 16–14 reprinted with permission from Bisque et al., *Earth Science Patterns in Our Environment,* Prentice-Hall, Inc. Figure 16–13 courtesy USDA Soil Conservation Service. Figures 17–7, 17–9, 17–10, 17–13, and 17–14 courtesy U.S. Dept. of Commerce, NOAA. Figure 17–15 © 1977 by the New York Times Company, reprinted by permission.

PRENTICE-HALL INTERNATIONAL, INC., *London*
PRENTICE-HALL OF AUSTRALIA PTY. LIMITED, *Sydney*
PRENTICE-HALL OF CANADA, LTD., *Toronto*
PRENTICE-HALL OF INDIA PRIVATE LIMITED, *New Delhi*
PRENTICE-HALL OF JAPAN, INC., *Tokyo*
PRENTICE-HALL OF SOUTHEAST ASIA PTE. LTD., *Singapore*
WHITEHALL BOOKS LIMITED, *Wellington, New Zealand*

TO THE CHILDREN

AND THE TEACHERS WHO TEACH THEM

Contents

2 Working with Children in Science 23

3 Teaching and the Nature of Science 60

PART TWO
Science and Children 181

13 Children and Animals 232

14 We Study Ourselves 265

Throughout the book activities are found within horizontal bars to set them apart from the regular text. Activities or suggestions for demonstrating a particular concept can also be found within the regular text. The list that follows includes not only the activities set off within horizontal bars, but also refers to activities incorporated in the regular text. Page references are provided for convenience in locating the activities.

In addition, at the end of each chapter in Part II, before the References, is a list of Related Activities from other chapters which may help teachers to demonstrate or illustrate concepts developed in that chapter. We hope that this cross-referencing of activities will make the book more useful to the reader and broaden the experiences which may be developed in conjunction with a particular chapter.

CHAPTER 1

CHAPTER 2

CHAPTER 3

CHAPTER 4

CHAPTER 5

CHAPTER 6

CHAPTER 7

CHAPTER 8

CHAPTER 9

CHAPTER 10

CHAPTER 11

CHAPTER 12

CHAPTER 13

CHAPTER 14

CHAPTER 17

CHAPTER 18

Preface

Experiences in science are important for the physical and intellectual development of children. In science experiences children have a chance to handle, manipulate, and experiment with magnets and wire, air and water, and many other concrete materials and objects. Science is one of the areas in which children deal with more than words and abstract symbols; they have firsthand experiences with the materials and activities to which words and symbols refer. In science children can begin to have the experiences that are the foundations of logical thought. It is now becoming increasingly evident that these kinds of experiences early in life are critical for optimum intellectual development.

Science experiences are for all children. Science learning should start in home and community, continue in nursery school and kindergarten, and be an integral part of the entire elementary school program. Blind, deaf, and other handicapped children can gain special benefits from science experiences. Also, such experiences may be especially important for those who are not doing as well in other areas of the school program; there are children who flower when given a chance to work with concrete materials. But science is for all, and almost all children delight in it.

Science experiences can be developed in a variety of settings. While special attention is given to science experiences that are developed in schools, this book is unique in also offering a wide range of science experiences for nurseries, day care centers and in cities. Some of these experiences are introduced into the literature of elementary school science for the first time.

This book is designed for use by teachers and those preparing to teach, teacher aides, and others who work with children. A great variety of science activities are suggested. The activities are described in clear language. In many cases diagrams and other illustrations are provided to show how science activities can be initiated. A deliberate attempt has been made to select science experiences that can be developed using materials often found in the home, school, and local community. No specialized expertise in science is needed to develop these activities.

The chapters in this book that deal with different areas of science, such as plants and animals, the earth, and air and the atmosphere, open with short, but to-the-point, discussions of background content. The emphasis is on some of the most important ideas that are related to the science experiences suggested. These discussions of science content are designed to help you as you work with children.

Are you tired of trying science activities that don't work? All of the activities suggested in this book have been carefully tested—most of them in a wide variety of situations. Over the years we have had the privilege of working with literally thousands of teachers, parents, and teacher aides. We have benefited from their reactions and criticisms, and we are indebted to them. We are confident that this testing has made this a book that you will find useful and reliable.

Ultimately, it is the children who have guided us in devising these activities. It is their interests, their responses, their spontaneity that is at the heart of any successful science experience. Try some of the activities described in this book with children. We are confident that both you and they will find them rewarding.

HOW THIS BOOK CAN BE USED

This is a comprehensive treatment of the teaching of elementary school science. It has several special features and it can be used in a variety of ways:

Preservice Elementary Science Teaching Methods. This book is designed to serve as a text for preservice elementary science teaching methods. It contains a tested model of a competency-based teacher education program in elementary school science.

In-Service Teacher Education Program. The book incorporates new professional and scientific materials. Entire chapters are devoted to science and mathematical development, science and reading and language development, and science and the development of logical and formal operations.

A Reference Book. Background information in science and a wide variety of science activities are included. Science content related to concepts and activities developed in the elementary school has been selected.

Dealing with Children Who Are Often Overlooked. Research and development materials for use with blind, deaf, and the educable mentally handicapped have been summarized. A special chapter on science for children in cities calls attention to unique opportunities for science experiences in urban settings. Special attention is given to science in early-childhood education.

Ideas for "Going Further." Some children want to "go further" in science. Many of the science activities suggested can be undertaken by individual children. The final chapter describes a number of investigations in science that children can undertake.

This book can be an important source of ideas and background information for the young teacher entering the profession. But it has also been especially designed to be of use to the teacher in service. We hope it will become worn with use over many years.

ACKNOWLEDGMENTS

Children have taught us! We have worked with hundreds of children. Their questions, some of which we could not answer, have led us to explore further. We continue to be amazed at how seemingly naive questions from curious, inquiring children can trigger new ideas in those who work with them.

We have also had the marvelous opportunity of working with literally thousands of idealistic, dedicated teachers and future teachers. Their concern for the quality of the educational experiences of their students have been an inspiration to us. Their continuing insistence on the "practical" and to "remember the children" have been important influences on our work.

We are indebted to parents who have reacted to their childrens' science experiences. This was an essential ingredient in the development of the science activities. We wish to thank the many parents who gave us permission to use photographs of their children.

Fellow professionals in elementary science education have contributed more than they may think. We have benefited from informative discussions in the Council for Elementary Science International (CESI). Our colleagues in the Science Curriculum Improvement Study (SCIS) and Science—A Process Approach (SAPA) helped us find new ways of doing things. The challenging discussions with scientists and science educators at the New York Academy of Sciences helped us to probe deeper. Certainly, we have benefited from the sharp, critical discussions that are endemic in the Department of Science Education at Teachers College, Columbia University.

We cannot possibly name all who have helped us in this work, but we wish to mention a few. Elizabeth Meng, Robert Walrich, and Doris Withers helped us with chapters in which they have special expertise. Lorraine Riccobono-Mahony helped us with the diagrammatic recipe. Mitchell Batoff, Bonnie Brownstein, Albert Carr, Morsley Giddings, Lester Mills, Catherine Namuddu, Herma Perkins, and John Youngpeter are among the elementary science specialists with whom we have discussed various ideas in this book.

Thomas Caynon and Thomas Jacobson helped us prepare the manuscript, and Barbara Tholfsen helped us with some of the illustrations.

Robert Sickles, Marina Harrison and other editors at Prentice-Hall provided moral support in addition to technical expertise.

As always we are indebted to our wives, Carol and Rose. They made it all possible.

WILLARD J. JACOBSON
ABBY BARRY BERGMAN

PART ONE

Children and Science

Experiences in science can make significant contributions to the development of children. Children can be helped to "see where they have only looked before." Through science experiences even very young children can sharpen their perceptual development. Science in the elementary school can enhance the possibility that children can achieve the logical, formal intellectual operations that are critical to optimum intellectual development. Science can also provide the meaningful contexts in which reading, language, and mathematical development can occur.

In science, both how we teach and what we teach are important. Stated negatively, science should not be taught unscientifically. There should be openness to new ideas, concern for the possible consequences of proposed actions, and willingness to submit ideas to empirical tests. We should attempt to make our work with children in science consistent with the nature and structure of science.

If variety is the spice of life, then it surely is the "salt and pepper" of teaching. Even the most exciting content can become dull if it is approached in the same way day after dreary day. Science is especially conducive to the use of a variety of approaches to teaching. Experimentation, investigation, demonstration, field trips, reading, discussion, and projects are among the approaches to teaching that imaginative teachers can use with children, who almost always have an innate drive to explore and find out.

Science is for all children. Young children and older children, handicapped and unhandicapped, slow learners and fast learners, in the country and in the city. Sometimes our work with children has to be adapted to their needs and interests, but all children need and deserve the advantage of rewarding experiences in science.

CHAPTER 1

Why Science for Children?

"I wonder if an earthworm can move in both directions?"
"What would happen if a plant were turned upside down?"
"Where are the stars in the daytime?"
"Where is the sun at night?"
"How did the world begin?"
"Where did I come from?"

These are a very few of the hundreds—nay, thousands—of questions that children have asked us. Are they not like the questions that every child asks sooner or later? We have all wondered about the world in which we live. In part, *Science for Children* is the further exploration of many such questions.

A child is born, and it cries to help start the breathing mechanism that will keep it alive. But look at a baby soon after birth and it already seems to have its own individuality. It has embarked on its great adventure. This is the adventure of learning to understand itself and to interpret what happens in its environment. It will continue this adventure as it squishes the sand at the beach through its tiny toes, stares awed at the wonders of the nighttime sky, peers through a school microscope at a cell and what's in it, and perhaps someday uses the electron microscope, the linear accelerator, and other sophisticated tools of science to help us all to gain a better understanding of the world in which we live.

Lives there a soul with a heart so dead
That never to himself has said
I want to know!
I want to know about the birds that leave in the fall.
I want to know about the trees that grow so tall.
I want to know about the stars that twinkle in the sky.
I want to know about thee, you, and I.
I want to know!

Science for Children is part of our answer to every child's quest to know.

But science is also humanity's quest to know. Science is the investigation and interpretation of events in the natural, physical environment and within our bodies. Scientists have studied the stars above our heads and the rocks at our feet. They have probed the atom and peered into the living cell. To a large extent they are asking the same kinds of questions as the curious child. They have learned much over the millennia, and all of us can benefit from it. No child needs to or can duplicate the work of the thousands of scientists who have given us so much. For example, no child has to learn again which plants or exotic chemicals are poisonous. No child has to find out how the planets move or where houseflies come from. The child can make use of what has already been learned. One of the purposes of teaching science to children is to help each child learn how to tap the knowledge that has been accumulated and distilled over the ages. In Isaac Newton's words, the child too can "stand on the shoulders of giants."

Parents and teachers have special opportunities and responsibilities to help children explore and find out. From the very beginning of a child's life, parents can encourage and support children as they explore their environment. Later, answering questions about the commonplace, finding time to explain, planning special experiences, and encouraging exploration and inquiry are among the ways that parents can continue to help children in their explorations and investigations. Teachers have special responsibilities to children in the study of science. In addition to helping and supporting them as they explore and investigate, they have the responsibility for helping children in their intellectual development. They can help children gain the experiences with concrete materials that give meaning to words and numbers. They can plan the science experiments and investigations that now appear to be essential if children are to acquire the ability to engage in formal, logical thought. They can help children learn how to learn. We are among the many who believe that a child's intellectual development is stunted without a wide variety of experiences in science. In this book there are descriptions of some of the most essential of these experiences.

We want our children to be curious, to wonder, and to be eager to find out more about the world in which we live. Very young children are very curious and eager to explore. Too many, however, seem to lose much of this beautiful curiosity as they grow older and move up through elementary school grades and on into the secondary school. This curiosity should be nourished, and perhaps the best way to do it is to be curious with children. A very simple, but extremely important, contribution that parents and teachers can make is to provide a model of a curious, inquisitive human being that children can see and possibly emulate. For years we have asked teachers and future teachers to try "I'm curious!" In this assignment they, along with a group of children, are asked to identify some object or phenomenon about which they are curious. It may be the spots that look like algae in well-trampled sidewalks or the second fruit that they find in a navel orange. Then, they and the children use all the ways and resources that they can muster to find out more about the phenomenon. Try "I'm curious!" See if you can nourish a sense of wonder and curiosity among the children with whom you work.

"Why science for children?" From decades of experience with science for children in homes and schools, it is now possible to be more explicit about the purposes of science for children. With these experiences, lives can be enriched. Without these experiences, children's growth will be stunted. The following are some of the answers to "Why science for children?" with a few selected examples of what this means for our work with children.

DEVELOPING A VIEW OF THE WORLD

While science is one way in which we seek to interpret and understand the world in which we live, it is also a way in which each individual investigates and interprets. Thus, science is a matter of both the individual and humanity. For the individual this quest to know begins very early, and parents and teachers can help each child along the path of discovery. Children are curious, and the child explores and learns through such experiences as watching an earthworm move, connecting a bell to a dry cell, keeping a record of the changes that take place in a terrarium, and comparing shells found at the beach with fossils seen in rocks. But children can also learn from those who have inquired before. An important dimension of every child's science education is the opportunity to relate his or her developing view of the world with that which has been developed in the past.

In science some aspects of our view of the world have been distilled into broad generalizations that summarize a great deal of experience and usually do not change very much with the passage of time. For example, one of the great generalizations of science is the first law of thermodynamics: "In a closed system the total amount of matter and energy remains constant." It is sometimes stated another way: "Matter and energy can be neither created nor destroyed." A little kindergartener put it quite succinctly when she said, "What you really mean is that 'You can't get somethin' for nothin'." Knowledge of the most important of these broad generalizations of science is extremely useful as we strive to investigate and interpret the phenomena of our world. Part of everyone's science education is to deepen our understandings of these broad generalizations of science.

Children wonder about where they are, where they came from, and what is to become of them. Surely these are healthy wonderings, and most people through the ages have probably given thought to these profound questions. Some of our deepest philosophical discussions and our finest literature essentially deal with these questions that children ask. Children have a right to wonder and to begin their exploration of the profound questions that they raise. Certainly, they have a right to become aware of some of what we have learned in the sciences and the meaning of these facts for them.

"We live in a world of science and technology." These words creep easily into our thoughts, conversations, and writings. But children are also surrounded by anti-science, and some aspects of anti-science can be harmful. The forces of anti-science may suggest that we should order our lives according to the configurations of the planets, the periodic appearances of zodiacal constellations, or the claims of those who will not offer evidence to substantiate their claims. Through their experiences in science, children should develop a view of the world that will help them evaluate claims and counterclaims and compare other views of the world with those that have been developed in the sciences.

Part of developing a view of the world is to learn of how we fit into this vast universe.

Susan Polansky
1234 Wideview Drive
Uphill, Pennsylvania
United States
North America
Western Hemisphere
Planet Earth
Solar System
Milky Way Galaxy
Universe

This might be the complete address of a child, and it does locate her in the universe as we know it. This is a vast universe, and we have explored very little of it. We live on a medium-sized planet of a medium-sized star.

But although there may be many others, it is the only place in the universe where we know (as of 1979) that there is intelligent life. Some may feel that the immensity of the universe tends to diminish us and our planet, but others see the exploration and investigation of the vast reaches of the universe as a challenge worthy of humanity. While it is a challenge for humanity, it is also a challenge for each individual. The curious, wondering child is embarking on such a quest. Parents and teachers can encourage, support, and guide the child in these explorations. The child, those who help the child, and humanity in general are enhanced by the never-ending quest to find out more about the universe.

Children look at themselves and the myriad of creatures that they see about them. There is great variety among these animals and plants. They differ in size and shape, color and organization. Throughout their studies children will learn the significance of this great variety in the world of life. But there are also many ways in which living organisms are alike. All living organisms have ways of getting food and, probably, ways of sensing the immediate environment. There are relationships between living organisms, and our inquiring children will learn more and more about these relationships.

There is also a relationship between living organisms and their environment—the atmosphere; the rocks and soil; lakes, rivers, and oceans. These elements of the environment make life possible. But humanity makes drastic changes in the environment. The human population has been growing very rapidly, and with our technology we have been mining more minerals, pumping more oil, cultivating more land, and pouring more wastes into the atmosphere, rivers, lakes, and oceans. It is critically important for children to become keenly aware of the interrelationships to be found in our environment.

DEVELOPING SCIENCE PROCESS SKILLS

Certain basic process skills can be developed by children, and a command of these skills is an important asset in later inquiry-centered science studies.[1] Mathematical and reading skills are among the abilities that are important in inquiry. Because it provides a meaningful setting for the use of mathematical and reading skills and because children have so many experiences with the concrete materials of science that are represented by language symbols, science instruction is one of the more effective mediums for the development of reading and mathematical

[1]Robert Gagne has suggested that such competencies are prerequisites for effective enquiry and that this should be emphasized at the earliest levels of instruction. Robert Gagne, "The Learning Requirements for Enquiry," *Journal of Research in Science Teaching,* 1, no. 2 (1963).

skills. In addition to these more general skills, certain specific competencies are essential if children are to learn how to inquire into questions and problems. The following are some of the most important of these process skills together with illustrations of what they can mean for our work with children.

Observation

Careful *observation* is important in many science investigations. It requires attention to detail, but it also requires the ability to see "the forest as well as the trees." In science with young children, attention is given to the observation of physical properties of objects and systems because physical properties are basically more fundamental and less ephemeral than the functions of objects. As their powers of observation become more sophisticated, children observe how objects and systems interact and seek evidence for those interactions.[2] Particularly with young children, attention is given to observing with the senses of seeing, hearing, smelling, tasting, and feeling.

A group of first-grade children had been observing the changes taking place in an aquarium. In the aquarium were guppies, snails, plants, and sand. They noticed that some "black stuff" was being formed on the bottom of the aquarium. "What is this black stuff?" "What is it like?" The children gathered some from the bottom of the aquarium. They examined it with their magnifying glasses. Smelled it. Felt it. With the help of the teacher, they wrote quite a comprehensive description of the "black stuff" on an experience chart that was used in further investigations. (See Figure 1 – 1.)

Sorting

Basically, *sorting* is grouping objects on the basis of some property. In the simplest but most important sorting, decisions are based on whether objects do or do not have a particular property. Technically, this is called a yes-or-no or *binary* sort, and it is the basis for many classification systems. Many computer systems also operate on a binary sort. The ability to sort is a prerequisite for classification.

A group of children in a large urban school wanted to collect and display materials that do or do not give off odors. They collected a variety of materials. Into one pile they sorted all the materials that three children agreed gave off some kind of odor. Into the other pile they sorted

[2]"Material Objects," "Interaction and Systems," and "Organisms" are three units of the Science Curriculum Improvement Study (SCIS) program designed for use with young children. They emphasize the observation of physical properties and evidences of interaction. *Science Curriculum Improvement Study Elementary Science Program* (Chicago: Rand McNally).

Figure 1–1 What is the "black stuff" that forms at the bottom of an aquarium?

the materials that they agreed did not give off any odors. Their basis for sorting was "odor or no odor." Later they were able to assemble a display of materials that gave off various kinds of odors.

Classification

Classification is the systematic arranging of objects and materials into a series of groups. For example, plants and animals are often classified into phyla, classes, orders, families, genera, and species. Usually materials and objects are classified on the basis of their observed physical properties. For example, minerals are usually classified on the basis of such physical properties as color, hardness, cleavage, streak, and luster. Systems of classification are designed to help us identify relationships between materials and objects.

Children can develop useful classification systems. A group of third-graders collected a variety of leaves from their schoolground. The teacher asked, "By what property could we sort the leaves into two almost equal piles?" The children had had some experience with binary sorting, and one of them suggested, "Into one pile we could put all the leaves that are like needles, and into the other pile we could put all the others. By what property could we sort each of these two piles into two almost equal piles?" Eventually the children developed the classification scheme shown in Figure 1–2. These third-graders had actually learned how to develop a useful system of classification.

Serial Ordering

Serial ordering is the ranking of materials and objects on the basis of the extent to which they have certain properties. For example, perfumes can be serially ordered by the relative sweetness of their smell. The very

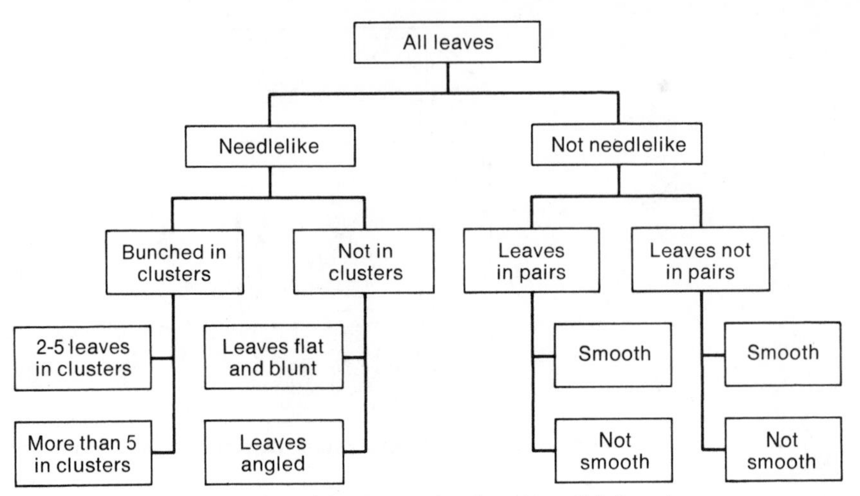

Figure 1–2 A system for classifying leaves developed by a third grade.

significant Periodic Table of the Chemical Elements was developed by Dmitri Mendeleev and others by arranging the known chemical elements in serial order on the basis of comparative weight, progressing from the lightest to the heaviest. It was noticed that certain properties, such as chemical activity, appeared periodically in the list. On the basis of these observations and the serial ordering, Mendeleev and others devised the Periodic Table, one of the most important developments in the science of chemistry and one of humanity's greatest intellectual achievements.

Children can have experiences in serial ordering with such familiar objects as peanuts. Give each child eight or more peanuts. Have them describe some of the properties of the peanuts. Then have them select some property that all of the peanuts have to some degree. Have them arrange the peanuts in order on the basis of the extent to which they have this property. If the peanuts are placed on small slips of paper and numbered, a record can be kept of the serial ordering. Then ask the children to order the peanuts on the basis of other properties. Then comes the critical question: "Do you see any relationships between the properties of the peanuts?" One youngster found an interesting relationship: The darker the color of the peanuts, the louder the rattling sound inside the peanut when it was shaken.

Operational Definition

In science it is important to be precise, and one of the steps toward precision is *operational definition*. In literary works definitions are usually made in terms of other terms. The precision of these definitions and our ability to use them to communicate precisely are limited by the precision with which we define all of the other terms. Operational definitions, on

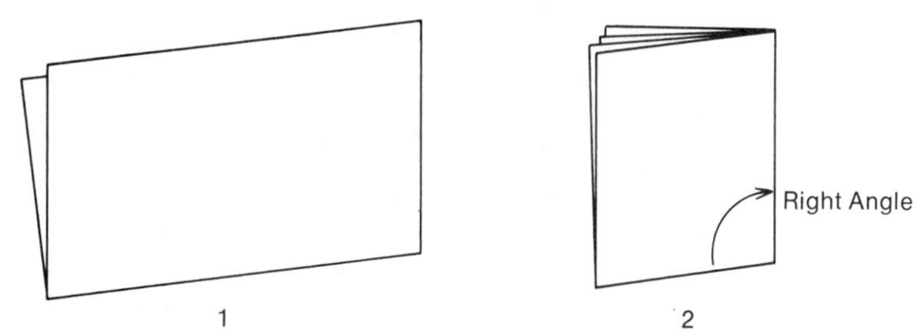

Figure 1–3 A right angle is the angle made when a piece of paper is folded in half one way and doubled over the other way. Other angles can be compared to this right angle.

the other hand, are stated in terms of the operations carried out in the process of defining. The meter, for example, is a standard unit of length, and it is defined operationally in terms of the wavelength of the orange–red light emitted by a krypton 86 lamp. This enables people to carry out the specified operations to obtain and communicate the precise meaning of the definition.

To give children experience in operational definition, a teacher gave them the operational definitions for "right angle" and "yellow." A right angle is the angle made when a piece of paper is first folded in half one way and then doubled over the other way. The resulting creases in the middle of the page make right angles. (See Figure 1–3.) The color yellow can be operationally defined as the color of a ripe lemon. The colors of other objects can be compared to those of a ripe lemon to see whether we would consider them to be yellow.

Communication

Science is a public undertaking. Scientists discuss their work informally with their colleagues, testing their ideas and getting help with their problems. They communicate the results of their investigations through scientific journals and at science meetings. Scientific work has to stand the test of criticism by one's peers. Many scientists also believe that scientists have a responsibility to communicate the meaning of their work in such a way that it can be understood by the educated public. Thus, communication is an important process in science as well as in many other areas of modern life.

Children should have opportunities to discuss their investigations with others, communicate their results to their peers, and test their ideas in public discussion. Every Friday afternoon in a primary school in Hertfordshire, England, the children assemble for a science colloquium. Some of the children present the science investigations that they have conducted. Usually they have planned their presentations with the help of teachers or parents to try to make it as effective a presentation as possible. After the presentation there are questions from other children. The day we attended there were presentations of investigations in the field of magnetism. One girl had made a survey of what objects in the school had been magnetized in one way or another. The questions were penetrating and the discussions helpful. The colloquium resembled many other scientific meetings that we had attended; certainly, these children were having a splendid experience in scientific communication.

Prediction

Prediction is the process of using what we know to predict what may happen. In some sciences, such as meteorology and seismology, one of the major reasons for gathering information is to use it to make predictions. In other sciences the information provides a basis for experimentation.

In traditional education children too often lack opportunities to use what they have learned. What they learn, then, seems to have little purpose or meaning for them. However, if they are asked to make predictions they need to use what they have learned to make the predictions.

A group of sixth-grade children set up a weather station. (See Figure 1–4.)

WEATHER RECORD AND FORECAST

Date ____________________

Cloud type ____________________

Temperature ____________________

Barometric pressure and direction ____________________

Wind speed and direction ____________________

Relative humidity ____________________

Precipitation amount and type ____________________

Radio reports of weather ____________________

Weather prediction ____________________

Weather that occurred ____________________

Figure 1–4 Children use what they have learned about weather to make predictions.

They recorded temperature ranges, relative humidity, precipitation, barometric pressures, cloud types, wind speed and direction, and radio reports of the weather in a city to the west of them. Each morning they "published" their weather record and predictions for the remainder of the day. Soon they noticed that high cirrus clouds, a "falling barometer," and reports of storms from the neighboring city almost always presaged stormy weather for their school. They were justly proud that the accuracy of their predictions was generally greater than that of the predictions published in their local newspaper and those reported over the radio station in the nearby city. They were making good use of all that they had learned about weather. They were learning how to use some of what they had learned in science.

DEVELOPING INTELLECTUAL POWER[3]

Our minds are our most important assets. Young children have the potential to develop intellectual power. Dimensions of this intellectual power are the ability to interpret data and suggest patterns to explain data, to suggest models to explain observations, to view from different frames of reference, to use symbols and abstract ideas, to suggest possible consequences for proposed actions, and to learn how to learn from experience. Whether or not children develop these abilities depends to a great extent on the kinds of educational experiences they have. Certain kinds of science experiences are especially appropriate for the development of some of these intellectual powers. In fact, they may be essential for the development of these abilities. In this book a great deal of attention is given to the development of intellectual powers.

Children can develop the ability to suggest patterns and apply them to data. Children can make observations and measurements, but the data will have little meaning unless patterns are suggested and applied. Some children periodically cut strips of paper that were as long as a plant was tall. By arranging these in order as a histogram (see Figure 1–5.), they could see a pattern in the plant's growth. Other children, by keeping a record of the changes in a tree that they had "adopted," gained a better picture of the seasonal changes in trees. (See Figure 1–6.) The actual measurements of a plant and records of changes in a tree have little meaning until the data are interpreted and patterns are sug-

[3]Much of the discussion in science education of the development of intellectual power through science experiences is based on the work of Jean Piaget and others who have studied the intellectual development of children. For more extended discussion see Herbert Ginsberg, *Piaget's Theory of Intellectual Development* (Englewood Cliffs, N.J.: Prentice-Hall, 1969) and Celia Stendler Lavatelli, *Piaget's Theory Applied to an Early Childhood Curriculum* (Cambridge, Mass.: American Science and Engineering, 1970).

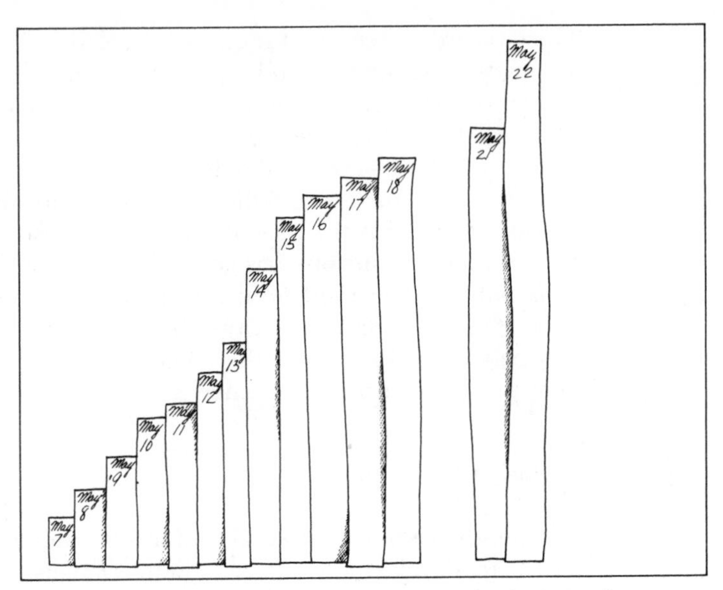

Figure 1–5 A paper strip histogram showing a pattern in plant growth.

gested. The ability to interpret data and suggest patterns is important in many life situations.

In science children can learn to suggest models to explain observations that are made. No scientist has actually "seen" the inside of an atom, but atomic models have been developed to explain observations that have been made. Similarly, models are suggested for the solar system, crystal structures, the nucleus of the cell. Some children were given a sealed "black box" containing a variety of objects. They could not look into the box. But by tipping, shaking, listening, smelling, and thinking they tried to develop a mental model of what was inside the box. (See Figure 1–7.)

We sometimes see things differently if we view them from different frames of reference. It is useful, for example, to imagine what the solar system would look like if viewed from the sun rather than from the earth or how "stationary" objects would appear from a moving platform. Children may try to imagine how the earth would seem to move if viewed from the moon. Or they can imagine how they would appear to a doll passing by on a moving toy truck. Or they may even be asked to "try to put yourself in other people's shoes. How do we appear to them?" (See Figure 1–8.)

The ability to use symbols and abstract ideas increases intellectual power. In the sciences some of the greatest advances have resulted from the manipulation of abstract ideas. The young child tends to operate in a world of concrete objects. One of the critical contributions of science

OUR TREE DIARY

March 25	We examine buds on our tree.	**Sept. 10**	Tree seems in fine shape after summer. Find seeds of tree on ground. We fertilize tree again.
April 10	We put a metal sheet around the trunk of the tree to protect it from dogs.	**Oct. 2**	Some leaves are beginning to change color.
April 17	We put fertilizer into small holes around the tree.	**Oct. 4**	Many leaves are now yellow.
April 19	A branch breaks off in strong wind. We trim it.	**Oct. 7**	Many leaves are now orange.
April 21	Leaves are coming.	**Oct. 15**	Leaves continue to fall. We bring some leaves into our classroom.
April 24	Leaves are now out.		
May 10	We examine the tree's flowers.		
May 16	Petals of flowers are falling.	**Nov. 5**	Almost all leaves are now off. We see buds for next year's leaves.
June 5	We say good-bye to tree for summer.		

Figure 1–6 A diary of a tree.

experiences is that children have a wide variety of experiences with many concrete objects. They learn to use symbols to represent these concrete objects. Thus, in *Science for Children* symbols usually have concrete referents. These are important experiences that should precede situations in which symbols represent only other symbols. Very early, children can also gain experience in using concrete materials to gain an understanding of abstract ideas. For example, when asked to point "down" very young children will point toward the ground. They are

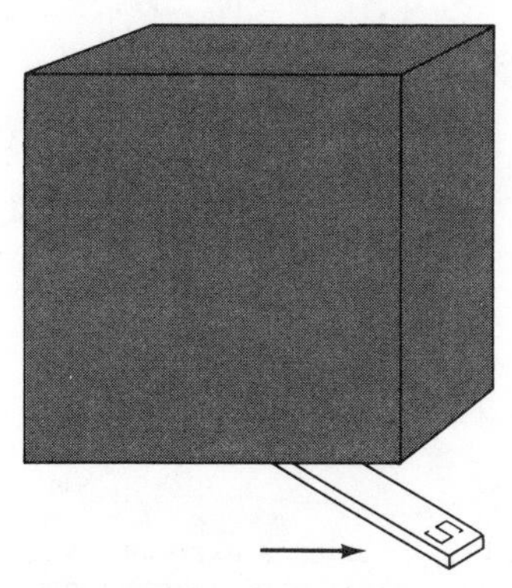

Figure 1–7 A magnet may be pulled along the bottom of a "black box" to try to detect magnetic materials inside.

Figure 1–8 To the doll, do we seem to be moving?

shown a globe, and it is suggested that the direction "down" is toward the center of the earth. "But how about the children in Australia. What is the direction of 'down' for them?" (See Figure 1–9.)

One of the most important intellectual powers is the ability and predisposition to suggest possible consequences of proposed actions. "What may happen if we try it?" In some life situations asking such a question can often prevent dire consequences. Many science experiences call for children to suggest possible answers to questions or problems—usually called "hypotheses." They are then asked to deduce possible consequences of these hypotheses—usually called "designing an experiment." Often they then have an opportunity to carry out the experiment to see whether the consequences they predicted actually do occur. After finding that an electric charge was generated on a comb when it was rubbed with wool, some children asked, "Would an electric charge be generated if we rubbed the rubber rod with plastic? Would it be the same kind of charge?" On the basis of their previous studies of electrici-

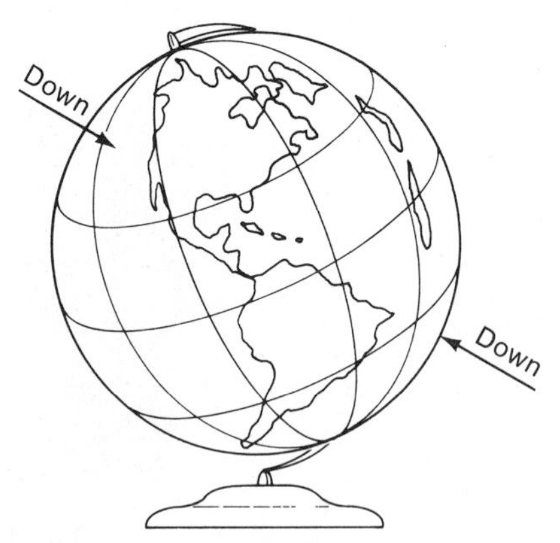

Figure 1–9 In science experiences, concrete materials are used to gain an understanding of abstract ideas.

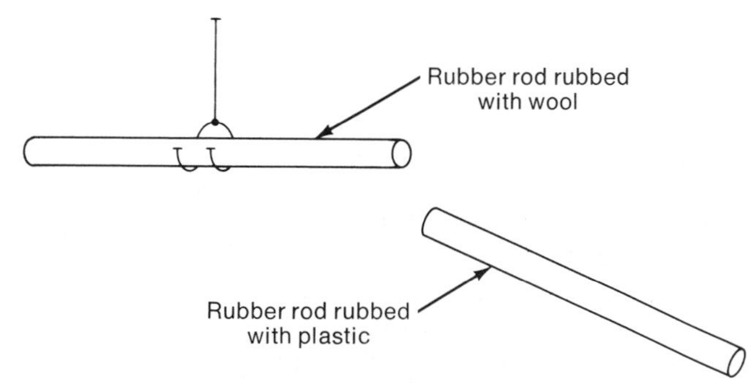

Figure 1–10 What will happen if a rubber rod rubbed with plastic is brought near a suspended rubber rod that has been rubbed with wool?

ty, they suggested that if it were the same kind of charge, it should repel a rubber rod that had been rubbed with wool. They designed an experiment to test this hypothesis. (See Figure 1 – 10.)

An essential intellectual ability is the ability to learn from experience. In the sciences this may take the form of systematic observations and carefully designed experiments. Kepler, for example, was able to base his laws on the very careful and systematic observations made by Tycho Brahe. Many of the most important ideas in the sciences have been tested experimentally to see whether the consequences deduced from ideas actually occur. Some children wondered whether hot water will freeze sooner in a refrigerator than cold water. One of the children hypothesized, as many believe, that hot water will freeze sooner. The children were eager to try it. But the teacher working with them began to ask some pertinent questions: "Will it make any difference whether you have more of one kind of water than the other?" "Will the kind of container make a difference?" "Will the place in the refrigerator where you place the container make a difference?" "How can we determine when water freezes?" After taking such questions into consideration, the children were able to suggest other factors that should be controlled. If they had not controlled these factors, they would not have learned whether hot water really does freeze sooner than cold water. They were learning how to learn from experience.

DEVELOPING SCIENTIFIC LITERACY

Scientific literacy is defined as the ability to understand and discuss developments in science and technology as they are communicated via newspapers, magazines, television, and other public media. To be scientifically literate is to be able to participate in the explorations and new developments taking place in science and technology and to understand and be able to discuss some of the results of these investigations and their possible implications. We are exploring the solar system with ever more sophisticated space vehicles and the vast regions far beyond the solar system with large optical and radio telescopes. We are probing deeper into the atom, trying to get a clearer understanding of the nature of matter itself. Although we all use energy, we continue to struggle for a better understanding of what energy really is. In a myriad of ways we try to gain a better understanding of the nature of life and how it has evolved on this planet. These are great intellectual adventures, and every youngster has a right to the preparation that will make it possible for him or her to participate in some way in these adventures throughout a lifetime.

Scientific literacy is closely related to the ability to read, view, and discuss ideas. Naturally, growth in the ability to read and discuss contributes to the development of scientific literacy. Of equal importance, the development of scientific literacy contributes to the ability to discuss and read. After all, reading and discussion are enhanced if there is something interesting and stimulating to read and discuss. For many children science and technology provide this interesting and stimulating content.

If science is the investigation and interpretation of events in the natural, physical environment and within our bodies, then every individual should have a right to participate in this quest. Few will become professional scientists devoting most of their time and energy to this quest. But all should be able to keep informed about the nature of these explorations and have the background to interpret many of the findings. Surely this will enrich the lives of our children, and certainly it should be a significant part of what it means to be "educated."

There are many ways in which children can keep informed about what is happening in science. There are journals and magazines for children that regularly report developments in science and technology.[4] Some encyclopedia yearbooks contain annual reviews of developments in science and technology. There are newspaper columns on science and occasional radio and television reports on developments in science and technology. Some schools have found it useful to hold regular colloquiums at which new developments in science and technology are discussed

[4]Among the journals children can read that report developments in science and technology are the following:

The Curious Naturalist (South Lincoln, Mass.: Massachusetts Audubon Society)

Current Science and My Weekly Reader (Columbus, Ohio: American Education Press)

Ranger Rick's Nature Magazine (Washington, D.C.: National Wildlife Federation)

and their possible implications explored. We hope children will carry their interest in "what's happening" in science and technology with them throughout their lives.

ACHIEVING OPTIMUM PHYSICAL HEALTH

"Health is a state of complete physical, mental and social well-being and not merely an absence of disease or infirmity." This is the definition of optimum health offered by the World Health Organization. Obviously, for each individual to achieve optimum health requires education, and part of this education should be through science.

There are several approaches to the study of the body and health education, and children should have the benefit of all of them. Through science children gain a better understanding of their bodies and what is required to keep them in good health. They learn, for example, about the nutritional requirements of the body and how these can be obtained. They also learn some of the ways in which harmful diseases are transmitted and some of the ways in which links in these chains of transmission can be broken. In the approach to health education through science, emphasis is placed on the broad generalizations that can be used to interpret many phenomena. These generalizations summarize a great deal of specific information and provide a base for delving deeper into questions concerning health and the human body. For example, most communicable diseases are transmitted from one human being to another. An understanding of this broad generalization can help a child understand the reasons for certain actions that are taken to prevent the spread of disease. Also, as they deepen their understandings of these broad generalizations children become better able to evaluate the claims and counterclaims that bombard all of us through the various media. Without some understanding of the broad generalizations that have been developed in the sciences, the individual has no defense against spurious pleas and extravagant claims.

In the approach to the study of the human body and health education through science, emphasis is placed on general understanding. In other approaches the emphasis may be on "do's and don'ts," but in science the "whys" are of central importance. For example, children may study the energy relationships between the body and its environment. The body cannot take in more or less energy than it uses for an extended period without there being some effect on the body. While recognizing that there are significant differences among people in how energy is utilized and how much is needed to meet bodily functions, children can gain a general understanding of the energy relationships in the body. In many other areas as well, the approach to the study of the body and health through science can broaden and deepen understanding.

THE CITIZEN AND SCIENCE EDUCATION

Although this fact is often overlooked, our children are citizens. Because they have more of a future than we do, they have a greater stake in what we do and don't do.

First, reflect on the events and changes that have taken place in your lifetime. You have probably taken part in humanity's first ventures into space and may have watched as the first man set foot on our satellite, the moon. You have heard and read of the findings of our probes to other planets and our search for life elsewhere. You have read and heard about developments in science and technology that may revolutionize the way we live. But during your lifetime, and depending on how long you have lived, the population of our planet may have more than doubled. Weapons have been constructed that, if they ever are used, might destroy all people, perhaps all life, on the face of the earth. We have used a large share of the fossil energy that is stored in the earth as coal, oil, and natural gas. The environment has been greatly altered, and radically different methods of communication and transportation have been developed. In a sense the world has become smaller and people have been drawn closer together. Some suggest we are at the edge of catastrophe. Others suggest we are on the verge of a future of unparalleled prosperity, with peace and plenty for all. Obviously, your life and your future have been drastically affected by developments in science and technology.

Now think of the lives that today's children will lead. They will live well into the twenty-first century. What further explorations will they have a chance to take part in? What revolutionary discoveries will impinge upon their lives? During their lifetimes the population of Earth will almost certainly double and perhaps double again. Our environment and its resources will almost certainly be placed under great stress. The peoples of the earth are likely to become even more interdependent, and the problems of one city, state, nation, or region are quite likely to have an impact on people everywhere. Genetic engineering may be a boon or a bane. The creation of living tissue in a test tube will have profound practical as well as philosophical implications. Beyond this, our children will have to deal with developments that we cannot even foresee. They may have opportunities we never had and problems that make our difficulties seem piddling. Certainly, our children deserve an education that will prepare them to profit from the possibilities and deal with them. Many of the problems are unprecedented, and no one has previously had the experience of dealing with them—the Earth will never before have had as large a population as it will have during the lifetimes of our children. Many of the problems have no final solutions, so that we and our children can only hope to achieve satisfactory resolutions; for example, because of the widespread knowledge of how to make nuclear weapons and the general availability of fissionable materials, the danger of nuclear conflict can be diminished but not completely eliminated. Many of our problems and those our children will face can be resolved only through the cooperation of many people, and this makes it critically important that all children receive an education that equips them to analyze and work with others in the resolution of these problems—the solutions to most of our environmental problems, such as the desecration of our parks or the pollution of our hydrosphere and atmosphere, cannot be dictated from above but require informed cooperation.

It may seem trite, but it is indubitably true, that our children's future depends critically on their educational opportunities. H. G. Wells said, "Human history becomes more and more a race between education and catastrophe." For our children's sake, let us hope that the education that we give them will ensure that the race will be won and that they will continue to be the winners.

WHY SCIENCE FOR CHILDREN?

Through science experiences children gradually build a more comprehensive and sophisticated view of themselves and the world in which they live. In developing this world view, they have a chance to become informed about and consider some of the broad generalizations that have been developed in the sciences. Children master science process skills, such as observing and measuring, that many people believe are essential for effective investigation of the questions and problems of science. They will have the greater intellectual power that comes from having learned how to use abstract ideas and how to learn from experience. They will have experienced another approach to the understandings that are needed for optimum health. They will have gained some of the scientific literacy that will help them take part in some of the great explorations and adventures that we hope will take place during their lifetime. Our children will live in a world of science and technology, and the science experiences they have as children will help equip them to analyze and, we hope, deal with some of the difficult problems they will surely face.

But there is another answer to the question "Why science for children?" This answer cannot really be put into words. It is to be found in the faces of children. You will see it in their bubbling excitement as they encounter something new. When you perceive their deep engrossment as they experiment with magnets or watch an egg hatch, you will have more of an answer. You may have a chance to see a child who hasn't been touched by much of what goes on in school seemingly come to life as she becomes involved in a science experience. When you see the face of a "dull" child brighten with interest and understanding, you will have even more of an answer. Given a little chance, most children enjoy and learn from the kinds of science experiences described in this book. If you don't believe it, try it! Your children have much to gain.

REFERENCES

ENNEVER, LEN AND WYNNE HARLEN. *With Objectives in Mind.* London: Macdonald Educational, 1972. An interesting and practical approach to objectives for elementary school science developed for the British *Science 5/13 Project.*

JACOBSON, WILLARD J. AND HAROLD E. TANNENBAUM. *Modern Elementary School Science.* New York: Teachers College Press, 1961. A comprehensive outline of directions for elementary school science.

KESSEN, WILLIAM. "Statement of Purposes and Objectives of Science Education in School." In *Science—A Process Approach, Commentary for Teachers,* pp. 3–7. Lexington, Mass.: Ginn, 1970. A statement of purposes and objectives developed to give direction to the AAAS' *Science—A Process Approach Elementary Science Program.*

Working with Children in Science

CHAPTER 2

The children who come into your classroom on an August or September morning may seem to have endless energy, strong drives, pixie-like mischievousness, and either a superficial opposition to learning or an indomitable desire to find out. But regardless of their individual characteristics, almost all of them will have a keen interest in finding out more about themselves and the world in which they live. *This is science.* Science is often popular with children because it not only allows but encourages their natural curiosity—it sanctions the freedom to question, to explore, to investigate, to find out, and to wonder.

What is the nature of these children who come into our classrooms? Certainly, each child is different, and we should be cognizant of these differences. This is one of the reasons that we will place considerable stress on individualized instruction in science. However, there are important similarities among children. For example, almost all children pass through somewhat the same stages of growth and maturation. We should also be aware of these similarities as we develop science experiences with children.

STAGES OF INTELLECTUAL DEVELOPMENT

Almost all children pass through certain stages of intellectual development. Many students of the intellectual development of children believe that most individuals pass through four more or less distinct stages of intellectual development:

1. *The Sensory–Motor Stage.* The first stage, approximately the first eighteen months of life, has been called the sensory-motor stage. Perhaps the most profound discovery of this period is that there is an environment; there is something outside of "me." The infant begins to develop some concept of objects. At first, objects exist only when they are perceived. Later the infant will retain some notion of an object and will try to find it. This young infant is also beginning to develop some concept of space. For example, he begins to learn to adjust the movements of his hands as he grasps for objects. He also begins to develop some notion of causality. Basically, the infant is learning about his orientation in space.

There are many students of child psychology who believe that the sensory-motor stage is of critical importance in both the physical and the intellectual development of children. Apparently, it is extremely important that infants have the loving care of parents at this time if they are to take the physical and intellectual steps that are expected at this stage. Infants in some orphanages, where individual loving care is not possible, often learn to walk two years later than those reared under more fortunate circumstances.

There are, perhaps, some environmental considerations that parents can control in order to provide a more stimulating atmosphere for the child in infancy. Attractive, colorful mobiles suspended in a crib entertain the infant as she spends a great deal of time lying on her back or side. The sides of a crib or carriage may also be decorated with colorful pictures or toys so that additional stimulation will be provided as the infant looks to the side of her world.

Objects for grasping, squeezing, poking, and prying are all of interest to the child in the sensory-motor stage. Materials of varying texture and consistency can be introduced, as well as objects that make sounds and noises. No physical environment, though, no matter how stimulating, can ever replace the critical need for warm and loving human interaction.

2. *The Preoperational Stage.* The second stage, from approximately 18 months to 6 or 7 years of age, has been called the preoperational stage. In this stage children begin to use the symbols of language. With these symbols they are able to think and probably reconstruct many of the experiences they had during the sensory-motor stage. This stage is called preoperational because the child cannot yet act or operate on the

basis of experience with objects or events. Apparently, the child's thought processes are not yet "reversible." The child has a mental image of an object but cannot use this image to act or operate on the object.

The basic behavior that defines the preoperational stage is the young child's inability to "conserve." The following task shows what is meant by conservation:

Conservation task. A child is given two similar tall containers and asked to fill both so that they contain equal amounts of water. When the contents of one of the containers are poured into a shallow tray such as a cake pan, the preoperational child will usually say that the other container has more water.

In the preoperational stage children should have abundant opportunities for firsthand, primary experiences with the objects of the environment. The child should have many and various opportunities to handle and play with many kinds of objects in the environment. The varied activity-oriented nursery school and kindergarten programs are consistent with this view of intellectual development. A science program should emphasize a large number and wide variety of concrete, primary encounters with material objects of the environment, even though, of course, the child cannot be expected to conceive sophisticated connections between these objects.

3. *The Stage of Concrete Operations.* The third stage is called the stage of concrete operations. Usually children from the age of 6 or 7 to about age 11 are in the stage of concrete operations; therefore, this includes most of the children in the elementary school. Children enter this stage when they are able to "conserve." For example, when equal amounts of water are put into two tall cylinders and then the water in one of the cylinders is poured into a flat pan, children in the stage of concrete operations will say that there is an equal amount of water in each of the containers. (See Figure 2–1.)

Children in the stage of concrete operations can operate on objects but not on verbal statements. They can describe the physical properties of objects, classify them, and arrange them in order on the basis of some property, and are also beginning to develop more sophisticated concepts of space, time, and number. They can deal with objects in quite sophisticated ways, but they have difficulty stating hypotheses and making logical deductions from those hypotheses.

In the stage of concrete operations, children should continue to have many and far-ranging firsthand experiences with concrete objects. But now they can go a step further: They can operate with these material objects. They can group objects into various systems and use classification schemes. They can use materials in experiments, investigations,

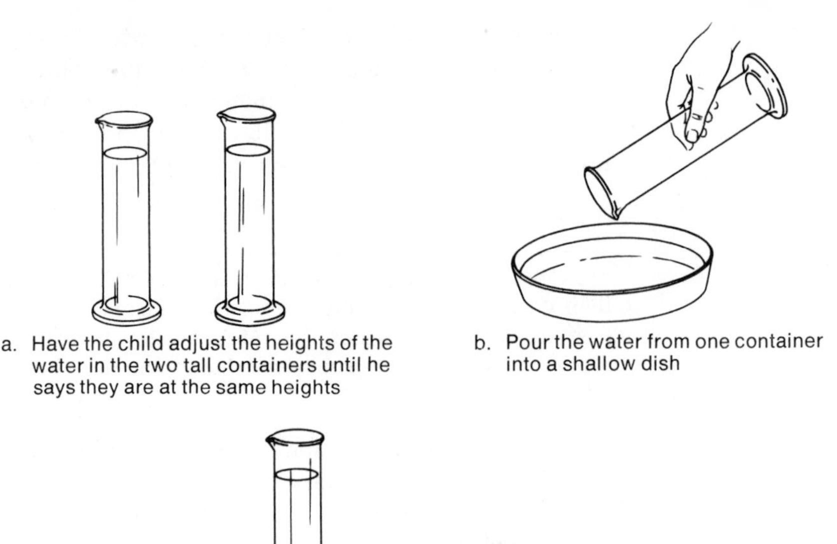

a. Have the child adjust the heights of the water in the two tall containers until he says they are at the same heights

b. Pour the water from one container into a shallow dish

c. Ask, "Now which container has the most water?"

Figure 2–1 A conservation task.

and projects. Children should be encouraged and helped to verbalize about the material objects and the operations they carry out with them. They should have the opportunity to practice suggesting hypotheses. Activity and language need to be brought together. In fact, it may be that a wide range of concrete operations is essential for future verbal development; it apparently is essential if the child is to move into the fourth stage of intellectual development.

4. *The Stage of Formal Operations.* The fourth stage, called the stage of formal operations, can begin at about 12 years of age, although some children in the upper elementary school achieve this level. (See Chapter 8 for a more extensive discussion of experiences that can lead to formal operations.) Children can now reason about hypotheses and ideas as well as objects. In a sense this might be called the stage of "abstraction." Children in this stage do not have to experience concrete objects but can deal with the abstractions that represent these objects or the actions that they have taken with these objects. Children, or for that matter adults, may revert to the concrete stage of operations when confronted with new and unfamiliar situations, but they have the potential to move into the more efficient and effective formal mode of operations.

Children in the stage of formal operations are able to use general laws. They can carry out logical operations. For example, they can use a law such as Archimedes' principle (an object placed in a fluid is pushed up by a force equal to the weight of the fluid it displaces) to predict whether or not an object will sink.

In the stage of formal operations, the individual can think of hypothetical situations that are not necessarily related to specific, real situations. It has been suggested that this partially explains the adolescent proclivity for social reform. Adolescents are exercising their formal power to hypothesize better worlds.

Some of the characteristics of intellectual behavior in the formal stage of operations, as contrasted with those of behavior in the stage of concrete operations, can be seen in the way an experiment in floating is handled.

Flotation task. A large container of water is made available, and a child is asked to classify a variety of objects, such as pebbles, blocks of wood, nails, toothpicks, bottle caps, and hollow metal cylinders, according to whether or not they float. After the child has classified them, he or she is asked to summarize his observations and to look for a law. (See Figure 2–2.)

Early in the stage of concrete operations, children can classify the objects as (1) objects that float, (2) objects that sink, and (3) objects (such as the bottle cap and hollow cylinder) that sink or float depending on how they are put into the water. They may explain their classification by saying that "light objects float" and "heavy objects sink." However, they encounter difficulty when they find that a large piece of wood is heavier than a small nail, yet floats. They may not yet have operational use of such concepts as density and specific gravity.

Adolescents in the formal stage of operation can handle this problem in more sophisticated ways. They can make hypotheses about whether or not an object will float and reject those hypotheses without necessarily carrying out the concrete operations. Adolescents can view various possibilities. They can arrive at such concepts as specific gravity and density (the same volume

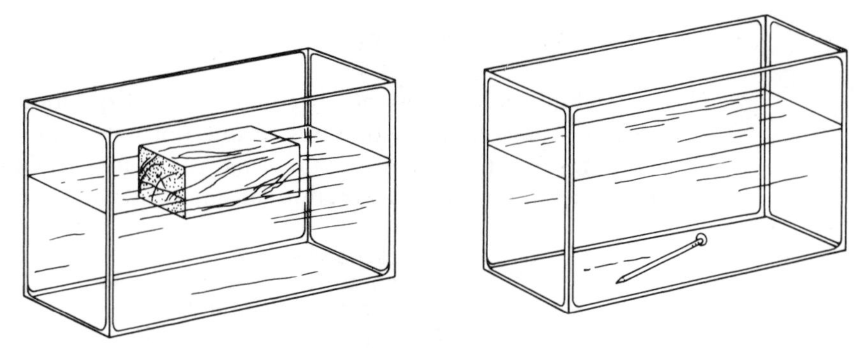

Figure 2–2 Which objects will float? Which will sink?

of water would not be as heavy as the nail). Adolescents also tend to prove things logically and to take all factors into consideration. They may also be able to carry out mental experiments.

IMPLICATIONS OF DEVELOPMENTAL STAGES

The nature of child development has important implications for our work with children in science. The following are some suggestions based on our understanding of child development.

Direct Experiences

Children learn as they operate with objects and ideas. Three different approaches are often used in science instruction: "chalk and talk," teacher or pupil demonstration, and laboratory experience. In "chalk and talk" there is almost complete dependence on the manipulation of symbols. This approach can be very effective when the individuals have already had rich experience in concrete operations but is of limited value in the elementary school. Again, in demonstrations children have an opportunity to observe, but they have limited opportunities to manipulate materials and equipment. There is an important place for demonstration in elementary school science, but only after children have had many firsthand experiences with the equipment and materials used in the demonstrations. It is in the laboratory experience itself, in which children actually manipulate materials and equipment, that elementary school science must function if the children are eventually to move on to the formal stage of operations.

In the laboratory approach the classroom becomes the laboratory. Each child pushes and pulls, sets up equipment, experiments with materials, and carries out the classic proposal "Let's try it and see!"

Wondering and Questioning

Children learn as they become aware of questions and problems and seek answers and solutions to questions and problems. The alert teacher is sensitive to the questions and problems that children raise and also poses questions that will stimulate and lead children into inquiry.

"Why is the sky blue?"
"Do fishes need air in order to live?"
"Where did I come from?"
"What happens when I grow?"
"What makes things stick to magnets?"

The sense of wonder and urge to question are important in the sciences. Some of the most successful scientists are said to have had a "childlike" inquisitiveness. And it is little wonder that this is so. If science is the investigation and interpretation of the environment, scientists will be led and guided by the problems they raise and the questions they ask. Probably some of the most important discoveries in the sciences have come about as a result of someone's unsatiated curiosity and persistent desire to find out.

Curiosity, inquisitiveness, and a sense of wonder are precious qualities that should be nurtured. But in many young children who come to school rich in these attributes, the sense of curiosity is deadened and the desire to find out stunted by the time they leave elementary school. In some cases this may be a result of maturation, but too often it is a result of the kinds of educational experiences children have had. Certainly, every effort should be made to have children's experiences in science nurture their curiosity and inquisitiveness rather than stultify them.

Science Concept Development over Time

Children's science experiences should build on one another. With each new experience it is helpful if the child has a related experience to which new learnings can be pinned. The process of concept development often takes place over extended periods.

Children's concepts, or pictures of themselves and the world in which they live, are limited by their stage of development. At one time, a child's concept of space may be limited to the boundaries of the local neighborhood. Twenty years later, the same individual may have some concept of the vastness of interstellar space. That person's concept of space has developed over a considerable time. Although science instruction is intended to help children develop concepts that are consistent with the broad generalizations of science, children need time to conceptualize from their experiences with the environment. For example, one young child had the concept that the water used in the home came from the wall.[1] It was only after a considerable time, during which he saw pipes in the wall and visited a water reservoir, that he developed a more sophisticated concept of the source of the water in the home. In some cases it may be necessary for the child to assimilate a wide variety of experiences before concepts can be developed.

Apparently, many children's concepts develop gradually. Careful studies of children's concept development show that children come back to the same ideas after a certain lapse of time and that each time they

[1]John Gabriel Navarra, *The Development of Scientific Concepts in a Young Child* (New York: Teachers College Press, 1955).

return to these concepts new information and insights refine and extend them. For example, when a group of children had their first experience with electricity some commented about the sparks they saw when one end of an electrical circuit was scratched on the other part of a dry cell. Months later, when the children were using flint and steel to make sparks to ignite some tinder, one child asked, "Is the spark that we get when we strike flint against steel the same kind of spark as that which we got from the dry cell?" (On a very sophisticated level, there is a relationship.) This youngster was connecting two observations separated by a considerable span of time. The sensitive teacher, who recognizes that children do not develop their scientific concepts "all at once," is able to help them interpret a variety of science experiences so that they can gradually refine and extend their scientific concepts.

An implication of this view is that teaching in science should be "open-ended." Rather than deal with a science generalization as a unit to be studied and completed, it is better to leave matters open-ended and come back later, when the concept can be expanded. Science lessons should not have conclusions in the sense that there will not be further exploration of the subject. Instead, children should have time for digestion and assimilation. Concepts are not closed at the end; they develop as they are nurtured and reconsidered.

Individual Differences in Development

There are important differences in the ways in which children move from one intellectual stage to another. Almost all children will move from the preoperational stage to the stage of concrete operations. The evidence and a great deal of experience indicate that there is no point in trying to hurry the transition between these two stages. However, the transition to logical, formal operations is dependent on the kinds of educational experiences children have, and for this transition science experiences are of special importance.

A major function of elementary school science is to help youngsters reach the stage of formal operations as they deal with the environment. At the level of formal operations, symbols are used. But if the symbols are to have meaning they must be related in some way to the concrete, physical environment. As more and more children come to live in urban and suburban environments, they may not have as many unplanned, direct primary experiences with the physical environment as children in rural environments have. It will become increasingly important, therefore, for children to have these experiences in the elementary science program—a program planned to include many firsthand experiences in touching, feeling, handling, smelling, hearing, and observing objects in the classroom and in the physical environment.

THE TEACHER AS A MODEL OF INVESTIGATIVE BEHAVIOR

The elementary school classroom teacher usually has a profound effect on the pupils under his or her guidance. To develop a curious, investigative attitude on the part of the children, the teacher must be seen as an inquisitive human being. He or she should be a model for the pupils in terms of seeking answers and using appropriate resources. When a problem is posed to a teacher, often the best way to deal with it is to conduct an investigation involving the pupils. Many questions that children ask cannot be answered because there are no answers.

To some people such questions may seem threatening: "I don't know the answers to all the questions my children will ask." No one does! Even the most learned scientist will not know the answers to all the questions that children will ask, and should not be expected to know them. This is not a reason to stifle the wondering and questioning of children. In fact, one of the best and most exciting ways of learning science is by working with children as they try to find the answers to some of their questions.

The teacher who uses materials and references or conducts small-scale investigations when questions are asked will be accomplishing more than just finding answers and solutions; he or she will also be demonstrating ways of finding out.

Changing Patterns of Behavior

Learning involves changing patterns of behavior. A child who studies the nutritional requirements of the human body and then changes her eating patterns has definitely learned. Learning is more than memorization. Youngsters may memorize the parts of a flower, but they will have *learned* the parts of a flower only when they are able to *interpret* the characteristics and functions of the parts of a flower in terms of the growth and development of a plant and are able to *use* their knowledge to propagate and care for plants. When children are able to use information, their patterns of behavior have changed and it can be said that they have learned.

In science it is what children actually learn that is important. Unfortunately, there often is a low correlation between what teachers plan and attempt to teach and what children actually learn. For example, in attempting to demonstrate how a thermostat in an aquarium heater operates, a teacher used a compound bar to illustrate electric circuits. (See p. 185.) When the compound bar was heated with a candle, it bent so that an electrical circuit was broken and a flashlight bulb ceased to light. When the candle was removed, the compound bar straightened out to close the circuit and the flashlight bulb again gave off light. When asked to explain what the experiment proved, one student said, "When there is light from the candle, the flashlight bulb doesn't have to shine any more, and it goes off. When we take the light from the candle away, the bulb lights up again." To the boy this seemed a sensible explanation. The teacher had intended that the demonstration would give the children a better understanding of electrical circuits. Since what children actually learn is often quite different from what teachers hope that they will learn, it is essential to determine what has been learned.

Behavioral Objectives

It is desirable to state what behavior is to be developed. If objectives are stated in behavioral terms, then there is some possibility of evaluation to see whether the objectives have been achieved. For example, in an exercise in which students are to learn how to define "density," it is suggested that they should be able to

1. demonstrate a procedure for finding the volume of a solid object by measuring the volume of water it displaces.
2. state an operational definition of density.
3. demonstrate the computation of the density of an object, given the known volume and known mass of the object.
4. distinguish between objects that will float in water and those that will sink in water on the basis of their density.

Note that each of these objectives is stated in terms of some operation that the child should learn to do. Upon completion of the exercise, it is possible to find out whether or not a student is able to do them.

Of course, the patterns of behavior that are developed should be desirable ones. An important function of the broad goals of elementary school science is to serve as a general guide as the kinds of changes in patterns of behavior to be achieved are formulated. (See Figure 2–3.)

APPROACHES TO TEACHING SCIENCE

We should use a wide variety of approaches to learning and teaching in science. Variety is the essence of good teaching. Teachers who use a variety of approaches to the teaching of science will have a classroom full of questioners seeking information and testing ideas – the essence of learning.

Approaches to teaching should be consistent with the nature of the scientific enterprise (see Chapter 3). An attempt should be made to check ideas empirically – "Does it work when it is tried?" The nature of the systems that are being studied should be considered. Hypotheses

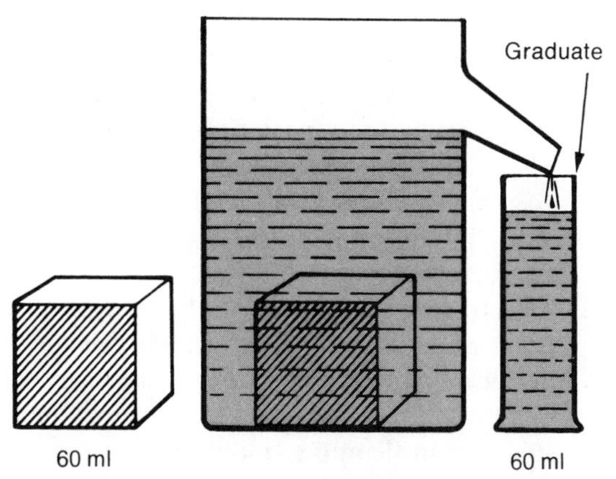

Figure 2–3 The volume of an object that sinks in water can be found by measuring the amount of water that is displaced. The mass of the object can be found with a platform balance. Then,

$$\text{Density} = \frac{\text{Mass}}{\text{Volume}}$$

should be used as intellectual tools in the study of questions and problems. At times, controlled experiments should be carried out. There should be ample opportunities for criticism and exchange of ideas, and ideas should be expressed as precisely as possible. Children should also gain some experience in using the cumulative dimension of science as a resource by relating their findings to those of others. It is important to keep such characteristics of the scientific enterprise in mind in working with children in science.

Each of the approaches to learning and teaching discussed in this chapter has advantages and disadvantages. They should be used when they are advantageous. None of these approaches is inherently difficult. However, the possibilities and cautions must be observed if each is to be used to its greatest advantage.

The Laboratory Approach

In the laboratory approach each child is directly involved in the activity, using materials, handling equipment, undertaking projects, conducting investigations, and carrying out experiments. The children may work as individuals or in small groups. The important feature of the laboratory approach is that each child has direct, primary experience with the materials and equipment. It is a "doing" experience.

Using the classroom as a laboratory. In the elementary school the classroom can become a laboratory. Desks can be pushed together to form flat surfaces. Rolling labs and table space along the perimeter of the room also make good surfaces on which children can work. Many of the materials that are needed can be found in the school or the local community. Children will need water and electricity, but for most elementary science experiences the source of electricity should be dry cells. While gas is very useful for science experimentation, it is not absolutely essential for most work in elementary school science. When water and other liquids are to be used, paper or fiberboard trays on which children can set up and carry out their experiments have proved helpful.

It is often desirable to have science materials assembled into small kits. For example, when studying magnets each child is given a small plastic bag containing two ceramic magnets, assorted pieces of metal and other materials, a compass, and a nail. With these materials each child is able to handle and explore the nature of magnets and to carry out the various experiments that are suggested. Often these kits can be used in more than one classroom, and when everyone is finished with the materials they can be stored for the coming year.

Often there may not be enough equipment of one kind for all the children to be engaged in the same science activity at the same time. Instead, different activities may be set up on different tables or parts of the room. For example, some children may be experimenting with magnets while others are using dry cells and still others are doing experiments using compasses.

The teacher's role in a laboratory approach. The teacher's role is as critical in the laboratory approach as in any other approach. To a certain extent the teacher will have to explain the nature of the activities to be undertaken. For example, in the work with magnets the teacher should explain that the children are to find out as much as they can about the properties of the magnets. However, care should be taken not to give too full an explanation; the children should not be robbed of all the possibilities of discovery. The teacher may also have to demonstrate how to use certain kinds of equipment. In the study of the magnets, for example, the teacher should demonstrate how a compass can be used to indicate whether or not an object is a magnet. While the children are experimenting, the teacher moves from child to child and from group to group asking questions, leading children to observations that they might otherwise not have made, unobtrusively helping some youngster over an obstacle that seems to have blocked his or her progress, and helping all children derive further meanings from the experience.

It has been said that "if laboratory work is worth doing, it is worth discussing." Certainly, with children in elementary school science there should be discussions in which children can compare their findings, explore the meanings and implications of their work, and suggest other lines of investigation that may be of interest and value.

It is important to recognize that laboratory work means activity on the part of the youngsters, and inevitably some noise. Children learn when they are actively engaged in a learning activity, when they are acting and reacting in situations. The laboratory is a place where this can happen. And the laboratory approach to teaching science has great potential in the elementary school.

Cooperative Investigations

In a sense cooperative investigation is a laboratory approach in which a group or the entire class works together. Often what is to be investigated is demonstrated by the teacher and then discussed by the entire class. Questions are asked; ways of finding answers are proposed; hypotheses are suggested; operations are carried out by the teacher or one or two pupils; results are obtained; and the implications and significance of the results are discussed.

The case of the sliding towel. As an example of a cooperative investigation, a teacher hung a towel across a rod and asked, "How far can we pull one side of the towel down before the whole towel will slide off?" The children were eager to try it, but the teacher insisted that they plan the experiment. "How are you going to pull the towel—from one corner or in the center?" "How are you going to find out how far you were able to pull it?" The children decided to pull the towel straight down at the center and to have a tape measure alongside the towel so that they could note how short the shortest end was before it started to slide.

After several trials the teacher asked, "How could we arrange the towel and the rod so that one side of the towel could be pulled down farther before it would start sliding?" There were many suggestions, ranging from "We should try to pull the towel down very slowly" to "We could wet the towel."

The children and the teacher worked together in developing the cooperative investigation. The children learned from each other as well as from the teacher. Some experience in cooperative investigations is often helpful before children undertake investigations individually. This particular investigation became known as "The Case of the Sliding Towel."

Obviously, children cannot be expected to have a sophisticated understanding of experimental procedures, but if they are involved in cooperative investigation they can learn some of these procedures as they work with others. They will then be better able to undertake investigations and devise experiments on their own.

Demonstrations

In a demonstration a teacher or pupil shows how some scientific apparatus works or the operational meaning of some scientific generalization. Some demonstrations are carried out to show pupils how to use a particular tool or piece of equipment. For example, a microscope is an important tool for the study of small objects and organisms. However, there are certain procedures that have to be followed in using a microscope if the pupils are to derive any benefit from its use. Certain cautions must be observed to avoid damaging microscopes, slides, and specimens. One of the most effective ways of teaching children the use of a piece of equipment such as a microscope is through demonstration.

Another common use of demonstration is to show the operational meaning of scientific generalizations. One of the most important generalizations in science is Newton's third law, "For every action there is an equal and opposite reaction." But what is the operational meaning of this generalization? A test tube can be suspended horizontally from two wires. A little water can be placed in the test tube and a cork fitted loosely into the neck of the tube. When the water in the test tube is heated, it changes into steam and forces the cork out. The cork goes in one direction, but the suspended test tube moves in the opposite direction. (See Figure 2–4.) This is a physical demonstration of an operational meaning of Newton's third law. For elementary school youngsters who are at the stage of concrete operations, such demonstrations are needed for them to derive meaning from generalized statements.

The following points should be kept in mind when preparing a demonstration:

1. *The demonstration should be as simple as possible.* The operational meaning of a generalization will be obscured if the equipment and apparatus used are complicated. Also, children often become so enamored of complicated setups that the central purpose of the demonstration becomes lost.
2. *The demonstration should be visible to all the children.* Use large equipment and apparatus for demonstrations. If this can't be done, the children should work in small groups so that all can see.

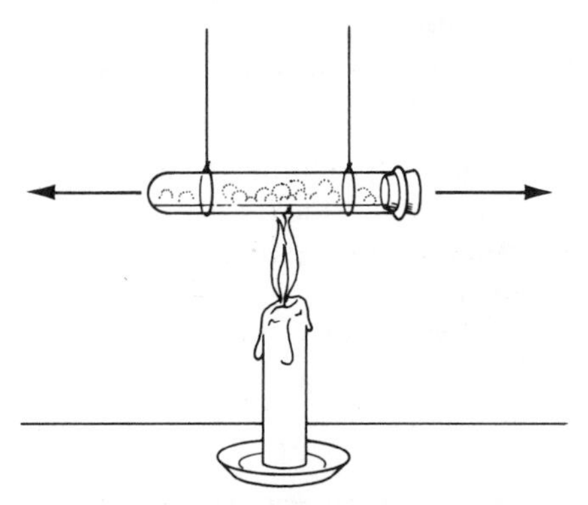

Figure 2–4 When the water in the test tube is heated, the steam forces the cork in one direction and the test tube in the other.

3. *All parts of the demonstration should be visible to the children.* If the most important parts of a demonstration are hidden in a "black box," children may very well attribute magical powers to the black box or to the demonstrator. In either case the value of the demonstration is diminished.
4. *Children should be encouraged to ask questions about the demonstration.* In many cases the demonstration can be altered to find answers to the children's questions. The demonstration then evolves into a cooperative investigation.

Investigations

In a sense investigations might be called "student research"—they are studies of questions and problems to which the answers are not known. In carrying out science investigations, students have some of the same kinds of experiences that scientists have as they explore the unknown. Students use the various processes of science, and they find that these are not discrete processes. They learn, instead, that these processes are components of inquiry. Also, they experience some of the frustrations of things not working the way they are "supposed to" and the exhilaration of finding ways of overcoming obstacles and arriving at tentative answers to knotty questions.

Obviously, children must tackle questions to which answers are not known but lie within their grasp. Examples of specific investigations are provided in later chapters. (For a list of investigations that children can undertake, see Chapter 20.)

PLANNING FOR SCIENCE INSTRUCTION

Planning for science instruction involves determining who will teach, where science instruction will occur, under what conditions it will occur, what supplies will be needed, what material will be covered, and when it will happen. Some schools have science specialists, but even where such specialists exist the classroom teacher may want to become involved in the pupils' science activities, follow up the work covered by the specialist, or at least coordinate other learnings with the science work. In most schools, however, science instruction is primarily the responsibility of the classroom teacher. It is often helpful for teachers on a grade level or in a school division to plan their science teaching cooperatively. Where conditions permit, some teachers have found it helpful to team teach in science, bringing classes together for important demonstrations, films, and other activities. A variety of teaching and learning configurations are possible and might, with careful planning, be attempted.

Where will science instruction occur? Some schools do have separate elementary science labs, but most often the classroom is the site of science instruction. In either case flexible desk and seating arrangements are highly desirable. Teachers may want to have all of the desks pushed together to form a large table or a "U-shape" for discussions, demonstrations, reports, colloquia, and so forth. At other times, when activity stations are being used, the desks ought to be separated, with materials and equipment set up at each desk or cluster. For demonstrations children may be grouped around a demonstration table or seated on the floor or at desks in a variety of arrangements. Teachers will often want to use outdoor areas. Planning for these spaces or areas is necessary as well as making sure they will be available. For further treatment of field study see pages 46–48.

The acquisition, storage, and distribution of materials must be carefully thought out. As teachers plan for science instruction, lists should be made detailing the materials that will be needed. It should not simply be assumed that "someone" in the school building will have a particular item. If the school does not have a science storage room, some items may have to be ordered. This may take several weeks or months, so advance planning is essential. Many materials for science teaching can be purchased in grocery, hardware, or department stores. A letter may be sent home with the children asking for specific items likely to be readily available in the home, although local school policies should be checked regarding such requests.

Once the materials for science teaching have been obtained, storing them becomes the next concern. Teachers have found that keeping all of the materials that will be used in a particular unit together is very helpful. Often the materials that are to be used for a day's lesson can be placed on trays or in shoe boxes. This saves time in that individual items do not have to be distributed. Youngsters can aid in the preparation of these trays or boxes. Orderly procedures for the distribution and collection of materials should be established. Again, pupils should be involved and should take responsibility for this operation. Finally, just where in the room or school building the materials will be stored should be decided in advance.

When science instruction will occur is another important consideration. Units should be planned so that they will be seasonally appropriate, particularly if they involve weather, planting, fieldwork, and the like. The time of day at which science instruction occurs is also a consideration. Often science gets pushed to the end of the school day. This practice might convey some of the teacher's feelings about the value and importance of science education. Science should be considered a basic area, not a minor subject.

The specific work to be covered is another issue. Collaborative planning with other teachers or administrators can be most helpful here. State or local guides or syllabuses give suggestions as to which topics should be covered and in which grades. A science textbook or commercial program often directs the content and processes in a curriculum. This does not mean that teachers no longer need to plan. Activities described in this book can be coordinated with and used in conjunction with textbooks and other programs. Children's interests should also serve as input for science activities. The interests of children are generally apparent from their discussions, the objects they bring to school, or the things they construct in their hobbies. Capitalizing on children's interests has built-in motivational value. A class with which the authors have worked became so involved with pumpkins one October that a month's activities ensued as children investigated the external and internal features of a pumpkin, made pumpkin pie, germinated and roasted pumpkin seeds, drew pictures, and took part in other related activities. This spontaneous interest in the pumpkin helped promote considerable science and language arts learning.

Realistic goals for what work to cover should be set. Overplanning is often wise for beginning teachers, but attempts to cover too much work in a given time span should be avoided so that flexibility can be maintained. The activities covered in this book are not arranged by grade level; most of them can be adapted to a variety of grade levels. Care should be taken, though, to make sure that the concepts involved are not too difficult for young children (in kindergarten and first grade) to deal with cognitively. When working with young children, concrete concepts should be conveyed; abstract material is usually not within their realm of understanding. For example, young children probably should not investigate the field lines of magnets, but it is entirely appropriate for them to sort a collection of objects into two piles based on whether or not each item is attracted to a magnet.

In planning science lessons the objectives and procedures should be carefully thought out and clearly stated. Even though lessons seldom proceed exactly as planned, it is important to know in advance the expected outcomes. The essential elements of lesson and unit plans follow.

The Lesson Plan

A plan for a single teaching lesson may be detailed or sketchy, depending on the teacher's needs. Single lessons should be planned in advance but cannot realistically be too detailed if they are to cover a 30- or 40-minute science period. However, there are some elements that should be included in all lesson plans: objectives, materials, procedures, assessment, and evaluation.

Objectives should be stated clearly and derived from the concepts or processes to be developed. For example, a content objective might state that "at the completion of this lesson, the pupils will be able to identify the common property of all materials attracted to a magnet." A process objective might be "At the conclusion of the experience, the pupils will be able to *predict* whether or not an object will be attracted to a magnet by looking at it and feeling it."

The *materials* needed for the lesson should be listed so that they can be secured in advance. This list should be as specific as possible (e.g., including paper clips, scissors, paper cups, etc.) in order to avoid a last-minute search for materials.

Procedures should be listed in the sequence in which they are to occur. For example:

1. Distribute the trays of bar magnets and paper clips to each child.
2. Let the children "play" with the materials for three or four minutes.
3. Ask the pupils to let as many paper clips hang from each pole of the magnet as it will hold.
4. Ask the children to record their results.
5. Ask each child how many paper clips were suspended from the north pole, the south pole, or the middle of the magnet.
6. Produce a bar graph of the average results on the chalkboard.
7. Ask the pupils which part of the magnet seems to be the strongest. Why?
8. Have two pupils collect the materials.

Assessments should be built into the lesson to find out how well the pupils have fulfilled the demands of the task(s). This may be accomplished by noting pupils' answers to questions asked, teacher's observations, completion of procedures, and generalizations and conclusions drawn by the youngsters.

Evaluation is also important. If the objectives of the lesson were stated clearly, then determining how well they were fulfilled should definitely aid in evaluating the lesson. Often, having a pupil repeat a procedure or demonstrate a concept can help in evaluating the effectiveness of the experience. Paper-and-pencil tests can be useful when definitions, procedures, concepts, or the thoughts involved in problem solving are to be tapped.

The Science Unit

Most science activities, particularly in intermediate and upper elementary grades, are not isolated experiences but parts of a larger set of learning episodes called a unit. A *unit* is a wide variety of related ideas, concepts, activities, experiences, materials, and resources for teaching and learning a single topic. Often learnings from other subject areas are integrated. Unit instruction helps children obtain a fuller appreciation of concepts than can be acquired through single, isolated lessons because the children become intensely involved in a wide range of interrelated learning experiences. The unit also helps teachers individualize science instruction through the use of related learning centers (see pp. 44–46), learning contracts, committee work, assigned reports, interviews, and individually assigned experiences or excursions.

Teachers often ask how much work should be covered in a unit. There is no one answer. Any topic selected for a unit should be dealt with to a sufficient degree so that the major concepts of the topic are investigated. The length of a unit, though, more often depends on the time available, the intensity of pupil and teacher interest, and the richness of concepts in the topic. Some units may require a week to complete; others may require three or four months. A short unit may involve studying different kinds of apples, making applesauce, investigating apple seeds, and visiting an apple orchard. A longer unit might cover several aspects of magnetism and electricity and require months to complete.

Units are built by first analyzing the skills, concepts, and processes to be developed in terms of which ones should be dealt with first. The experiences in a unit are built upon one another, with the simpler ideas presented first. As this hierarchy is constructed, the planner must select the experiences and skills that are prerequisite to an understanding or appreciation of the next step. The content chapters of this book are arranged in this way. The sequence of concepts discussed builds from simpler ideas to more complex ones.

The materials in a unit must be planned for well in advance so that they can be obtained before they are to be used. Trips or excursions often require several weeks' notice. Books and resource guides should also be acquired in advance.

The unit plan contains several components that distinguish it from a lesson plan. In some respects it may be thought of as a series of lesson plans with a unifying structure. There is no one way to organize a science unit. The plan will depend on the subject to be studied, the experience of the teacher, and the needs of the pupils. There are, however, common elements in almost all unit plans.

Broad goals. All unit plans should include some broad goals or general problems. These goals or problems should be general enough to include the nature and scope of the entire unit. The stated problem for a unit might be "How is our daily weather determined by the interaction of the sun, air, and water?" Obviously, in order to find answers to this problem the children will have to deal with concepts and engage in activities involving the sun and its energy, the nature and movement of air, air pressure, the nature and shape of the earth, the tilt of the earth's axis, planetary wind belts, the water cycle, the atmosphere, and many other associated subtopics. Stated as a broad goal, the purpose might be: "To investigate the many interactions of air, water and the sun and how they determine our daily weather."

Importance. The importance of the unit should be considered. How does the problem affect the daily lives of the youngsters? How will involvement in this unit improve the pupils' skills in other areas? How will successful completion of the unit prepare the children for future learning? How much interest have pupils demonstrated in the topic?

Skills and concepts to be covered. A list of the skills and concepts to be covered in the unit should be generated. The concepts should be fairly specific. For example:

> Warm air is less dense than cool air because its molecules are farther apart.
>
> Air rises as it is warmed.
>
> Air descends as it is cooled.
>
> Air is warmed at the equator.

Process skills to be developed in the unit should also be listed. Examples might include the following:

> The children will *demonstrate* the effects of evaporation by fanning a water-streaked chalkboard.
>
> The pupils will *predict* the results as a bottle of warm water interacts with a bottle of cold water when the bottles' mouths are placed together.
>
> The children will *construct* a greenhouse using plastic pans, soil, tape, and a thermometer, and *demonstrate* how a greenhouse traps heat.

All of the major process skills and concepts to be developed in the unit should be listed.

Materials. A materials list is essential in a unit plan, and as mentioned earlier, many of the items will have to be secured in advance. Each activity has to be carefully analyzed in terms of all of the anticipated supplies. Perishable and live materials may have to be acquired at the last minute.

Motivation. An initiating activity or motivation should be included to capture the children's interest and generate enthusiasm. This might take the form of a fascinating demonstration or display, a trip to a local resource, or an intriguing story, photo, or object. In a motivating activity the importance or relevance of the unit should be related to the children's lives and interests.

Activities and experiences. These are the heart of the unit. The experiences should be listed sequentially. Each activity should fulfill the demands of one or more of the skills or concepts listed previously. Beginning teachers may wish to organize this part of the unit plan as a series of lesson plans similar to those described in the preceding section. Each activity should contain specific objectives, procedures, new vocabulary, and a built-in assessment.

Community resources. A list of community resources (i.e., places to visit, people to invite to the school, exhibits, etc.) should be compiled. The children may be involved in the generation of this list.

Ancillary activities. Related activities in which some of the pupils may participate should be considered. Activities involving preparation of bulletin boards, displays, collections, and food are often included in science units.

Bibliography. The unit plan should contain a bibliography of teacher resources, children's books (on a variety of reading levels), pamphlets, brochures, and other printed material. Also included can be posters, study prints and audio-visual materials such as films, filmstrips, single-concept film loops, and slide collections.

Culminating activity. A single activity should be planned to conclude the unit. It should be a natural outgrowth of the children's work in the unit and should be designed to help summarize the major concepts covered. Examples of culminating activities include assembly programs produced by the children, displays, panel discussions, report presentations, and dramatizations.

Evaluating the unit. The effectiveness of the total unit can be evaluated in a variety of ways. The assessments associated with the individual activities should give clues as to how well the children retained the concepts and performed the processes set as initial objectives. The enthusiasm generated and the accuracy with which the children proceeded are other clues. Individual and committee reports and presentations are still other means of judging the effectiveness of the unit. How complete were the children's reports in terms of accuracy and comprehension? Tests are often used to evaluate retention of concepts but should not be composed of questions requiring only factual recall. Tests should ask children to analyze and evaluate situations or procedures, select appropriate data, transform information in some way, or solve a problem.

Working with science units requires careful and extensive planning, but it offers many advantages compared with the use of a series of single, somewhat isolated lessons or experiences. Units tend to stimulate intense involvement. Pupils seem to be constantly on the lookout for related materials when they are involved in a unit. Making notes on the success or pitfalls of individual aspects of units serves to improve them and enables them to be recycled for use in future years.

Individualizing the Science Program

An effective means of individualizing the science program is the use of classroom learning centers, activity stations, or task cards. These techniques help the teacher organize instruction so that children can learn on their own and at their own pace. A learning center (interest center, activity center) may be no more than a few task cards with specific instructions for independent investigation. It can be materials set out on a table with specific procedures to follow, or some apparatus or equipment with which the child must conduct an experiment. The common characteristics of all learning centers is that they are self-directing, promote independent learning, and help meet the individual strengths and needs of all the pupils in a class.

Learning centers can take several forms. They may be the total program or approach for an entire unit. For example, a teacher may be conducting a month-long unit on air and its properties. In order to convey the basic concepts involved, ten self-contained activities for the pupils are set up. One or two of the activities are basic, and all the children must complete them. Of the remaining eight activities, the pupils may choose five that they wish to complete. As each child completes an activity, he or she must submit evidence to the teacher to demonstrate his or her learning and involvement. The teacher may conduct a summary of findings and a discussion after the children have completed the unit. Here is an example of a task card:

AIR STUDY *Card #2*

In this activity you will try to find out if air takes up space.

Materials you will need: a plastic bag
a twist tie
a book.

1. Try to trap air from the room in your bag.
2. Are there any places in the room that you think do not contain air? If so, try to collect air in them.
3. Trap some air in your bag and close it with the twist tie as quickly as you can. Try not to let any air out.
4. Try to balance the book on the bag full of air.

Now answer these questions on a separate sheet of paper and give it to your teacher.

a. Was there any place in the room where you did not find air?
b. When you balanced the book on the bag, what was between the table top and the bottom of the book?
c. What do these activities tell you about air?

These activities may be approached and the questions answered in a variety of ways, depending on the individual child.

Activity centers may be designed to supplement the regular instructional program in order to extend and reinforce skills already covered. For example, if a class unit on metrics has been taught, the teacher might want to have the pupils apply their newly learned skills by measuring and keeping records of the length of each of five objects displayed on a table. The children express their answers to the nearest centimeter and submit their results to the teacher.

Learning centers may be initiated by teacher or pupil. An example of a child-initiated activity center may occur the day a pupil brings a pet to school. The youngster may devise ten questions based on specific observations of the pet that the other children must answer when they have a turn to observe.

Ingredients of a well-designed learning center. A well-designed learning center ought to have all of the following components:

1. A variety of interesting materials, including commercially prepared items as well as teacher-made objects. There ought to be manipulative as well as paper-and-pencil materials.
2. Clear instructions for self-directed activity must be provided. When younger children are involved, instructions are often given in rebus (pictoral) form. Many of the activities suggested in this book can be used in this way.
3. If the activities are to be self-evaluated, there must be a clear way for pupils to do so. Children should have no difficulty checking their own work.
4. There should be some provision for multilevel activities so that youngsters in a class of varying interests and abilities can select stimulating, challenging activities.

The teacher should have the following very clearly in mind when planning activity centers:

1. Objectives must be clear and observable. Anticipated outcomes, for the most part, should be specified.
2. Sufficient materials and resources to complete the demands of the task should be available to the pupils.
3. An evaluative mechanism must be planned in advance so that the teacher will know whether the demands of the activity have been met and to what degree. Usually teachers have pupils submit some completed work or answered questions as a result of their investigations. If work is to be submitted, the teacher should have developed checklists so that he or she may note which pupils have completed which tasks.

Classroom learning centers can provide endless enjoyment to both pupils and teachers while fostering independence and meeting individual needs.

Field Experiences

Field experiences may be broadly defined as experiences that children have outside the classroom. The possibilities for field experiences are limitless. They may include visits to different parts of the school building, studies carried out on the school grounds, projects undertaken in outdoor laboratories, strolls down a nature trail, trips to a farm or a manufacturing plant, visits to laboratories and research facilities, and carefully planned school camping experiences. (See Chapter 10 for field trips within a school and schoolgrounds and Chapter 11 for field trips within a city.) While some field experiences may involve extended and carefully planned field trips, many take the form of short forays to visit a site near at hand or to interview someone who is easily accessible.

Field experiences can provide the following benefits:

1. Firsthand experiences with materials that cannot be brought into the classroom.
2. Study of plants, animals, and technology in natural settings. This makes it possible to study interrelationships among various elements of the environment.
3. Study of problems related to science and technology in their natural settings.
4. On-the-spot interviews with people in their occupational environment. This helps children get a better understanding of the nature of their work.
5. Situations in which the applications of science can be studied.

On many field trips the collection of specimens may be planned. The children should begin to understand conservation considerations. Obviously, the laws designed to protect rare species and property owners should be respected and obeyed. In other cases the basic conservation rule to follow is that *materials and specimens can be collected if this results in no significant change in the environment.* A beaker of sand collected from the beach makes no significant difference. However, the collection of a lone flower from a roadside will make a significant difference and should be avoided.

It is important to recognize that field experiences may take many forms. In some cases, the teacher may take the entire class on a trip. In others, the field experience may be undertaken by a single child, who is encouraged to "go look at ______ and see it firsthand after school." In still other cases, a small group may undertake an investigation in the field.

The following are suggestions for planning and conducting field experiences with small and large groups of children:

1. Have the children think through the purposes of the field experience.
   a. What kinds of information will be obtained?
   b. What kinds of questions should be asked?
   c. How important is this information? How can it be used?
   d. Will the field experience benefit others as well as us? A community-service project, such as a conservation project, can be an important field experience.
2. Reconnoiter the site before the visit.
   a. What observations should be stressed?
   b. Where should the group make stops?
   c. What precautions need to be taken?
   d. How should the resource people be briefed so that the children will gain the maximum benefit from the experience?
3. Organize carefully for the field experience.
   a. Clear your plans with the school administration. Most school systems have standard policies regarding field experiences. These should be adhered to.
   b. Obtain parental permission. Parents have a right to know when their children leave school, and it is usually wise to obtain written permission on simple duplicated forms.
   c. Arrange for transportation. Many school systems have school buses or other modes of transportation that can be used for trips.
   d. Invite other adults, such as parents and other teachers, to accompany the class on the trip.
   e. Make advance arrangements for visits to the park, museum, factory, or farm.
4. Brief the group for the field experience.
   a. Suggest appropriate dress.
   b. Discuss responsibilities for science equipment, first-aid kits, maps,

cameras, tape recorders, and notebooks that may be taken on the field trip.

c. Present an overall picture of the field experience using maps and charts.
d. Discuss behavioral and safety standards required. Some orientation to conservation aspects of the trip may also be necessary.
e. Discuss and further clarify the purposes of the trip.

5. Make the field experience as profitable as possible for all the children.
   a. Help the students keep the purposes of the field experience in mind. In this way they may "see where they have only looked before."
   b. Encourage children to make observations that may not have been anticipated in the previous discussion. At times "freelance browsing" can be quite fruitful.
   c. Make certain that all members of the group hear explanations and discussions. At times it may be necessary to interpret the remarks of experts and relate them to the children's previous experiences.
   d. Check that the planned information, specimens, pictures, recordings, and notes have been obtained.
6. Back in the classroom it is important to help children derive meaning from their field experience. Have them do the following:
   a. Analyze the data collected.
   b. Discuss the meaning and implications of various observations.
   c. Evaluate the field experience and discuss how future experiences can be made more effective.
   d. Write thank-you letters, when appropriate.

Using Science Resources

There should be no artificial limitations on children as they study, investigate, and inquire. In fact, they should learn how to use all the resources available to them. A standard dictum in teaching is to *always use the resources and procedures that are best suited to the goals to be achieved.* The following are some of the resources that should be used in an effective elementary school program:

1. *Trade Books.* There are now many excellent library books dealing with a wide range of scientific topics at various levels of reading. These are sources of information, suggestions for experiments and investigations, and places where children can check their findings against those of others. Trade books are especially important sources of information about subjects ranging from the interior of the earth to the reaches of outer space that children cannot deal with firsthand. Some of the trade books you may wish to consider are listed at the end of the science process and content chapters of this book.
2. *Textbooks.* Textbooks are common sources of information and suggestions available to the children in a class. Like trade books, textbooks should be used within the context of inquiry. They can be a handy reference book and a guide for laboratory and field experiences. They can also provide a broad general outline of a program

in science. See Appendix II for a list of elementary school science programs.

3. *Films.* Films can convey motion when this is important to an explanation, as in the explanation of the electric motor or the slow motions of plants as they grow and orient themselves toward light. Films can also be used to develop attitudes and feelings. It is difficult, for example, to view a film such as the classic "The River" without experiencing deep feelings and perhaps changes in attitude.
4. *Filmstrips.* These are excellent mechanisms for bringing a coordinated series of pictures into the classroom. An important advantage of filmstrips is that they can be used at whatever pace the teacher deems desirable; sometimes a prolonged discussion can be carried out over one frame.
5. *Slides.* This is another excellent way of bringing pictures into the classroom. Of special importance are the slides brought into the classroom by the teacher, children, and parents. Many of these can deal with unique local features.
6. *Transparencies.* Transparencies shown with an overhead projector are a convenient way to show diagrams, sketches, and pictures to an entire class. Since transparencies are relatively easy to make, they can be designed by the teacher or the children to meet the peculiar instructional needs of a particular classroom.
7. *Television.* Through television, distant events and prominent personages can be brought into the classroom. The video tape recorder makes it possible to record television programs and to use them when they best fit into the instructional program.
8. *Telephone.* The telephone can be used to interview resource people. In a sense, it is another way of bringing the outside world into the classroom. The conference telephone arrangement is especially useful, since it makes it possible for everyone in a class to take part in the telephone conversation.

Projects

Projects, special undertakings of individual children or groups of children, may take the form of a special study of a selected topic, leading to a report. One group of children, for example, made a study of the moon in which they investigated children's literature dealing with the moon and presented a report of their findings. Other projects might be called construction projects. Many children have constructed electric motors and in the process learned a great deal about electricity. Sometimes display projects are made to depict some scientific principle or some procedure that is used in science. One project depicted the various ways in which archaeological artifacts can be dated.

It is important that a science project be a learning experience in science as well as an interesting construction experience. Constructing an electric motor without observing the scientific principles involved is of little value. An understanding of scientific principles, the most important learning outcome of a science project, is also useful when children encounter setbacks and problems in constructing and operating their projects. In later chapters a variety of projects in which children may become involved are described. (See Chapter 20 for some suggested projects.)

Colloquiums

Colloquiums are meetings at which children describe and discuss the results of an investigation, a study they have carried out, or a project they have completed. It is desirable that children have some help in organizing their report and planning their presentation. Other children should then have an opportunity to discuss the presentation.

Communication of ideas and the results of investigations is an essential component of the scientific enterprise. As they take part in colloquiums held in the classroom or on a schoolwide basis, children begin to experience an important dimension of the scientific enterprise.

Safety in the Science Program

A strong elementary school science program will necessarily involve the use of many learning materials for activities and investigations. It is important, within this program, to develop safe and desirable habits for the care, handling, cleaning, and sharing of science supplies and equipment. The classroom teacher is in an excellent position to help promote such positive habits. A list of dos and don'ts can be easily distributed and just as easily forgotten unless the pupils have had opportunities to work through the *reasons* for safety rules and precautions. For example, in one elementary school a rule was established that a pail of sand had to be on hand for any activity involving the use of fire or flame. But, when the children were asked if they knew the reason for the rule, few knew the answer. If we hope to use any form of fire in our classroom (even a candle flame), then we must help children to learn the nature of fire and how it can be controlled. The three things necessary for a fire are fuel, oxygen, and sufficient heat to keep the fire going. A fire can be extinguished by removing any or all of these necessary ingredients. Firefighters may pump great quantities of water onto a fire to reduce the kindling temperature and thereby eliminate the heat. Campers throw sand or dirt on a fire to deprive it of oxygen. As children are helped to think through the causes and effects of potentially dangerous situations, they will obtain a more lasting knowledge of what to do in hazardous circumstances. Children should consider in advance the possible hazards or inherent dangers of using a material, piece of equipment, or energy source.

Accidents which do occur should be turned into a learning experience in which the pupils are encouraged to examine the causes for accidents and how they might be prevented in the future. Children can keep an accident record or log, analyze newspaper reports of accidents, or even investigate possible safety hazards within their classroom or school and suggest corrections. The development of safe habits for living is primarily the development of attitudes—ones which require learning from many experiences, a positive example to emulate, and time.

There are some general areas of work in elementary school science which tend to be more hazardous than others and these are briefly discussed. Animals are a welcome addition to any elementary classroom; but, children must be taught to treat the animals gently and not to tease them. (See Chapter 13 for a detailed discussion of caring for animals in the classroom.) The use of glassware is generally inappropriate in elementary school science, but when its use is imperative, children should be helped to understand that glass can break under the slightest pressure and cause serious injury. Similarly, thermometers should be carefully distributed, stored, and handled. The use of microscopes should be reviewed with the pupils. A general procedure is never to focus downward (i.e. place the objective in its lowest position without touching the slide) before the child places his eye on the eyepiece and then he can turn the focus knob so that the objective is pulled upward. This will prevent the microscope slide from breaking, which can occur if the objective lens is pressed onto it.

Heat or energy sources can create safety problems in the classroom. In general a heat source without a flame (e.g., a hotplate) is safer than one with a flame. Where the use of a flame is essential, candles are probably best although they do not provide a great quantity of heat. Canned heat or alcohol lamps may also be used with proper precautions taken. Gas heat is not usually available in elementary classrooms and can cause many difficulties in terms of the flame and high degree of heat as well as the possibility of escaping gas. Batteries and dry cells should always be used in electricity projects. Standard house current is inadvisable since even momentary contact can be extremely harmful. It has been said that the greatest danger in handling a dry cell is the possibility of dropping it on your toe. If hand tools are used, the children should be instructed in their proper use and whenever chemicals are required, only the weakest solutions should be used. Strong, corrosive acids should not be used; lemon juice or vinegar usually demonstrate the properties of an acid just as well as stronger solutions.

Science is a natural area of the elementary school curriculum for the development of habits of safe living, but the promotion of careful behaviors applies to all areas of school and home life. Once an attitude of safe and careful investigation is established, with the reasons for safety rules clearly developed, it is likely to remain with the child for years to come.

EVALUATION IN SCIENCE

Evaluation in elementary school science can take several forms, depending on the purposes of the evaluation. Generally, however, the aims of evaluation are to determine children's growth and development in science and to provide feedback for teachers and administrators for the purpose of improving science content, methodology, and programing.

Goals, Behavioral Objectives, and Evaluational Activities[2]

Effective evaluation is in terms of broad goals and more specific objectives, and is carried out through evaluational activities. *Goals* are broad statements of the directions in which we wish children to grow. For example, one of our goals in elementary school science is to help children develop a view of the universe that is generally consistent with the view that has been developed in the sciences. *Objectives* are more specific and precise statements, usually written in behavioral terms. *Behavioral objectives* include words—such as *describe, state, construct, locate, design, suggest, relate, classify, sort,* and *operate*—that indicate what a child should have learned to do as a result of an educational experience. *Evaluational activities* are designed to ascertain whether children have achieved the desired behaviors.

Thus, in order to evaluate it is important to delineate the broad, general goals of the program, goals like those discussed in this book. (See pages 5–22.) Then the specific behaviors that are desired should be described. Evaluational activities should be designed to determine whether the children have achieved these behaviors. The following are broad, general goals for instruction in science, together with samples of behavioral objectives and evaluational activities under each of the goals.

Goal: To build a world view that is generally consistent with the world view that has been developed in the sciences.

[2]In the development of this approach to goals, objectives, and evaluation, we have been influenced by our work in the development of the elementary science program "Science—A Process Approach." This program features behavioral objectives, competency measures, and appraisal devices to determine whether the behavioral objectives have been achieved.

Objective: 1. Children should be able to state their location or address on Planet Earth and in the solar system, galaxy, and universe. (See page 6.)
Evaluational Activity:
Have the children write their address in the universe.
2. Children should be able to describe ways in which human beings are similar to and different from other animals.
Evaluational Activity:
Show the children a picture of a chimpanzee and ask them to describe three ways in which they are like the chimpanzee and three ways in which they are different.

Goal: To develop science process skills.
Objective: 1. Given a set of objects having a variety of properties, the children should be able to sort the objects on the basis of their properties and state the basis on which they have been sorted.
Evaluational Activity:
Give each child twelve peanuts and ask the children to sort the peanuts into sets of peanuts so that all the peanuts in a set have a common property. Ask them to state how all the peanuts in a set are alike.
2. After making a series of observations of an object that cannot be opened, the children should be able to suggest a model of what is inside the object.
Evaluational Activity:
Place a piece of chalk inside a shoe box and tape the cover shut. After carrying out whatever operations they wish on the box, ask the children to suggest what is inside the box.

Goal: To develop intellectual power.
Objective: 1. Children should be able to describe how things would appear from a different frame of reference.
Evaluational Activity:
Place a small doll and a marble on a roller skate and push it across the classroom floor. Ask the children to describe how the marble would appear to the doll.
2. Children should be able to design an experiment to test an idea or hypothesis.
Evaluational Activity:
The children have found that a rubber comb that has been rubbed with wool will repel another comb that has been rubbed with wool. Ask the children to design an experiment to see whether a rubber comb rubbed with plastic has the same static electric charge as one rubbed with wool.

Goal: To develop scientific literacy
Objective: 1. Given an article on science, children should be able to read it and state in their own words the key ideas in the article.

Evaluational Activity:
Have the children read a section of a science textbook that is at their reading level and ask them to state the main ideas that they have read.

Goal: To develop an understanding of the human body and the relationships between human beings and their environment.

Objective:
1. Children should be able to locate and identify in a manikin the major organs that make up the respiratory system.
   Evaluational Activity:
   Have the children find and point to the major organs of the respiratory system in a manikin.
2. Children should select foods that will make a nutritionally sound diet for a day on the basis of the four basic food groups.
   Evaluational Activity:
   Have the children keep a record of the kinds and amounts of foods eaten during a day. Have them analyze this record into the four basic food groups.
3. Children should be able to predict some of the possible consequences of making various alterations in an ecosystem.
   Evaluational Activity:
   Show the children a picture of some natural setting (such as a forest) or take them to some natural spot (such as a hillside) and ask them to predict the possible consequences of such changes as cutting trees or planting more trees, removing the grass or planting more grass, digging drainage ditches or building a pond.

Goal: To prepare for effective citizenship.

Objective: Children should be able to identify and state the science concepts or generalizations involved in a public issue.
Evaluational Activity:
After studying the growth of the human population, ask the children to state two basic science concepts that are involved in human population growth.

Other Sources of Information for Evaluation

Evaluation is an important dimension of teaching. Teachers can gain valuable insights into children's degree of understanding and skill through such informal procedures as the following.

Answers to *questions* asked of pupils during a lesson or learning episode about what they observe or discover often provide teachers with clues as to whether the concepts developed are being grasped and related to other concepts. Teachers should make wise use of questioning techniques (see pages 36, 104) to check on the clarity of a demonstration or discussion.

Brief performance assessments can be "built into" science lessons. For example, in a class discussing the requirements of a complete circuit each child (or group of children) can be given a bulb, a socket, a screwdriver, a wire cutter, wire, and a battery and then asked to make the light work. By observing and noting the success of the children's efforts, the teacher can check to see whether the children were able to apply their knowledge about circuits or replicate a procedure. Children may also be called upon to demonstrate a specific science technique, such as focusing a microscope, using an eyedropper, or handling a piece of equipment.

Science projects are often assigned in which a pupil must construct a model, conduct an investigation (see Chapter 20), interview a local scientist, or prepare evidence supporting a viewpoint (or viewpoints) on an issue. Completed projects can be brought to school and readily evaluated by the teacher. Findings or experiments can be reported to the class.

Reports are another means of evaluating science learnings. Children can become actively involved in researching a topic, conducting related investigations, and delivering an oral or written report or demonstrating an experiment or operation. In projects and reports the child's efforts and involvement are generally easily discernible and provide a tangible means of evaluating vocabulary usage, organization of thoughts, concept development, and application of scientific concepts.

Tests can have some significance for evaluation in elementary school science. They can help teachers discern whether children have retained the material covered or can apply or evaluate new knowledge. However, the limitations of formal tests should also be noted. Testing is a "point-in-time" sampling of knowledge acquisition or usage; tests do not necessarily indicate what children can do to demonstrate a concept or procedure. In addition, tests often measure a child's ability to read. If the reading content of a question is beyond a child's general reading level, his or her science knowledge might not be tapped. Also, essay questions are probably best used only when children can organize their thoughts in sentences and adequately express themselves in writing.

Test items should be clear, concise, and easy to understand. Any test ought to combine basic recall and recognition items with questions that call forth higher levels of mental operation such as application, interpretation, analysis, prediction, inference, and evaluation of information.

Objective test items are usually in the form of fill-ins, true or false, matching, multiple choice, or labeling. An example of a multiple-choice item that calls upon children's interpretive skills follows:

A farmer reported that rocks "seemed to grow in his fields." The rocks probably appeared as a result of

a. meteorites that exploded as they entered the atmosphere.
b. frost heaving the rocks up from below.
c. the rocks being washed down from above by running water.
d. the soil that once covered the rocks being washed away.
e. the rocks being blown there by strong winds.

Essay questions for elementary school children should be clear and concise and call for relatively brief answers. In evaluating responses it is important to focus on the quality of the ideas expressed and the general organization rather than being unduly influenced by such factors as neatness and penmanship. Here are two examples of essay questions appropriate for upper-grade elementary school children:

1. Why is it difficult to make a model of the solar system in the classroom?
2. We can collect a tremendous amount of information about the atmosphere. Yet we cannot predict the weather with 100% accuracy. Suggest some reasons for this.

Tests should be viewed as a learning experience, and children should be given opportunities to go over their errors. The tests should be returned as soon as possible so that the material covered is still fresh in the children's minds. Testing should not be used as a threat or punishment and should not be a frightening experience. Particularly, in elementary school, tests should be viewed as helpful tools for both teacher and pupil; this can serve to build positive attitudes toward testing, from which the children will benefit as they move from one grade to the next.

Keeping records of growth in science is important in the evaluative process. Growth takes place over time, and therefore it becomes necessary to keep records of children's behavior. These can then be compared with specific learning objectives. In order to systematically record information, teachers devise record books, individual science folders, or record sheets in which they can check off objectives or skills achieved, reports or projects completed, test results, and so forth.

Anecdotal records with a specific focus can be kept to note behavioral changes in children's approaches or attitudes toward science. For example, a teacher may see and record a child destroying a plant or some other living organism. If, after a discussion of this attitude with the child, the child expresses concern for the welfare of certain organisms and uses what she or he has learned to help maintain the living organisms, this would probably be evidence of an attitudinal change. Comparison of these two anecdotal records can help the teacher assess pupil growth.

Observational checklists may also be used as a device in record keeping. Each time certain behaviors are observed, they can be checked. For example, we can observe and check children's handling of science materials in relation to specific safety behaviors.

Other areas of importance in evaluating science teaching and learning can be readily detected by the alert teacher. These include the degree of pupil involvement and enthusiasm, the interests developed, and the joy of learning. These important behaviors and attitudes are not only pertinent to the success of teaching and learning a given topic but also significant in the children's future learning.

How Can Your Teaching of Science Be Improved?

Even the best teachers can improve their teaching. The following questions may be suggestive of ways in which you can improve your work with children in science. If you answer "no" to any of the following questions, you may wish to consider that aspect of your teaching and in some cases seek ways of implementing the ideas suggested by your self-evaluation.

1. Do all the children in your class have frequent opportunities to handle and manipulate science materials and equipment?
2. Are the children asked to assume some of the responsibility for the distribution, collection, care, and storage of the science equipment and materials?
3. Have your children had an opportunity to gain some experience with each of the following kinds of approaches to science learning?
   a. laboratory experimentation
   b. individual or small-group investigations
   c. cooperative investigations
   d. demonstrations
   e. science projects
   f. field trips
   g. reading to obtain needed information
   h. use of community resources such as museums and nature centers
4. Do the children have an opportunity to report the results of their investigations and projects to their class?
5. Do you ask some divergent questions (i.e., questions to which you do not know the answer, questions that may lead children to consider and explore further implications of the science that they are studying)?
6. Do you consider a child's response to questions, even though it may seem implausible or unrelated to the question, before turning to some other child for his or her response?
7. Do you help children try to derive meaning from the activities in which they engage?
8. Do you have generalized goals, such as those discussed in Chapter 1, toward which you direct your work with children in science?
9. Do you have specific behavioral objectives for some of the science activities that you undertake?

10. Do you make a systematic attempt to ascertain the extent to which the children achieve those behavioral objectives?
11. Do you keep some kind of a record of each child's achievement and growth in science?
12. Are some of the children's science activities undertaken as a result of questions that they have raised?
13. Are children sometimes led to consider the methods and processes that they have used to deal with a question or problem?
14. Are children sometimes asked to use what they have learned in science to try to "think through" what may happen when they try an experiment?
15. Are children helped to learn how to locate and evaluate sources of information?
16. Do you systematically review some of the new science materials for children that are becoming available?
17. Do you read articles and books on science, view science programs on television, and take advantage of other opportunities to become better informed in science?

How Can Your School's Science Program Be Improved?

All educational programs can be improved. The following questions have been used by many teachers, principals, and science curriculum coordinators to suggest ways in which programs can be improved. A "no" or "maybe" answer to any of the following questions may indicate areas for improvement.

1. Is there a planned program for elementary school science?
2. Is there sufficient flexibility in the planned program so that the interests and concerns of children can be explored? Can you give an example of how such an interest has been used as a basis for a science activity?
3. Are there specific behavioral objectives for instruction?
4. Are testing procedures or instruments available to help teachers ascertain the extent to which behavioral objectives have been achieved?
5. Is it possible for teachers to obtain needed equipment and materials quickly?
6. Are enough materials and equipment available so that each youngster can take part in laboratory experiments?
7. Are such audio-visual resources as films, filmstrips, recordings, slides, overhead projections, and television easily available to the teacher?
8. Are enough textual materials, preferably of several kinds, available so that all the youngsters in the classroom can have a common reading experience?
9. Are there a variety of science books at different reading levels available in the classroom for children to consult in their science studies?
10. Is a variety of science books available in the school library?

11. Is at least one children's encyclopedia available in the school library?
12. Does the school have a systematic arrangement for the utilization of such community educational resources as museums, planetariums, zoos, botanical gardens, nature centers, and aquariums?
13. Can the teacher easily and quickly arrange for transportation for field trips?
14. Does the school system have a way of monitoring the ongoing science program in order to identify problems and difficulties and suggest ways of improving the program?
15. Is sufficient time provided in the science program? Many elementary school educators have suggested that about one-fifth of the time in the elementary school program should be devoted to science.
16. Are professional books and curriculum materials in elementary school science available for use by teachers?
17. Is there a planned program of in-service education to help teachers develop proficiency in science?

REFERENCES

See Appendix 3 for a complete list of professional books in elementary school science. Here we annotate a few of the books that are especially relevant to this chapter.

GEGA, PETER C. *Science in Elementary Education,* 3rd ed. New York: Wiley, 1977. A popular generalized text for prospective and practicing elementary school teachers.

GOLDBERG, LAZAR, *Children and Science,* New York: Scribner's, 1970. An inspirational book for teachers that can stimulate teacher and pupil enthusiasm for science. Human values and respect for children are conveyed.

HURD, PAUL DEHART AND JAMES JOSEPH GALLAGHER. *New Directions in Elementary Science Teaching,* Belmont, Calif.: Wadsworth, 1968. A detailed summary of major elementary school science projects and curricula. Includes a section on problems and trends in elementary school science.

MAGER, ROBERT F. *Preparing Instructional Objectives,* Belmont, Calif.: Fearon, 1972. A programmed guide to the basic components of instructional objectives.

PILTZ, ALBERT AND ROBERT SUND. *Creative Teaching of Science in the Elementary School,* 2nd ed. Boston: Allyn and Bacon, 1974. A clear treatment of creative science teaching, including child development, teaching methods, and examples of activities.

VICTOR, EDWARD. *Science for the Elementary School,* 3rd ed. New York: Macmillan, 1975. A popular content and methods book for pre- and in-service teachers in elementary school science.

VICTOR, EDWARD AND MARJORIE S. LERNER. *Readings in Science Education for the Elementary School,* 2d ed. New York: Macmillan, 1967. A book of selected readings covering diverse topics in elementary school science.

CHAPTER 3

Teaching and the Nature of Science

"How should science be taught in our elementary schools?" This question was asked at a school board meeting at which various school programs had been described. An elementary school teacher had just given a very fine description of some of the kinds of science activities that had been developed in that particular school system. The board members had been impressed, but one of them raised this fundamental question. The teacher replied, "We try to teach all subjects in ways that are consistent with the nature and structure of the subject. It is especially important that we teach science in ways that are consistent with the nature and structure of science."

No one could argue with this statement. As a board member put it, "Science should be taught in a scientific way. Certainly, it should not be taught unscientifically." Accordingly, the remainder of this chapter deals with the nature and structure of science.

Science is the investigation and interpretation of events in the natural, physical environment and within our bodies. It is an important human enterprise. Scientists are specially prepared to undertake such investigations and interpretations, and they devote much of their professional life to it. However, science is more than what scientists do. It is important to recognize that everyone, including young children, can engage in science. They do so when they investigate and interpret events in the natural, physical environment and within their own bodies. Through science education they should grow in the ability to undertake such investigations and interpretations.

While science can be a complicated human enterprise, it is important that children begin to gain some understanding of it. In science instruction it is often useful to employ analogies to help explain that which is complicated. In explaining the nature and structure of science, we have found it useful to compare it to the nature and structure of a building. (See Figure 3 – 1.)

The three basic components of this view of the nature and structure of science are the assumptions, methods and processes, and broad generalizations of science. In the framework of our analogy, the assumptions are the foundations on which the structure is built. The methods and processes are the horizontal beams. In the structure of science, work goes on, using these methods and processes. The vertical piers are the broad generalizations of science. The methods and processes of science are used to clarify and extend these broad generalizations. Occasionally a revolutionary new generalization is generated. This is akin to adding a new vertical pier to the structure.

Various dimensions of the structure of science will be discussed in this chapter, with examples from the history of science and with demonstrations and experiments that can be used to clarify the structure of science. To make somewhat abstract ideas more concrete, it is suggested

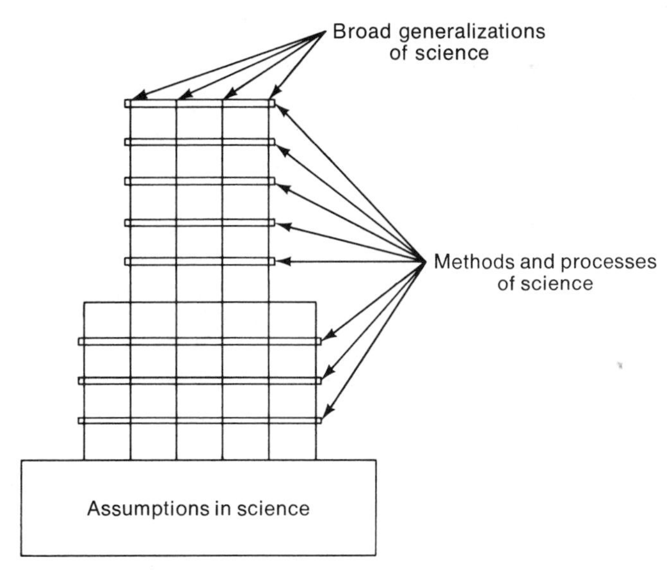

Figure 3–1 The nature and structure of science can be compared to the structure of a building.

that both examples from this history of science and science activities be considered for use with children.

ASSUMPTIONS IN SCIENCE

Assumptions are ideas or matters that we take for granted without necessarily having complete evidence to support them. In everyday life, for example, we usually assume that most people with whom we associate want to do that which is right even though this assumption may at times be expressed in strange ways. We do not have evidence for this assumption for all the people we meet, but it is generally useful to make such an assumption.

An assumption made in the sciences is that various phenomena that we encounter can be explained in some rational, consistent way. The thunder rumbles; a boom may shatter windows; a conspicuous mark may appear on a stone wall; or an airplane may disappear on a flight over the ocean. We assume that there are rational, consistent explanations for these phenomena and we proceed to investigate and seek better explanations. The very structure of science is based on this assumption.

During an electrical storm, you will often hear the loud crackling sounds of static on your AM radio. In 1930 Karl Jansky was assigned the task of studying atmospheric interference with radio and telephone communications. The static from lightning discharges were readily understood—we can induce such static with other electrical discharges. But Jansky was puzzled by the continuing static that remained even when there was no lightning or other electrical activity. What was the source of this continuing interference with radio and telephone communication?

Jansky *assumed* that there was some rational, consistent explanation for this interference. He built a directional antenna with which he could determine the direction from which the interference came. Surprisingly, he found that the interference was most intense when the directional antenna was pointed in the direction of the core of the Milky Way Galaxy. Jansky had discovered that there are radio waves emanating from regions in outer space that are picked up by antennas on earth. Since that time these radio waves have been studied very intensively, and through this radio astronomy we have learned a great deal more about the nature of the universe. Jansky made the assumption, as other scientists do, that there was a natural explanation for the radio and telephone interference that he was studying. Working on the basis of this assumption, he made a discovery that opened up a new field of science: radio astronomy.

What has happened to the streak of water? A child had wet his finger and made a wet streak on the chalkboard. (See pp. 69, 70 for further extensions of this activity.) Many hypotheses are usually suggested, but one child said, "It just disappeared." This kind of response provides an opportunity to consider and discuss a basic assumption in science. In science we cannot be satisfied with a suggested explanation like "It just disappeared." We assume that there is a rational, consistent explanation for phenomena that occur. What would happen to the structure of science if we were to accept such explanations as "It just disappeared"?

METHODS AND PROCESSES OF SCIENCE

The methods and processes of science are another dimension of the structure of science. They are among our most powerful intellectual tools. These methods are characterized by careful analysis to clarify the nature of the question or problem to be investigated. Children are acting scientifically when they try to clarify the meaning of the questions they ask. It is not possible to study all of the universe at once. Systems within the universe must be defined and studied. Children define systems when they decide what is and is not pertinent to the study of a problem. Available research must be consulted to see what others have found out about a question or problem. Children do this when they consult science books in the school library or query adults for information or opinions. Possible answers are suggested, and these hypotheses become tools for the investigation of the problem. Experiments and observations are carried out in order to get the information that will help determine whether a hypothesis should be accepted or rejected; the acceptability of a hypothesis depends on what actually happens when it is tested, not what we wish would happen. The answers we obtain are checked with the answers that others have reported in books and papers. In our elementary schools children should begin to have experience with using these methods.

Scientists do not use these methods and approaches all the time, but they tend to use them when they are operating scientifically. Usually these procedures are used when the scientist is operating in his or her area of specialization. A physicist is more likely to operate scientifically when dealing with questions and problems in physics than when dealing with difficulties at home or in business, although he or she may use them under those circumstances as well. The methods and approaches of science are among the most powerful intellectual tools we have developed, and it is especially important that children begin to gain some understanding of these methods.

Empirical Tests

In science the primary test of an idea is an empirical one: "Does it work when it is tried?" Whenever possible, ideas in the sciences are checked by direct observation or experimentation. Usually it is desirable that the observations and experiments be checked by a number of qualified scientists.

In the latter part of the seventeenth century, it was commonly believed that dead bodies, filth, or any sort of decayed matter engendered worms. To test this hypothesis the Italian scientist Francesco Redi placed three dead snakes in a box to decay. Worms soon appeared and began to devour the meat. The worms were of different sizes, which led Redi to believe that they had been born on different days. After the meat had been devoured, the worms escaped through a hole in the box. In order to find out what happened to the worms, Redi repeated his experiment using a box in which all openings were closed. He noticed the worms changed into egg-shaped objects that we now know as *pupae*. After a number of days, adult flies emerged from the pupae. Then Redi had an important idea: "Perhaps all worms found in meat come from flies and these worms, in turn, develop into new flies!"

Redi went on to test the idea that all worms found in meat came from flies by conducting a *controlled experiment*. He placed various kinds of meat in eight wide-mouthed flasks. Four of these flasks were sealed so that no flies could get near them, while the four remaining flasks were left open. Worms soon appeared in the open flasks, and flies were seen entering and leaving at will. No worms appeared in the closed flasks. Through careful observation and controlled experimentation, Redi obtained empirical evidence to test his hypothesis. The work of Redi and others laid the foundation for the scientific generalization that living things come from other living things.

This insistence on empirical tests differentiates science from several other areas of human endeavor. For an idea to be consistent with widely held dogma or pervasive belief is not an adequate test in the sciences. Similarly, majority vote is of little consequence. To test an idea in the sciences is to subject it to the rugged, demanding check of "Does the idea work when it is tried?"

Which object will strike the ground first? Children should have an opportunity to consider various kinds of tests of ideas and to experience carrying out empirical tests. For example, hold a heavy steel ball and a light cork of about the same size at the same height above the table. "If they are both dropped at the same time, which will strike the table first?" The teacher may suggest that, on the basis of greater age, college degrees, and other qualifications, his or her judgment is that the heavy ball will strike the table first. (This is reliance on authority to test an idea.) "Is this a reliable test?"

Then the teacher may ask for a vote. (This is a majority test.) "Is this a reliable test?" Finally, the empirical test of actually dropping the ball and the cork is tried.

Systems Are Selected for Study

In dealing with a question or problem in the sciences, the entire universe is not investigated and analyzed. Instead, systems of objects and structures believed to be pertinent to the question are identified. For some kinds of problems, the solar system may be the object of study. Or the system may be limited to the chemicals in a test tube. Identification of the pertinent system, an almost automatic response on the part of the trained scientist, makes it possible to deal with the problem with some hope of success.

William Harvey identified a system as he studied the problem of how blood moves in the body. It had long been known that the heart moves blood, but Harvey studied valves in the heart and realized that blood could not flow through the heart in both directions. By measuring the capacity of the heart, he was able to calculate the amount of blood that went through it at each beat. He found that more blood passed through the heart in one hour than was contained in the entire body. The blood must circulate from the heart throughout the body and back to the heart.

Harvey studied the flow of blood in the heart, arteries, and veins. He did not devote his attention to the brain, the nervous system, or the skeletal system. Instead, he identified the parts of the body that were most pertinent to the problem he was studying.

The identification of systems for study has led to specialization within the sciences. Specialization, in turn, sometimes leads to difficulty of communication between specialists, who often use special languages developed within their specialties. Many scientific advances, however, are a result of this specialization, which has allowed scientists to devote their energies to the study of particular systems.

Children should practice defining systems. Drop a drop of ink into a glass of water. Ask the children to observe carefully what happens and report their observations. Then have them note the objects they did not mention. Did they mention the clock on the wall? Was the bird that flew by the classroom window part of the system? Were the children in the adjoining room part of the system? The children, like scientists, will identify a system of objects that can reasonably seem to be related to the phenomenon under study. This is often done almost automatically by scientists. However, some important advances have been made by changing the nature of the system that is being considered.

Using and Testing Hypotheses

Hypotheses, or suggested answers, are used as tools in the investigation of questions and problems. Observations, experiments, and investigations are carried out to test hypotheses.

Charles Darwin is reputed to have said, "How odd it is that anyone should not see that all observation must be for or against some view, if it is to be of any service."

An automobile mechanic uses hypotheses to investigate why an automobile will not start. Obviously, she cannot check the entire automobile all at once. Instead, she suggests possible answers and then checks these answers, one by one. For example, she may say, "Perhaps there is no gasoline in the tank" or "Perhaps there is dirt in the fuel line." Such suggested answers—hypotheses—help the mechanic discover what is wrong with the automobile.

The scientist Charles Nicolle often visited a hospital in Tunis where there were patients suffering from many diseases, including typhus. It was well known that while typhus was very contagious outside the hospital, it seldom, if ever, was spread from one patient to another within the hospital. Why? One day Nicolle noticed at the entrance to the hospital the body of a man who had died of typhus. What was the difference between the dead man and the patients inside? The patients had been stripped of their clothing, shaved, and washed. Typhus, then, must be carried by something on the outside of the body. The body louse? Once having gained this idea—the hypothesis—Nicolle was able to prove that typhus was transmitted by the body louse and to show how this deadly disease could be controlled.

To formulate a hypothesis children must use knowledge that they have derived from previous experiences. If thinking is using what we know to deal with questions and problems, then this is what children engage in as they strive to suggest useful hypotheses. Actually, one of the great faults with much of our science instruction has been that children seldom use—and thus do not see the value of—anything that they have learned. However, when they are urged to suggest hypotheses they have to call upon their past experiences and use what they have learned. They are having practice in thinking and, perhaps, learning how to do it.

Germinating seeds in different ways. A fourth-grade class had been germinating seeds, and the seeds had been arranged in such a way as to allow them to find out if it made any difference whether the seeds were "right side up, upside-down, sideways, or 'kitty-corner.' " With little surprise to anyone, the children found that the orientation of the seeds had no effect on their germination. But then one child asked, "What would happen if a full-grown plant were turned upside-down?"

Like many children, these fourth-graders were eager to carry out the empirical test. Their first response was, "Let's try it!" But the teacher insisted that they first suggest hypotheses. "What do you think will happen when you try it?" The responses to this question helped prepare the children to observe the empirical test.

"The stem of the plant will bend and begin to grow up."

"The plant will grow straight down."

"The plant will die."

Controlled Experiments

Controlled experimental tests are an important means of investigation in the sciences. In the classic type of controlled experiment, all the factors but one are controlled. Any changes that take place must be due to the variable factor. For example, when a new kind of seed is being tested, plants from the new seed are compared with plants from other varieties. However, it is important that all varieties be grown under the same soil, water, sunlight, and cultivation conditions. If all such factors are controlled, any differences in yield must be due to the variety of seed used.

The control is an important factor if the experiment is to be a scientific one. A controlled experiment helps show what factors are causally related to the phenomena being studied.

Will a dark-colored surface absorb more heat than a shiny surface? Children should have many experiences in setting up controlled experiments. For example, one group of youngsters wished to find out whether a dull, dark-colored surface would absorb more heat than a shiny surface. They blackened one test tube by holding it in the flame of a candle. Then they put equal amounts of sand into the black test tube and a clear test tube. Thermometers were inserted into the sand in both test tubes and the tubes were supported in front of a goosenecked lamp. All conditions for both test tubes were the same (distance from lamp, amount and kind of sand, etc.) except for the color of the tubes. Any difference in the temperature of the sand must have been due to the color of the tube.

Science Has a Cumulative Dimension

Children need not start learning science where the cavemen started. They can make use of the knowledge acquired by those who have preceded them. For example, Michael Faraday discovered how to use magnets to generate electricity. He could not have made this discovery if the following had not already been achieved:

Volta—Developed the voltaic cell, which is a source of electric current.

Oersted—Demonstrated that an electric current flowing through a conductor will affect the magnetic needle of a compass.

Arago—Showed how an electromagnet could be made.

Ampere—Showed that two adjacent wires will be affected when electric currents are sent through them.

Unknown—Showed how electric conductors could be insulated.

Meteors and meteorites. Children can crawl up onto "the shoulders of giants" by using the books and other science materials available to them. A group of fourth-graders were studying meteors and meteorites. Although one of the children had a relic of what was supposed to have been a meteorite, there is a limited amount of information that can be gained from an examination of a part of one meteorite. To augment this information the children scoured all the available books in the school and public libraries. Unlike our early ancestors, who regarded meteors with ignorant awe and fearful superstition, because of their research these fourth-graders were able to view meteors with scientific understanding.

Checking Ideas and Findings

Ideas and findings in science are criticized and checked by others who are competent in the field. Imaginative new ideas provide the breakthroughs that lead to new scientific generalizations. But before these ideas can be accepted they must be subjected to the critical scrutiny of other scientists. When Copernicus published his theory that the planets revolve around the sun, the idea was subjected to very severe criticism. Now, however, the heliocentric theory, with some modifications, is generally accepted by all astronomers. This series of checks and balances is known as the public dimension of science. It was not sufficient that Copernicus publish his theory; the theory had to be examined, criticized, and finally accepted by others competent in the field.

Experimental findings must also be checked by other scientists—the results of an experiment must be accompanied by descriptions of the experiment that are sufficiently detailed so that other scientists can repeat and check the experiments. The German scientists Hahn and Strassman reported that uranium 235, when bombarded with atomic particles, split into such elements as barium and krypton. It was quickly realized that if uranium actually fissioned in this way a great deal of energy would be involved, and that this energy might be released in an explosion. When the fissioning of uranium 235 was reported at a scientific meeting in the United States, it is reported that the scientists didn't wait for the close of the meeting before rushing to their laboratories to begin checking the findings.

"No one can work as a scientist in complete isolation from other scientists." Ideas and findings are communicated to fellow scientists to be checked and criticized. Obviously, this means that freedom of communication is essential in science and that restrictions on freedom of speech and publication inhibit the continued progress of science. This is one of the reasons why scientists are often in the forefront in the battle to protect such basic liberties as freedom of speech and press and the right to hold unorthodox opinions.

Children should have a chance to communicate their ideas and the results of their experiments to other children for discussion and criticism. One way of doing this is to hold colloquiums at which children describe their work – whether it be a study on how to grow plants in various kinds of soil or an investigation of various ways of building telegraph sets – to their schoolmates. The children have to prepare their demonstrations carefully and be ready to answer the probing questions of their colleagues. Usually the teacher helps bring out the most important points connected with the demonstrations. This is one way for the children to learn a great deal of science in the traditional sense and, of equal importance, to gain a more profound understanding of the "public dimension" of science.

New Approaches to Problems

One of the important aspects of science is the attempt to view questions or problems in new and different ways. Irving Langmuir, for example, improved the electric light bulb by approaching the problem in a radically new way.

The first light bulbs were made by putting a carbonized filament inside a glass bulb and evacuating the air from within the bulb. Subsequent attempts to improve the light bulb often took the form of trying to pump a greater fraction of the air out of the bulb. Langmuir considered other ways in which the filament could be kept from burning. Rather than pumping air out of the bulb, why not pump into the bulb an inert gas that will not support burning? This was done, and a better light bulb was eventually constructed. Langmuir succeeded because he viewed the problem in a radically new way.

Viewing questions and problems in new ways is difficult:

> In this connection it is not irrelevant to note that, of all forms of mental activity, the most difficult to induce even in the minds of the young, who may be presumed not to have lost their flexibility, is the art of handling the same bundle of data as before, but placing them in a new system of relations with one another by giving them a different framework, all of which virtually means putting on a different kind of thinking-cap for the moment.[1]

[1]Herbert Butterfield, *The Origins of Modern Science* (New York: Macmillan, 1957), p. 1.

The idealized experiment is another important scientific tool. Galileo used such an idealized experiment to achieve a better understanding of motion. If a cart is pushed along a level road, it will continue to roll for a short distance after the pushing has stopped. If the road is very smooth and the bearings well lubricated, the cart will roll farther before it stops. What would happen if the road were perfectly smooth and there were no friction? The cart would continue to roll forever. By using an idealized experiment like this one, Galileo helped achieve a better understanding of the nature of motion. Later Isaac Newton stated this finding as one of his laws: "A body at rest tends to remain at rest and a body in motion will tend to remain in motion in a straight line unless acted upon by an outside force."

Taking a point of view directly opposite to the prevailing point of view, as Langmuir did, and the idealized experiment of Galileo, are two approaches that we can help our children use. For example, we can ask children to try to envision what the sky must look like at noon to children on the opposite side of the earth. Or, "What would our city be like if there were no air pollution?"

Quantitative and Precise Statements

In science an attempt is made to express ideas and findings as precisely as possible. Whenever possible, ideas and findings are stated in mathematical terms.

For example, to answer the question of how many seeds in a package actually germinate, one kind of answer would be "many or most," "few or not many." However, it is more precise to state the actual number or, even better, the fraction or percentage of the seeds that have germinated. These more precise statements can be communicated to and checked by others. Statements such as "many" or "few" are imprecise and depend on the judgment of the observer or experimenter; obviously, someone may consider "many" to be "few," and vice versa. However, when ideas and results are stated quantitatively there is greater precision and less room for confusion. Also, more precise statements are more useful. For example, a precise statement of germination rates will help determine how many seeds should be planted.

New developments in science are sometimes a result of precise statement and measurement. For example, according to the atomic theory of the time, the atomic weights of various elements should be in whole numbers. However, very careful measurements of the atomic weights of elements indicated that many elements apparently had atomic weights that differed slightly from whole numbers. How could this be explained? The measurements were so accurate that the discrepancy could not be explained by lack of precision of measurement. Investigations of these discrepancies led to the discovery of isotopes of elements. Each isotope has an atomic weight that is a whole number. However, in the form in which we study them elements are usually composed of a mixture of various isotopes of the element; the composite weight of these isotopes usually approximates, but is not exactly, a whole number. In this case precision of measurements led to a discovery of tremendous importance.

Using measurement tools. Children should be encouraged to be as precise as possible in discussing their ideas and reporting their findings. They should learn how to use rulers and measuring tapes, clocks and watches, scales and balances, measuring cups and graduates. As they use these tools for measurement, they will learn the more precise language of science and the importance of precision. More effective ways of reporting can be learned. Graphs, for example, are useful for seeing and showing relationships among data.

THE BROAD GENERALIZATIONS OF SCIENCE

The broad generalizations of science may be considered the upright pillars of the structure of science. One of the major aims of working with children is to help them gain a clearer concept of various broad generalizations of science so that they can use them in their lives.

Science generalizations are the "big ideas" of science. They are based on the cumulative experience of the scientific community. The law of gravitation—"all matter anywhere in the universe is attracted to all other matter in the universe by a force that is proportional to the product of the masses of the objects and inversely proportional to the square of the distance separating the objects"—is a well-known example of a scientific generalization. The child has seen objects fall to the earth; he or she has had experiences with this generalization.

Scientific generalizations are broad statements that help explain phenomena and serve as tools for the investigation of the new and the unknown. An apple falls to the ground; the baseball the outfielder throws to the catcher travels in a certain trajectory; and the planets move around the sun in certain paths. It is the law of gravitation, however, that helps explain the relationships among these phenomena.

The terms *concept* and *generalization* are sometimes used interchangeably. A *concept,* however, is the individual s view or mental image. For example, a child may have a concept of the distance to the moon, and this concept is peculiar to the individual. *Generalizations,* on the other hand, are statements based on the cumulative experiences of the scientific community.

In the elementary school, children must be helped to develop a more coherent understanding of the broad generalizations of science. The following are examples of important generalizations that should be included in the curriculum.

Conservation of Matter and Energy

The law of conservation of matter and energy states that the sum total of matter and energy in a closed system remains constant. Under ordinary conditions it can be stated that "matter cannot be created or destroyed" and "energy cannot be created or destroyed." Under unusual circumstances, however—in particle accelerators such as cyclotrons, where particles of matter are accelerated to velocities approaching the speed of light—the particles of matter increase in mass as energy is converted into matter. In nuclear reactions matter is converted into energy. Except under these extraordinary conditions, matter and energy can be neither created nor destroyed.

The law of conservation of matter and energy means, for example, that when two or more chemicals react in a test tube, in a burning match, or in our bodies, the mass of the products of the reaction will be the same as that of the reactants. A machine will not produce more energy than is put into it. In some ways the law of conservation of matter and energy indicates some of the limitations on what can be done. It can also be used, however, to predict what will happen when certain operations are carried out.

The law of conservation of matter and energy is often understood in an unsophisticated, but important, way by children when they say, "You can't get something for nothing." The law is used in a more precise way in balancing chemical equations or calculating the efficiency of machines.

The Second Law of Thermodynamics

The second law of thermodynamics states that heat energy can never be transferred spontaneously from a colder body to a hotter body. If some object, such as a bar of metal, is heated, the heat will flow from the hotter region to the colder one; it will not move spontaneously from the colder region to the hotter one. In refrigerators, air conditioners, and similar devices, heat is "pumped" from a colder region, such as the interior of a refrigerator, to a warmer region, but energy has to be supplied and work done. If a refrigerator is disconnected so that no energy is supplied to operate the compressor, the temperature inside the refrigerator will eventually become the same as that outside.

The science of thermodynamics is largely a study of the application of the second law of thermodynamics. Insulating materials are used in homes and schools to slow the inexorable transfer of heat from the warmer region to the colder one. On the other hand, heating systems are designed to accelerate the transfer of heat.

Transfer of heat. Children study this law of thermodynamics as they measure changes in temperature in solids, in liquids, and in the air around them and as they investigate ways in which heat is transferred in homes and throughout their environment.

Like the law of conservation of matter and energy, the second law of thermodynamics is a major pillar in the framework of science. It is a law of limitation in that it dictates certain operations that cannot be carried out. A spontaneous flow of heat energy from a cold region to a hot region will not occur. To accomplish this, energy must be supplied to the system. In analyzing energy systems, including those in which living matter is involved, the second law of thermodynamics is a valuable tool.

The Law of Mass Production

The law of mass production states that all living things tend to reproduce at such a rate that the population outstrips the food supply. The deer in some of our northern forests, for example, reproduce to the point at which they eat all available food. When there is insufficient food for all of the deer, many of them will die of starvation. With fewer mouths to feed, the food supply will once again become adequate. Reproduction, however, will continue at such a rate that again there will be an insufficient food supply.

This law of mass production apparently holds for all living things. Among plants there is a struggle for soil minerals, water, and sunlight. Usually the plant species that is best adapted to survive in a particular environment will be dominant in that environment. Among insects there is a very high reproductive rate. A large fraction of the insect eggs may not even hatch, but the number of eggs is usually so high that the small fraction that hatches usually leads to a total population of such a size that it has an inadequate food supply.

The law of mass production. Children can see many evidences of this mass production in the environment. The fluffy tuft of the dandelion contains hundreds of seeds from which new dandelion plants can sprout. Similarly, the female frog can deposit hundreds of eggs, which will become tadpoles and then adult frogs. Trees produce thousands of seeds that are dispersed in a variety of ways.

TEACHING SCIENCE: HOW AND WHAT

As we work with children, we must be concerned with both "how" we teach and "what" we teach. "What" children learn may be largely determined by "how" we teach. On the other hand, "how" we teach, especially in science, should be determined by "what" we are teaching. The teacher endeavors to gain a profound understanding of the "what" and resourcefulness and skill in the "how."

In science the "what" and "how" of teaching are intimately intermingled. Perhaps the most important goal in working with children in science is to help them acquire a better understanding of the various methods and processes of investigation. However, these understandings can probably be developed only through demonstration of various approaches and the use of methods that are characteristic of the sciences. The methods of teaching science must be consistent with the methods and processes of science—to teach science effectively, it must be taught scientifically.

The case of the candle and the jar. Certainly, one of the best ways to teach science scientifically is to encourage children to identify significant questions and problems in science and then work with them as they investigate these problems. Much of this work can take the form of cooperative investigations in which pupils and teacher work together to clarify questions and solve problems. For example, in one classroom a child demonstrated that when a large glass jar was placed over a burning candle, the candle eventually was extinguished. The child explained that there is a part of the air called oxygen that is necessary for the candle to burn. When much of this oxygen has been "used up," the candle can no longer burn. However, one of the children asked, "What would happen if the glass jar were raised so that some air could enter at the bottom? Would the candle go out?" The teacher asked him to clarify what he meant by his question. He placed the glass jar on two blocks so that air could enter the jar. Next the teacher asked the class to predict what would happen to the burning candle under these conditions. In other words, she asked the children to use what they knew about gases and the nature of burning to form a hypothesis. They then tried the experiment and tested their hypothesis. (See Figure 3–2)) After the experiment had been completed, the teacher called the children's attention to the various methods that they had used to investigate and try to answer a question. Through this cooperative investigation of a relatively simple, but important, question, the teacher demonstrated some of the characteristics of a scientific approach.

Teaching science scientifically should help children develop more sophisticated concepts of the broad generalizations of science. Through their investigation of the burning candle under a glass jar, the children

Figure 3–2 When the jar is raised so that the air can enter, will the candle go out?

should have developed a clearer understanding of the nature of burning. For example, they can learn that only a fraction of the air is utilized in the burning of the candle. By swishing limewater, which turns a milky color when exposed to carbon dioxide, around in the jar, they can discover that carbon dioxide, previously not present in very large amounts, was formed as a product of the process of burning. They can also discover that warm gases (in this case gases rich in carbon dioxide) tend to rise and cooler gases tend to descend. This is probably the principal reason that the candle will be extinguished even though the jar is raised so that fresh air could, but does not, enter through the neck of the jar. These are examples of important generalizations concerning the nature of burning and the behavior of gases. One of the important results of our teaching should be that our children gain a clearer concept of some of these major generalizations.

REFERENCES

BUTTERFIELD, HERBERT. *The Origins of Modern Science.* New York: Macmillan, 1957. This is a useful history of some important dimensions of modern science. Some of the statements and approaches to the history of science are especially useful in work with children.

JACOBSON, WILLARD J. ET AL. *Science: A Way of Knowing.* New York: American Book, 1969. A discussion of the nature of science, with associated science activities that can be used with older children.

KUHN, THOMAS S. *The Structure of Scientific Revolutions.* Chicago: University of Chicago Press, 1970. It is suggested that occasionally revolutionary new ideas occur in the sciences. The broad generalizations of science discussed in this chapter are examples of ideas that were revolutionary when they were first proposed.

SCHWAB, JOSEPH J. AND PAUL F. BRANDWEIN. *The Teaching of Science.* Cambridge, Mass.: Harvard University Press, 1967. Two leaders in science education discuss the nature of science and its implications for teaching.

TAYLOR, F. SHERWOOD. *An Illustrated History of Science.* New York: Praeger, 1959. Of special value to teachers are the many illustrations of key scientific experiments.

Science and Perceptual Development in Young Children

CHAPTER 4

Every day Paul walks through a beautiful park on his way to school. He sees nothing, hears nothing, smells nothing, touches nothing; he seems completely oblivious to his environment. Paula, on the other hand, observes the colors of the flowers and trees and notes the changes that take place from one day to the next. She hears the rustle of the leaves and the scratchings of the crickets. She kneels to catch the scent of the dandelion and compares it with that of the morning glory. She feels the sticky path of a snail and the sand of a small anthill. Which of these two youngsters will learn more from the environment? Which will enjoy life most?

The senses are as fundamental to the young child learning about the world as they are to the scientist studying chemical reactions in a laboratory. Just as the scientist must note color changes, odors, sounds, and textural changes, so too the young child learns by perceiving sights, sounds, odors, and textures in the immediate environment. A child's first impressions of physical reality are received through the senses. As sensory information is perceived, the brain acts on it to form thoughts or mental images. Perception is not a passive process—we do not just see, we look; we do not just hear, we listen. By relating current sensations to past impressions and experiences, children (and adults, for that matter) form their notions of the environment. Careful sensory observation is one of the basic skills of science work, just as knowing the sounds of the letters of the alphabet is one of the basic skills in learning to read. Through carefully planned science experiences, children can become more perceptive of the sights and sounds, odors and textures that surround them in their environments. As they become more perceptive, they enhance their ability to learn.

In the development of children's ability to perceive, the goal goes beyond the refinement of the child's direct sensory skills. We want to superimpose intellectual activity on the sensory experience so that children can recognize, differentiate, or operate in some way on the stimuli. For example, can children distinguish the smoother of two rocks? Can children compare the odor of the perfume they are sniffing to the odor of a lemon peel? These are examples of activities that develop the ability to perceive and can foster the employment of higher level thought while working with concrete materials. The following is a suggested hierarchy of skills to be developed while working with such materials:

1. Children should learn to identify and describe the physical characteristics of what they are perceiving—a lemon is yellow; it has a rough skin; it tastes sour, etc.
2. Children ought to be aware of the sense or senses they are using as they examine an object. If they are given an unfamiliar object to "play with," do they merely examine it with their eyes? Are other senses involved? Certainly they are receiving some information about the texture of the object if they hold it. If the object is dropped or moved along some surface, a sound is likely to be emitted.
3. Children should be encouraged to recognize similarities and differences among objects perceived. For example, two apples are similar because they are both red and are similar in form and taste, but they may be different sizes, different shades of red, etc. Two whistle sounds may have similar quality and sound when compared to a foghorn, but one whistle sound may be higher in pitch than the other.
4. Young children should also begin to develop the ability to order materials along some continuum. Objects or materials are ranked according to the extent to which they exhibit a particular property. Young children can arrange a series of cardboard dolls from shortest to tallest. A variety of sandpaper grades may be ranked from smoothest to roughest.
5. Materials may be sorted or classified according to various criteria. The very simplest form of sorting is binary (or yes–no) sorting. Buttons may be either black or not-black, two-holed or not-two-holed, rimmed or not-rimmed. Objects may have a distinct odor or no odor at all.
6. As children accomplish simple sorting, they should move on to more complex types of classification, e.g., sorting into three or more groups according to qualitative distinctions. A group of

shapes may be sorted by color, form, or size. Acorns, leaves, peanuts, buttons, and blocks may be sorted by kind or type.

These skills represent basic intellectual operations that can be promoted by utilizing materials that appeal to the senses. Activities may be devised for each of these skills, in each of the five senses. A program for young children has been developed using similar perceptual operations.[1] A diagram outlining the skeletal organization of this program appears in Figure 4–1. Each box, or cell, represents an instructional unit. For example, in the unit "Touching 5," activities are described in which pupils have to sort objects by texture according to a binary scheme.

	Content Areas				
Perceptual Skills	Seeing	Hearing	Smelling	Tasting	Touching
1. Identifying and describing the properties of the stimulus materials					
2. Identifying the sense or senses used in examining a given object					
3. Identifying similarities and/or differences in a set of stimulus materials					
4. Identifying stimuli on the basis of the extent to which they exhibit a particular property.					
5. Sorting objects according to a binary classificatory scheme					
6. Sorting objects into more than two groups according to qualitative distinction					

Fig. 4–1 Sequence of instruction for the Perceptually Oriented Preschool Science Program

[1]Abby Barry Bergman, "The Development and Implementation of a Perceptually Oriented Preschool Science Program," Doctoral dissertation, Teachers College, Columbia University, 1975.

In this section a variety of activities are described that can be used to help promote several intellectual operations. All of the activities involve materials that can easily be obtained at a grocery or hardware store.

Before attempting to direct the children's investigations with the materials, give them ample time to freely explore or "play with" the objects. Listen to their questions and direct them back to the children so that they can learn to seek their own answers and conduct investigations to find solutions. The activities are easy enough so as not to unduly frustrate children between the ages of 4 and 8, yet difficult enough to be challenging and interesting. Be alert to opportunities to enrich these experiences. Ways can be found to pose new questions and to adapt these activities to children of a variety of ages and interests. All of the activities that follow may be accomplished individually, in small groups, or with an entire class by calling on different youngsters to perform the required operations. A general guideline for all perceptual work with children is to first work with gross, easily discernible differences in materials and then move on to subtler, finer differences and distinctions.

IDENTIFYING AND DESCRIBING MATERIALS (PERCEPTUAL SKILL 1)

As children begin to work with materials, their first task should be to describe the materials they manipulate. This may be done with ever-increasing accuracy.

Shape characteristics. Prepare a variety of triangles, squares, rectangles, and circles from construction paper in different sizes and colors. Ask the children to describe each shape in terms of number of sides, number of "points," color, and so forth. Provide new terms when they are not part of the youngsters' vocabularies. Talk about the relative sizes of the shapes. Which ones are larger or smaller? In order to find out whether the children have retained the information, ask them to describe the shapes again a few days later.

Identifying sounds. Ask the children to look aside or turn around while you drop objects on the floor. Try items that are familiar—a fork, a pencil, a ball, and the like—and see if the children can identify the objects from the sounds they produce as they hit the floor.

Children can also be taken on "listening walks" in which they walk outdoors or in a school hallway and identify as many sounds as they can. A careful listener outdoors may hear birds, airplanes, automobiles, truck horns, and so forth.

Smelling objects. Collect a variety of objects that have distinct odors. They might include orange peels, perfume-soaked cotton balls, mothballs, or cloves. Introduce the objects one at a time to the children and ask them to describe the object, including its odor. The children's existing language patterns should be accepted initially without correction (e.g., an odor may be "yucky"). After the children have had time to explore the materials, it is appropriate to provide new terms to describe odors (e.g., perfume may "smell like flowers").

Introduce an odorous substance to the children a few days later and ask them to describe it. Do they include a comment about odor in their description?

A tasting party. Arrange a "tasting party." Give the youngsters samples of foods representing each of the basic kinds of taste: candy for sweet, a pretzel for salt, a lemon wedge for sour, and a piece of baking chocolate or coffee for bitter. Give the children a sample of each food to taste and ask them to describe its taste. After a while, help them with the appropriate terms for each sensation. Encourage them to talk about their preferences and why they have them.

A texture box. Obtain a cardboard box or a small carton and begin collecting scraps or small items that have interesting textures. Examples include aluminum foil, corduroy, velvet, wooden sticks, and so forth. Cut a hole in the top of the box large enough for a child's hand to pass through. Ask a child to reach into the box and, without looking, grab an object and describe it as well as he or she can. Encourage the child to try to identify the object before it is pulled out of the box. By changing the objects in the box, you can maintain children's interest in this activity for weeks or months. The children should begin to show greater facility for describing textures after they have played this game a number of times. (See Figure 4–2.)

IDENTIFYING THE SENSE OR SENSES USED IN EXAMINING AN OBJECT (PERCEPTUAL SKILL 2)

Children should develop an awareness of the organs or body parts responsible for each mode of sensory perception. Pupils should be encouraged to employ a variety of senses as they inspect objects and materials.

Figure 4–2 A child reaches into a box, describes a texture, and then pulls it out.

What's inside? Put three or four pennies in a milk container or small box. Have the youngsters try to guess what is inside of the box without opening it. After the children make their guesses, ask them what part (or parts) of their bodies helped them arrive at their guess. Certainly, they ought to mention their ears.

An orange is an orange. Give each child a small slice of orange and ask the children to write down as many observations of the orange as they can. There are probably dozens of observations that could be made. Next to each observation have the youngsters mention the sense or senses used in making the observation. How many senses did the children use? Did any of the observations deal with the odor or taste of the orange?

IDENTIFYING SIMILARITIES AND DIFFERENCES (PERCEPTUAL SKILL 3)

The ability to identify similarities and differences in materials is a basic information organization skill. In order to detect such distinctions, the children must concentrate on the property the objects have in common before they can distinguish the differences. For example, given a set of various-sized red triangles, the children must first realize that all of the shapes are triangles and that all of them are red; then, in looking for differences among the shapes, they can focus on the size differences.

Before attempting any of the following activities with youngsters, it is important to determine whether or not they understand the meanings of the terms *same* and *different*. You may conduct a simple test to check this. Give a child three blocks, two of the same color and the third of another color. Ask the child if all of the blocks are the same. If the response is no, then ask what is different about them. If the child does not mention any difference, explain the use of the terms *same* and *different*.

Finding similarities and differences in coins. Give the children some pocket change to play with for a while. Then give each child two coins of the same denomination but of different dates and degree of wear. Ask the children to carefully examine the coins and ask them what is the same about the coins. Next ask whether there are any differences between the coins. There may be some slight color differences as well as differences in degree of wear and date. Older children may notice different mint marks.

Hearing similarities and differences in sounds. Obtain six small boxes (small milk containers from the school lunchroom will do) or 35 mm film cans. Put some rice into two of the containers, some lima beans into two others, and two pennies into the remaining two. Staple the milk containers at the top or seal the boxes. Give a group of children the six containers and ask them to try to find the pairs (i.e., sets of two containers filled with the same objects) by listening to the sounds produced by each as it is shaken.

Repeat the activity using different materials as fillers. Subtler differences in sounds may also be introduced.

Finding similarities and differences in odors. Obtain small boxes or containers as in the preceding activity, but this time pierce holes in the containers with a nail. Fill two of the containers with cloves, two with perfume-soaked cotton balls, and two with cinnamon, pepper, or some other spice. Ask the children to match the containers by smelling each one. This "game" may be repeated over and over again using different odor-producing substances. Young children generally enjoy using their sense of smell.

A texture matching game. Obtain a cardboard box and cut a hole in the top of it just large enough for a child's hand to pass through. Collect a variety of texture samples (e.g., corduroy, velvet, sandpaper, aluminum foil, etc.) and cut each sample in half. Put one half-sample of each texture in the box and then place the remaining halves, one at a time, in a child's hand as he or she reaches into the box with the other hand to try to find the same texture. When the child finds a similar sample, he or she should pull it out of the box and check visually to see whether the sample appears to be similar. Interest in this activity can be maintained by changing the materials. Some interesting textures are flannel, fur, waxed paper, cotton quilting, and burlap. (See Figure 4–3.)

IDENTIFYING THE EXTENT TO WHICH PROPERTIES ARE EXHIBITED (PERCEPTUAL SKILL 4)

Characteristics or properties of materials are expressed to greater or lesser extents. One doll may be longer than another; one piece of sand-

Figure 4–3 A child reaches into a texture box with her right hand in order to match the texture she is holding in her left hand.

paper may be coarser than another; one soft drink may be sweeter than another. Identifying the extent to which certain characteristics are exhibited helps children establish relationships among objects. The activities that follow also help develop new vocabulary and expression, particularly with regard to comparative adjectives (e.g., *rougher, harder, longer, sweeter,* etc.).

Finding the longer of two objects. Collect a variety of object pairs that are essentially the same except for their length. Some possible examples are pencils, drinking straws, rulers, and the like. Give the object pairs to a group of youngsters and ask them to identify the member of each pair that is longer. If the children cannot detect differences in length by visual inspection, have them place the objects with their ends down on a table or other surface to compare length. (See Figure 4–4.)

Identifying the heavier of two objects. Try to find a variety of objects that are similar in dimension or shape but different in weight. Such materials may be made by filling same-sized boxes, milk containers, medicine vials, 35 mm film cans, or other containers with objects of different density. For example, one container may be filled with paper and another with pebbles. Have the child determine by hefting (lifting the objects, one in each hand, at

Figure 4–4 A child compares the length of two pencils by placing their ends on a table top.

the same time) which container is heavier. This activity becomes increasingly difficult if the objects to be hefted vary in size or overall dimension, since the relationship between weight and volume can be deceiving. For example, hefting a 1 lb fishing weight and a 1 lb bag of cereal may become confusing, since cereal will undoubtedly take up more space than the fishing weight. Similar-sized rocks may be hefted and the actual weights checked on a balance. (See pages 125–126 for the construction of a simple equal-arm balance.)

Identifying the sweeter of two drinks. Prepare two samples of water or tea for children to taste. Do not sweeten one sample, but put some sugar in the other. Offer the two samples to the children, giving them the unsweetened one first. Ask them to select the sample of tea or water that is sweeter. (A group of children may sample the same cup of liquid if each child is given a drinking straw.)

Identifying the rougher of two objects. Collect a variety of texture samples (e.g., plastic, rocks, flannel, burlap, etc.). Present the materials to the children in pairs and ask them to identify the smoother or the rougher of two objects. Two rocks may be compared for their smoothness or roughness. Children should have many opportunities to determine these comparative qualities of objects. As they practice such activities, their identifications should become more precise and accurate.

BINARY SORTING (PERCEPTUAL SKILL 5)

Classification is a basic skill in science, as it is in practically every subject area. The use of classification systems is an efficient and economical way of approaching and analyzing information. In the activities that follow, children use a particular type of classification called binary sorting. The children must divide objects into two, and only two, categories. (This is often called yes–no classification.) An object is either red or not-red; a texture can be soft or not-soft; a taste may be salty or not-salty; and so on. This type of sorting is basically perceptual in nature, yet it goes a step beyond purely perceptual sorting in that a mental judgment must be made; the children must consider each material and decide which group it belongs to.

These activities can be most exciting for children and teachers. It is suggested that you ask the children why they placed a particular object in a particular group, not only to find out whether they can recite the rule being used at the time but, more important, to get some clues as to how the children are approaching the task. The children may devise new criteria for grouping, and this should by all means be encouraged. (See pages 8–9 for additional sorting activities.)

Sorting buttons. Obtain an assortment of buttons. Spread them out on a table and ask a group of children to sort them into two groups according to various criteria (e.g., white and not-white, two-holed and not-two-holed, rimmed and not-rimmed, etc.). After the children have had several experiences with this activity, ask them to devise their own criteria for dividing the buttons into two groups.

Sorting by odor. Collect a variety of objects of which some have a distinct odor and others do not. Some examples are a piece of perfumed soap, a pipe cleaner, a clove, a wooden bead or block, a small ball of aluminum foil, a peppermint candy, and a small ball of clay. Discuss with the children the fact that some objects have an odor and others do not. Ask a child to sort the objects into two groups on the basis of whether or not each has an odor. Ask the children to name the objects in each group. Discuss the odors. Can the children devise other ways of sorting the objects?

SORTING OBJECTS INTO MORE THAN TWO GROUPS (PERCEPTUAL SKILL 6)

In the following activities children are involved in sorting objects by similarity of quality, often called sorting according to *qualitative distinction.* For example, given several geometric shapes, a child might place all of the circles in one group, all of the triangles in another group, all of the squares in a third group, and so on. Objects may be sorted by similarity of color, size, texture, taste, odor, and so forth.

Whereas in the previous activities the children were instructed to sort objects into two and only two groups, here they sort objects into more than two groups.

Sorting pocket change. A rather simple sorting activity is to merely ask the children to sort a pocketful of coins. The obvious categories would be pennies, nickels, dimes, and quarters, but other categories can be devised (e.g., color, degree of wear, etc.).

Sorting objects by sound. Six or eight small boxes, milk containers, or film cans can be filled with a variety of objects. For example, fill two with rice, two with paper, two with pennies, and so on. Have the children arrange the containers in groups according to similarity of sound.

Sorting liquids by taste. Prepare six small paper cups and fill two with lemonade, two with orange juice, and two with apple juice. A piece of paper may be used to cover each cup, or the liquids may be disguised with food coloring so the children cannot sort the liquids by color. Give each youngster a sip from each cup, using a straw, and have them arrange the cups in three groups according to similarity of taste. Discuss the tastes and see whether the children can identify them. Is the lemonade more sour than the orange juice?

Sorting textures. Obtain three shoe boxes or other boxes with similar dimensions. Line the bottom of one box with corduroy, the bottom of the second box with sandpaper, and the bottom of the third box with creased aluminum foil. The materials lining the boxes should be fixed in place with glue or paste. Cut a hole in the side of each box large enough for a child's hand to pass through yet small enough so as not to reveal the texture inside. (If the texture is visible, a small black cloth can be draped over the opening of the box.) Prepare three small samples of each texture for sorting into the boxes. The samples of aluminum foil should be mounted on cardboard.

Ask the children to touch the texture lining each of the boxes. Then have them describe their sensations as they feel the linings. Give a child one of the texture samples; ask him or her to feel it and then place it in the box that is lined with a material that feels the same as the sample he or she is holding. Repeat this procedure for each of the remaining samples.

CONCLUSION

In the activities described in this chapter, sensory materials are used in a variety of ways. Children are asked to make perceptual judgments about these materials. Involvement with concrete objects is a basic component of science work with young children. It is through such activities that the basic tools of science inquiry are developed. The establishment of these fundamental science skills helps develop sensory and perceptual readiness for more complex science investigations.

REFERENCES

BRUNER, JEROME. *The Process of Education.* Cambridge, Mass.: Harvard University Press, 1960. A classic statement on curriculum, teaching, and readiness for learning.

ELKIND, DAVID AND JOHN H. FLAVELL, EDS. *Studies in Cognitive Development.* New York: Oxford University Press, 1969. A collection of statements and research reports dealing with cognitive development, with considerable attention to perceptual learning.

GAGNÉ, ROBERT. *The Conditions of Learning.* New York: Holt, Rinehart and Winston, 1977. An essay expounding a behavioral theory of education. Learning is considered as a change in behavior, and the steps in this change are discussed.

GIBSON, ELEANOR. *The Principles of Perceptual Learning and Development.* Englewood Cliffs, N.J.: Prentice-Hall, 1969. A comprehensive work on perceptual development that takes account of several major theories. Primarily for specialists.

HYMES, JAMES L., JR. *Teaching the Child Under Six.* 2nd ed. Columbus, Ohio: Charles E. Merrill, 1974. A practical book for teachers of young children touching on many important issues.

PIAGET, JEAN. *Science of Education and the Psychology of the Child.* Translated by Derek Coltman. New York: Viking, 1971. In this volume Piaget addresses the issue of educational approaches, issues, and programs and criticizes many of these in light of his own theories.

Reading, Language Development, and Science

CHAPTER 5

Imagine a world without language! How would we communicate with one another? How could we share the knowledge we gain or the experiences we have had? While forms of communication are used by other species of animals, a highly developed language is unique to humans. Bees, dolphins, birds, apes, and other creatures do have ways of communicating, but in no other animal is communication as sophisticated or advanced as it is in the human.

Learning science is intimately associated with language development and learning to read. A basic, yet specific, vocabulary is helpful as children investigate science. Such terms as *longer* and *shorter, more than* and *less than, rougher* and *smoother* serve as a kind of shorthand for describing and comparing objects and events. Just as certain basic perceptual skills are necessary for a child to learn to read, so too there is a set of basic language skills that constitutes an integral part of science learning.

The relationship between science and language (either verbal or printed) is a mutually reinforcing one. Not only are certain linguistic underpinnings needed for science learning, but science experiences themselves promote language development and reading readiness skills in young children, for it is in science that children gain concrete experiences that form a basis for the use of language and symbols.

Language is a natural activity. Children seem to be able to learn a language even in the face of dramatic and severe handicaps. The average child acquires a vocabulary of some 2,000 words by the time he or she is 5 years old—indeed an amazing accomplishment.

Certain aspects of language development have important implications for the elementary school teacher. These include the ability to speak correctly, the acquisition of a conceptual vocabulary, the relationship between thought and language, and the acquisition of functional language skills. A brief discussion of each of these aspects of language development follows.

Speaking Correctly

The ability to speak correctly (i.e., according to grammatical rules) is an important area in the development of children's language competence. We often mistake poor language use for narrow experience with grammatical rules. When a young child says "foots" instead of "feet," he *has* internalized and used the proper rule for the formation of plurals, but he is not yet familiar with all of the exceptions. Through constant exposure to proper usage and instruction, the grammatical rules and their exceptions become part of the child's language.

Proper usage can also be aided by deliberate syntactical patterning while working with children. When showing a child a red ball, the teacher may say, "This is a large red ball." Then the child ought to be asked what the teacher is holding and encouraged to respond with a complete sentence. There are specific, commercially prepared readiness programs that attempt to structure syntactical patterning and language development in young children.[1]

Conceptual Language

The acquisition of a broad conceptual vocabulary is another important concern in language development. A *conceptual vocabulary* is defined here as a group of words that demonstrate a concept. Some examples are *over* and *under, in front of* and *behind, up* and *down, hot* and *cold, above*

[1]Lloyd M. Dunn and James O. Smith, *Peabody Language Development Kits* (Circle Pines, Minn.: American Guidance Service, 1967); Lloyd M. Dunn et al., *Peabody Early Experiences Kit* (Circle Pines, Minn.: American Guidance Service, 1976); Siegfried Englemann, Jean Osborn, and Therese Englemann, *Distar Language* (Chicago: Science Research Associates, 1976).

and *below, sweet* and *sour,* and so on. Children can learn such words in casual or informal discussion during play or small-group activity where free interchange is possible. The key here is that *real* objects ought to be used when demonstrating such conceptual words. For example, if the terms *in front of* and *behind* are under consideration, a doll and a toy house may be used as props. Place the doll in front of the house and say, "The doll is in front of the house." Put the doll in varying positions with respect to the house and, while doing so, use the appropriate terms. For the terms *hot* and *cold,* have the child place her hand in the stream of water running from a faucet and use the proper words as the temperature of the water is adjusted. The important aspect of such language training is that vocabulary should be introduced or used *while* the concept is being demonstrated, not before or after the demonstration.

The Relationship Between Thought and Language

The relationship between thought and language has been researched for years, and two major positions have emerged from this research. Vygotsky, a leading student of this subject, has argued that language and thought are intimately associated and that at about 4 or 5 years of age a child's internal speech controls and organizes his or her thought. Piaget has asserted that the origins of intellectual functioning are to be found in the child's actions and operations with objects, and not necessarily in language.[2] Although Piaget has maintained that language does not play as prominent a role in logical or intellectual development as Vygotsky believes it does, he has admitted that language is symptomatic of underlying thought (i.e., language is the thought or notion that is expressed). As ideas are verbalized, the ideas themselves tend to become clearer in our minds. Often when one searches for the right words to express an idea, the idea becomes clearer to us. Thought and language thus are constantly interacting with each other.

It is also important to note that language development and the development of logic and reasoning do *not* proceed at the same rate. In the young child verbal abilities often lag behind conceptual or logical abilities. The youngster may have internalized more than he or she is able to express in words. This can be confusing to those of us who work with youngsters because it is sometimes difficult to "get a handle" on what children know and can do. For example, in early science experiences many children are asked to choose two of three objects (e.g., blocks) that are the same color. Certainly, unless the child is severely colorblind, a 2- or 3-year-old is perceptually capable of identifying the

[2]Herbert Ginsburg and Sylvia Opper, *Piaget's Theory of Intellectual Development* (Englewood Cliffs, N.J.: Prentice-Hall, 1969), pp. 171–72.

blocks that are the same color. However, such an activity rarely tests perceptual acuity; rather, it is a language test. Does the child know the meaning of the word *same?* It is important to make sure children understand the terminology involved as they work with activities in which they are asked to cross out the picture in a series that "does not belong" or mark the pictures in an array that are "the same."

Functional Language Skills

A fourth area of concern in the young child's development of language is the acquisition of functional language skills. This is a crucial area of language development and the one that is most intimately allied with a young child's science experiences. Functional language skills, as defined here, are the abilities that demonstrate that a child can follow directions, perform operations with materials when given instructions, communicate the results of investigations, and so forth. In short, functional language skills are essential to children as they are guided through activities and investigations.

The development of functional language skills can be likened to the development of the process approach in science. The process approach, a deliberate instructional program designed to encourage a "scientific attitude" in dealing with physical reality, is basically the way in which a scientist works in a laboratory. On a sophisticated level this deals with the formulation of hypotheses, the outlining of a research procedure, the conducting of experiments and tests, the formulation and substantiation of conclusions. Part of this process presumes a curious nature and an inquiring mind, but these terms are often misunderstood, if not sloppy. Science educators have specified appropriate science process skills that children can learn at any level, from preschool to graduate. For the youngest child in school, some of these skills might include carefully observing an object, describing the characteristics of objects, sorting objects, following verbal instructions, following a laboratory procedure, recalling the sequence of actions in a procedure, and so forth. These are similar to important aspects of language work with children. In terms of following instructions, for example, a child may be asked to stand up, get the box that says "batteries," give one to each child, put the box back, and sit down. Children should initially follow one- and two-stage commands and gradually work up to following a relatively complex sequence of instructions.

In working with children in science, it is important to ascertain whether the youngsters understand what they are supposed to do. When asking children to perform a particular operation, the task may have to be rephrased several times until it is finally understood. For example, a child who is asked to sort a collection of objects may not know the meaning of the word *sort*. The question should be rephrased: "Put the objects into groups" or "Put all the things that are alike in some way in the same pile." It is through such activities and experiences that children grow in the ability to understand and use language. Individuals who work with young children should try to determine whether their terminology is clear.

As they develop functional language skills, children also need to acquire a bank of terms to use in accurately describing their observations. Information received through the senses often needs to be communicated. As pointed out earlier, use of such terms as *large* and *small*, *short* and *long*, and *rough* and *smooth* will aid the child in describing objects and phenomena. It is important for children to use as many real objects as possible when they demonstrate or describe materials or events.

Accurate description of sensations within the body is another major functional language skill. If a child is feeling ill and is asked to describe how and where "it hurts," it is most helpful if he or she has an adequate vocabulary for describing internal sensations. Teachers can check this ability by asking children to run in place for a minute and then having them describe how they feel after they stop. Imagine how parents or school nurses must feel when trying to accurately pinpoint a child's symptoms!

Communication of the results of experiments and investigations is another important language skill. Science is for humanity; it is for all citizens. As discussed in Chapter 1, scientists communicate their findings through informal discussions, in scientific journals, at colloquiums, and, it is hoped, through more widespread means so that important results reach the general public quickly. Just as the chemist in a laboratory records and describes the reactions produced when two chemicals are mixed, so too children in elementary school should describe their own experimental results. A bubbling reaction may fizz, smoke, or explode. The scientist notes these results. When a young child mixes flour and water to make play dough, he should be able to describe how the dough feels and how it changed from a powdery substance to a gooey mass.

It often seems that getting children to relate the details of such experiences is like pulling teeth. Describing an experience in the past may be quite difficult for a youngster whose vocabulary is limited. It is hard for us to know the frustration of a child who is hampered by an inadequate vocabulary. As adults, we can try to experience some of this frustration if we eat a piece of lemon and try to describe its taste without using the term *sour*. It is not an easy task. Often the only way to accomplish it is to use an analogy. For example, when asked to describe the odor of a mothball, one young child replied, "It smells like the stuff my mother cleans the bathroom with."

FOSTERING THE DEVELOPMENT OF LANGUAGE SKILLS

Language development is a slow, gradual process, but there are some specific things we can do to help children along. First and foremost, children should be provided with a wealth of experiences with concrete materials. This is expecially important in the home, nursery school, and kindergarten, but children's need to manipulate objects does not end there. Children should continue to have such experiences throughout their school years, and those experiences should be incorporated into as many learning episodes as possible. The materials children use should be labeled, and their experiences with those materials should be discussed. Elicit pupil descriptions of objects and events, initially accepting the child's existing vocabulary, no matter how unconventional it may seem. Children who have been provided with a fund of firsthand experiences generally have also developed the vocabulary necessary to describe such experiences.

Teachers can study the language patterns used by children. Pupils often respond to questions with one-word replies. If a teacher holds up a ball and asks, "What is this?" children will often respond, "Ball!" This is a fine opportunity to say, "Yes, it is a ball." Children who hear complete sentences used are more likely to use complete sentences in their own speech.

Introducing basic taste words. Give each child a small lemon wedge to taste. As the children are sucking on the lemon, elicit descriptions of its taste. This is an opportune moment to introduce the term *sour.* The same procedure may be employed while offering the children a potato chip *(salty),* a few grains of instant coffee or tea *(bitter),* and a piece of chocolate *(sweet).*

Learning body parts. Several games may be used to help children learn the names of body parts as they touch them (e.g., "Simon Says"). When working with older children, technical names may be used, such as *patella* for kneecap, *femur* for hipbone, and so forth. Again, the important aspect of this activity is introducing the names of body parts *as* the children touch them.

Object guessing game. Describe an object in words (basically adjectives) and ask the children to guess the object. For example, the teacher might say, "I see an object. It is in the back of the classroom. It is red. It is smaller than a breadbox. What is it?" The child who first guesses which object is being described may then describe another object in the room. Variations on this game involve letting the children know if they are "hot" or "cold" (ile., close to or far from the object).

Experience charts. Ask children to dictate the sequence of events in a story about the procedure for constructing a toy or piece of apparatus. Make an experience chart based on the children's dictated sentences large enough for the whole class to see. Some children may want to illustrate the chart. The story may be read and reread so as to review and reinforce new words, concepts, or procedures. (See Figure 5–1.)

Working with tactile sensations. Rub objects across the face (or arm) of a child. Try objects that display a variety of texture types, such as a cotton ball, a piece of sandpaper, a piece of crumpled aluminum foil, and the like. As the object is rubbed against the child's skin, ask him or her to describe its texture or the sensations he or she is feeling. As distinct materials are tried, appropriate terms such as *smooth* and *rough* may be introduced or reviewed. Children in the early elementary grades enjoy trying to guess letters of the alphabet as they are "printed" on their backs with a finger. It may be surprising to find that some youngsters who do not excel in academic areas are particulary keen in activities involving tactile sensations.

Two general guidelines for building vocabulary in the early elementary grades are the following: (1) Use real objects whenever possible and (2) try not to substitute pictures for real objects if the objects themselves are readily available. It is also important to read to children. This activity is often neglected after the second or third grade. A lively, fast-moving story or science book will capture the attention of most children in the elementary grades and help enrich their verbal imagery and expand their vocabulary. Have children make up their own stories or write down their own procedures for others to follow. Summaries of activities and investigations not only reinforce the concepts developed but also help children gain practice in expressing themselves in writing and increase their skill in sequencing events.

The tape recorder should not be overlooked as a tool for language development. Children may record summaries of investigations, conduct interviews on topics of civic importance, and demonstrate environ-

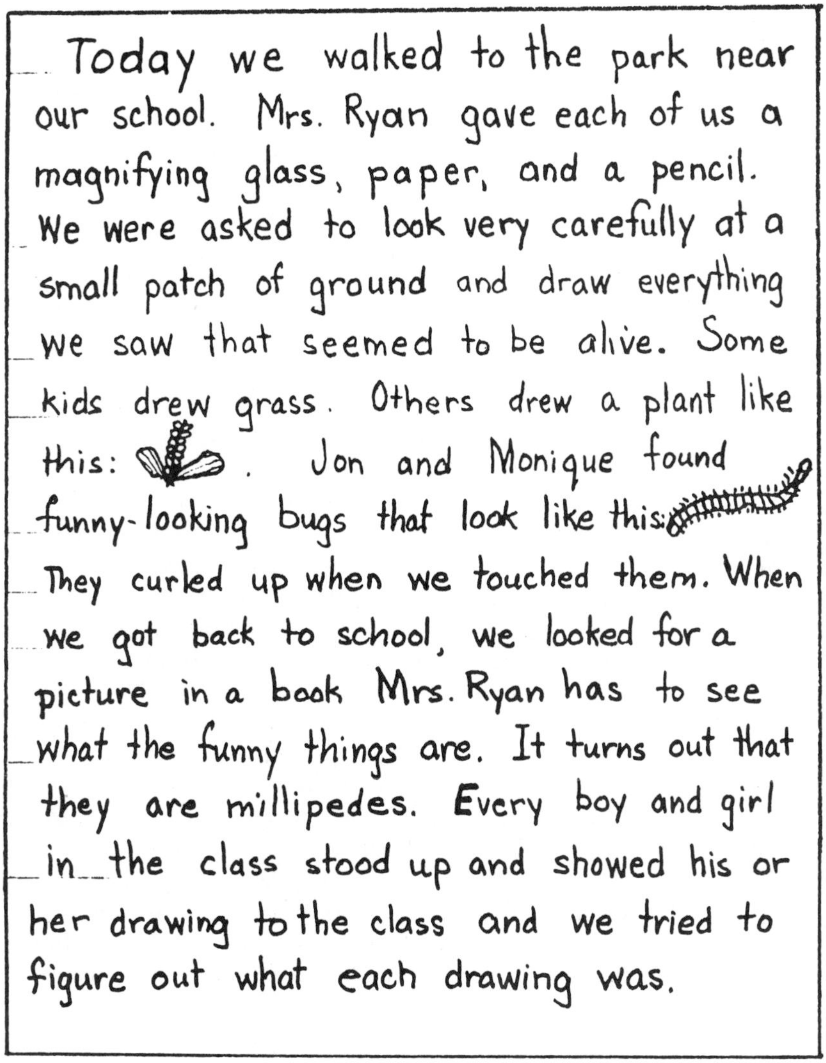

Today we walked to the park near our school. Mrs. Ryan gave each of us a magnifying glass, paper, and a pencil. We were asked to look very carefully at a small patch of ground and draw everything we saw that seemed to be alive. Some kids drew grass. Others drew a plant like this: . Jon and Monique found funny-looking bugs that look like this: They curled up when we touched them. When we got back to school, we looked for a picture in a book Mrs. Ryan has to see what the funny things are. It turns out that they are millipedes. Every boy and girl in the class stood up and showed his or her drawing to the class and we tried to figure out what each drawing was.

Figure 5–1 An experience chart.

mental sounds to their classmates. Household, outdoor, or in-school sounds may be recorded, played to a class, and identified by the pupils. Among the household sounds that may be recorded are a telephone ring, a vacuum cleaner, running water, footsteps, and the like. The tape recorder has also proven quite useful in helping children who read with difficulty. Tapes of laboratory procedures have been prepared for children's use.

Science experiences can be considered the core of a reading readiness program. *Reading readiness* is a term that is often used but less often defined. Most experts in early-childhood education would agree, though, that reading readiness is a state of preparedness to benefit from reading instruction with a minimum of stress and discordance. This definition assumes that there exists a set of skills and experiences that children ought to have before they begin formal reading instruction. These skills usually encompass the ability to distinguish among consonant and vowel sounds *(auditory discrimination),* the ability to discern fine similarities and differences in objects and symbols *(visual discrimination),* an adequate vocabulary of spoken words *(language preparedness),* and perhaps most important, a wealth of primary background experiences to help children attach meaning to what they begin to read.

Much of the discussion in previous sections of this book makes the point that perceptual and language development are integral parts of the process of learning to read. Taken together, they form a sound program for preparing children to read. Early science experiences are also essential for success in beginning reading. Such experiences often provide for free manipulation of concrete materials, which helps promote intellectual development in that these experiences serve as a basis for the formation of the mental images that are needed if children are to become intellectually equipped for instruction in reading.[3] Before a child can be expected to read the word *tree,* he or she ought to have seen, touched, and experienced a tree. Failure to provide many varied concrete experiences in early childhood can interfere with the development of reading comprehension, since some of the objects or events encountered in print may have no real meaning for the child.

Language development is closely allied with experiences in beginning reading, since reading is actually an extension, or symbolization, of language. As mentioned earlier, comprehension of written symbols grows out of personal experience. Once a child has had contact with the object represented by a written word, the object derives a real meaning for the child, a meaning that can then be attached to the written symbol. Labeling objects with both spoken and written words is essential. Children can make mini-exhibits of natural materials that they collect, and signs labeling the objects may be provided by the teacher or written by the child. The furniture and objects in an elementary school classroom are often labeled.

[3]John F. Newport, "Can Experiences in Science Promote Reading Readiness?" *The Elementary School Journal,* 69 (April 1969): 378.

Comparative terms should also be experienced before they can be meaningful in written form. Children need to have concrete experiences to demonstrate the meanings of terms such as *longer* and *shorter, over* and *under, in* and *on.* These associations are difficult to develop through words alone, or even through combinations of words and pictures, but can be built up through the combination of manipulations and verbalization. The close relationship between such concrete language and reading experiences and early science investigations is obvious, since manipulating and comparing environmental objects is at the heart of the science program for young schoolchildren.

Perceptual development is also intimately involved in the process of beginning reading. The relationship between the provision of perceptual training activities and success in reading is well documented. Children who are helped to make fine discriminations in objects and visual materials (as suggested in Chapter 4) will be well prepared to make the discriminations among letters of the alphabet (particularly the often-confusing lowercase *b, d, p,* and *q*) that are essential in beginning reading instruction. The ability to make auditory discriminations is also necessary if children are to hear the fine differences in initial and final consonants, and, later, the even more subtle differences in vowel sounds.

Perceptions that are remembered are categorized by the child to fit the ideas and patterns of thought that he or she already possesses. A group of impressions and perceptions becomes organized into concepts, and these, in turn, become parts of larger patterns of thought. Single perceptions, then, can be considered the units or building blocks of thought, and thoughts can be expressed through language or written symbols. The connection between perceptions and beginning reading, thus, is a close one and is worth the attention of teachers. The more children have perceived or done, the better prepared they are for reading.

In the upper elementary grades, as children become more fluent in oral reading, attention is usually directed toward the development of reading comprehension skills. These skills involve interpretation of material read by the child, and usually include the following:

Getting the main idea of a passage.
Interpreting data presented in material read.
Drawing conclusions from material in a selection.
Recalling and reconstructing the sequence of events in a selection.
Gathering facts and data to support conclusions and interpretations.
Analyzing, synthesizing, and evaluating information in a passage or selection.

Examination of these specific reading comprehension abilities reveals that they are very similar to science process skills. Evidently, working with children in science to observe, draw conclusions from experiments, and substantiate these conclusions with facts and data will help children develop the same skills in relation to reading.

These skills will seldom appear without formal or informal instruction; they are developed through questioning, practice, communication, and review. Teachers frequently ask pupils questions that tap these abilities as they check for comprehension of stories read in small groups or individually. Children are often asked to go back to the text of a story to find the details or facts that substantiate a particular interpretation or conclusion. Similarly in science, if pupils have been investigating the properties of air and a youngster claims during a discussion session that air is real because it has weight, she should be able to substantiate her claim. If, as part of the investigation of air, the youngster had carefully weighed a balloon or a volleyball that had been completely emptied and then re-inflated, she would be able to present sufficient evidence to support her claim.

Children can be asked to draw conclusions, reconstruct the sequence of events in an experiment, or support contentions as part of discussions, reviews, or reports to the class. Questioning by teacher or other pupils that probes their interpretations of findings helps children see their work in relation to other investigations and also helps sharpen their comprehension skills and general fund of knowledge.

Basic science activities not only are fundamental to the development of competence in science but can also aid in building the necessary foundations for language development, perceptual development, and the ability to read and interpret. The activities outlined in Chapters 4, 5, and 6 represent basic components of an early childhood science program; their relation to other branches of the elementary school curriculum suggests that they are interdisciplinary in nature. There is no substitute in a child's schooling for these firsthand experiences. They are extremely important if children are to grow intellectually and benefit from instruction.

REFERENCES

BREARLY, MOLLY, ED. *The Teaching of Young Children.* New York: Schocken Books, 1970. A practical book for classroom teachers that sets clear goals for primary school teaching. Based heavily on Piagetian theory.

GINSBURG, HERBERT AND SYLVIA OPPER. *Piaget's Theory of Intellectual Development.* Englewood Cliffs, N.J.: Prentice-Hall, 1969. A first reference in Piagetian psychology. A clear yet not oversimplified introduction to Piaget's major theories, with some mention of classroom applications.

HARRIS, ALBERT J. *Effective Teaching of Reading*, 2nd ed. New York: Longman, 1971. A reference for elementary school teachers on how to organize a classroom reading program.

ROSWELL, FLORENCE AND GLADYS NATCHEZ. *Reading Disability: Diagnosis and Treatment*, 3rd ed. New York: Basic Books, 1977. A professional book for both reading specialists and classroom teachers dealing with how to recognize and deal with a variety of reading problems.

RUDDELL, ROBERT B. *Reading–Language Instruction: Innovative Practices*. Englewood Cliffs, N.J.: Prentice-Hall, 1971. Some practical and ready-to-use ideas for helping children to read.

VYGOTSKY, LEV SEMENOVICH. *Thought and Language*. Edited and translated by Eugenia Hanfmann and Gertrude Vakar. Cambridge, Mass.: M.I.T. Press, 1962. A classic work on language development and the relationships between thought and language.

Science Experiences for Young Children

CHAPTER 6

One December morning Billy, a kindergartener with whom we had the pleasure of working, was experimenting with a simple balance. He would place an object in one pan of the balance and then count the number of paper clips needed in the other pan to balance the object. Billy had just balanced a small toy car and discovered that this operation required 38 clips. Next he wanted to balance a pair of scissors. He put the scissors in one pan and started counting out paper clips as he placed them in the other pan. Suddenly he cried out, "I don't have to do this again! The car weighs 38 clips so I'll put it on the other end." Then he proceeded to add clips to the pan with the car and counted "39, 40, 41, 42," and so on until the scissors were balanced. Billy used the information he had gained by balancing the toy car to find out how many *more* clips would be needed to balance the scissors. What a discovery for a 5-year-old! Such sudden realizations are among the most gratifying aspects of working with young children.

When we speak of young children in this chapter, we are referring to the children who have not yet entered the first grade. The majority of first-graders have had prior experiences in nursery schools, day care centers, play groups, Head Start programs, prekindergartens, and kindergartens.

Young children are natural investigators. Their curiosity about the world around them prompts activity and experimentation. This attitude can be fostered and preserved throughout childhood and later life if opportunities for investigation are provided and supported. Within reasonable limits, young children ought not to be thwarted in their explorations; they should be encouraged and helped to feel capable as learners and experimenters. They should feel that their world is full of life and phenomena to be studied.

The observations made by young children are delightfully fresh and clear. They are relatively unsophisticated, and their perceptions are largely free of prejudices about what should be seen or observed. Older children, when asked to relate what they are perceiving, often respond with the answer they think is expected of them. Younger children are more likely to describe their perceptions as well as their language abilities permit.

The youngest investigators also are not burdened by common misconceptions about science (e.g., "Science is hard"). When these youngsters are experimenting or investigating their world, they are unaware that they are engaging in science work, studying concepts and phenomena, or behaving in "scientific" ways.

The child's earliest years are years of rapid physical and mental growth. The first five years of life are a period of stimulation and learning—more learning may take place during this period than during any subsequent period. Positive environmental influences can have a profound effect on the young child's development. The problem facing teachers during this period is how to arrange the encounters children have with their surroundings in such a way as to effect optimum growth. There are no easy solutions to this perplexing problem; however, it appears that for the almost inexhaustible curiosity of youngsters to be maintained, their environment must be rich with stimulating materials. A setting full of objects to see, hear, poke, pry, sniff, taste, stretch, shake, and touch will provide endless interest for the young child. Manipulation of or action on objects will foster learning. In Piaget's words, "To know an object, to know an event, is not simply to look at it and make a mental copy or image of it. To know an object is to act on it. . . . An operation is thus the essence of knowledge."[1] If children are trusted to investigate, if they are permitted to explore, then they can operate on objects and learn. This requires more than an attitude of permissiveness or benign neglect; it requires the structuring of the child's surroundings to stimulate science learning and general mental growth.

[1]Jean Piaget, "Development and Learning," *Journal of Research in Science Teaching,* 2, no. 3 (1964):176.

It is essential that those who work with younger children provide them with science experiences; this point can hardly be overemphasized. Science experiences provide the conceptual bases on which later science competence can be built. Early science experiences help young children become more aware of their surroundings, observe, manipulate, discover. It may be that depriving youngsters of such experiences will limit their subsequent intellectual development. The more things a child has seen or done, the more he or she is likely to be interested in seeing or doing new things. Also, language skills can best be developed in a background of rich sensory experiences. Words are preceded by mental images, and these images are to a large part formed through experiences with objects in the environment. Therefore, early science experiences are also essential for the development of language skills.

In general, children should be provided with materials that stimulate their curiosity. Speaking with children as they investigate helps one assess how they are dealing with the materials or what they are thinking about them. Items that are unfamiliar to the children should be labeled. Have children seek answers to their own questions, and avoid being too quick to provide answers; it is better to help them find ways of answering questions for themselves. Often, with the youngest children, actions or activities must be repeated over and over again until a concept is internalized. Teachers of very young children usually need a great deal of patience, since the young child's attention span is short.

Children develop an individual learning style at an early age. Some prefer to be told what can be done with an object. Others prefer to try things on their own. Some children are most alert early in the morning. Others need a few hours to "wake up" and are most alert later in the day. Some children are visual learners; others prefer the auditory modality. Some children work best alone, while others work best in group situations. Some children are hesitant to explore new materials; others will "jump in." The children in any group will surely exhibit a wide range of tolerances and limits. Becoming familiar with individual learning styles can help teachers maintain realistic expectations in many learning situations.

As children grow older and their language becomes more sophisticated, the types of questions asked of them become very important. Activity may be motivated by the right question at the right time. Some questions can go to the heart of the concept or idea being investigated. Also, open-ended questions should be posed; the answers will require further investigation. Some questions help children pull together their findings and understand certain phenomena better. Above all, it is important for teachers and children to talk while they are investigating; this enriches the learning experience, brings children to higher levels of mastery, and often leads to further study.

SCIENCE AND YOUNG CHILDREN

As children have school experiences earlier and earlier in their lives, there is an ever-increasing need for teachers in nursery schools, day care centers, kindergartens, and the like to provide youngsters with appropriate science experiences. The children's encounters with materials and their environment can be carefully structured to promote maximum intellectual growth. A well-equipped schoolroom can provide countless opportunities for exploration and investigation. Youngsters should be able to do things with the materials they are given. They should make transformations (e.g., mixing flour, water, and salt to make dough). Children's questions about what they are doing should lead to other investigations; the natural curiosity of the young child can be maintained and preserved in this way. Science investigations can and should be an outgrowth of the child's daily work and play.

Curricula for young children are usually broad and flexible enough to include investigations of many aspects of the environment. These may draw upon topics from diverse areas of science. A science program for young children will generally include several basic components: unplanned, spontaneous events; the use of commercial science programs; planned but isolated science experiences; and the planned, in-depth unit study. (See pages 37–44.) A well-balanced science program for young children should combine all of these components.

Examples of unplanned, spontaneous events may be a snowfall, a pet brought to school, a windy day, or the experience some child had "last night." The teacher should be alert to such happenings as opportune times for teaching and learning. It is important to seize such moments and encourage children to observe, experience, investigate, and discuss. Activities may also develop spontaneously as a result of questions raised by pupils. If teachers use an investigative approach by responding, "Let's try it!" in answer to pupils' queries, they are providing a model of investigative behavior for the child to emulate. The more youngsters become accustomed to finding out by trying and doing, the more they are likely to internalize this process and seek answers to their own questions.

SAMPLE SCIENCE ACTIVITIES

Planned, but isolated, science experiences are quite common in nursery schools, prekindergartens, and day care centers. Examples of such activities include cooking experiences, short trips, or using a magnet or a magnifying lens. Such experiences are usually developed in a single session. Some examples of science experiences for young children follow. They are samples from the literally thousands that could be listed. Similar activities appear in later chapters of this book, but the purpose here is to demonstrate the simplicity and concrete nature of the content and approach in a few activities uniquely suited to young children.

Cooking with young children. Cooking is a marvelous experience for youngsters. All of the senses are stimulated, and the children actually witness physical and chemical transformations. Children should be involved in peeling, cutting, and slicing the raw materials; preparing the pots and pans; adding seasoning; stirring; and observing the substance at various intervals during the process. Children should also be involved in pouring, portioning, serving, and of course, eating the finished product. Of the many possible items for classroom cooking, the following are among the most popular; applesauce, popcorn, vegetable soup, mashed potatoes, relishes, cookies, oatmeal, butter, peanut butter, puddings, cereal, and candies. Recipes may be outlined diagrammatically to recount the sequence of events in cooking. (See Figure 6–1.)

Caring for and studying animals. Caring for a goldfish is a simple way to introduce children to animal needs. Goldfish do not need a heated tank. Any large glass container with a wide mouth will make a satisfactory home for a goldfish. The children can arrange bottom cover using sand or gravel and perhaps install a plant or an attractive rock. Let the water stand overnight. A goldfish should never be given more to eat than it will consume in ten minutes, but it should be fed each day. The swimming habits of the fish should be observed and discussed. The movements of the mouth, tail, fins, and gill covers should be observed. A special day for a group of youngsters could consist of preparing a container; walking to a dime store or a pet shop; buying a fish, food, and gravel; and returning to school to set up the fishbowl and accessories.

Care of larger pets can be approached in a similar fashion, but usually requires more elaborate preparations and care. Young children can learn to care for birds, gerbils, hamsters, guinea pigs, mice, rabbits, and the like. In any experience with pets, the children should be taught to treat the animal gently, to be careful in feeding and cleaning it, and to take an interest in its movements and habits. (See Chapter 13, for a more detailed discussion of securing and caring for classroom animals.)

Caring for and studying plants. Young children should be given responsibility for maintaining potted plants and studying plant growth, new shoots, and other changes. The needs of plants (i.e., air, water, soil, sun) should be discussed and could be studied
through various deprivations (e.g., noting differences in similar plants, one watered twice a week and the other watered twice a month).

Figure 6–1 A recipe in diagram form which the children may follow as the food is prepared.

Children can germinate seeds in clear plastic cups and observe root growth—lima beans are particularly good for this. Cut some heavy construction paper or a desk blotter and stand it up along the inside of the clear plastic cup. Place the beans between the cup and the paper and fill the cup with water. Observe the direction of root growth. Starting plants from cuttings, from carrot, beet, or pineapple tops, from sweet potatoes, and so forth is particularly interesting for the young child. Where possible, planting and caring for a vegetable or flower garden is a delightful experience for a child.

Youngsters can examine leaves and sort a variety of them. Collections of pine cones, acorns, nuts, bark, and other natural objects are of interest to children at this age. A careful examination of soil may also be stimulating. (See Chapter 12 for more extensive coverage of the plant kingdom and plant studies.)

Chicken bones. One of the more interesting activities for young children is to work with chicken bones. Have the children save chicken bones from home. Any meat left on them should be removed, and they should be washed and then dried thoroughly for a few days. The children can then sort the chicken bones by type or shape. They can arrange them in a variety of ways to form their own animals. The bones can also be pasted onto construction paper or cardboard to depict the skeletons of real or imaginary animals, with the rest of the creature's body drawn or painted around the bones.

The bones can also be pressed into clay or a tray of plaster to form an impression. (Make sure to coat the bones lightly with oil or petroleum jelly.) Such impressions may be made of single bones or an entire "skeleton." (See Figure 6–2.)

Self-exploration. Identification of parts of their own bodies can be an important activity for young children. More difficult body parts can be introduced (e.g., elbow, ankle, knee, hip, etc.) Children can look at themselves in a long mirror and describe the color of their hair, eyes, and articles of clothing. Games can be played (e.g., "Simon Says") in which children identify body parts. (See Chapter 14 for more explorations of the human body.)

Figure 6–2 Chicken bones may be pasted on paper to form strange new animals.

Explorations with light. Given a flashlight, young children can switch it on and off, shine it into a mirror, or shine it on walls and ceilings, doing so from various distances and angles. Block the light's path from various objects and ask the children to guess the object by looking at its shadow. Project the object's shadow from various vantage points. Hold a flashlight several feet from a blank wall and let the children make "animal shadows" by blocking the path of the light with their hands.

Water play. Few activities for young children are as rich with joy and opportunities for conceptual development as water play. The children should be given a large basin or plastic tub of water and a wide assortment of containers of various sizes and shapes as well as funnels, tubing, ladles, eggbeaters, and spoons. Water will pour easily from some containers (wide-mouthed ones), yet bubble, plop, and gurgle from others. Some short, squat containers may hold the same amount of water as some tall, slender ones. Discovering these relationships will help the child understand fundamental volumetric relationships.

For the fun of it, young children often delight in playing with a basin of soapy water and making mounds of bubbles with an eggbeater. Children enjoy painting with water and can see evaporative processes at work if they apply a thin film of water to a chalkboard or some other surface. (See Figure 6–3.)

Figure 6–3 Children enjoy waterplay in a kindergarten classroom.

Float and sink. An activity associated with water play is to test whether each of a variety of objects sinks or floats after it has been dropped into a basin of water. After each test the object may be placed in one of two groups or piles, depending on whether or not it sinks. It is also interesting to investigate materials that can be used to support other objects in the water. Cardboards, pie tins, and the like may be used to support pebbles, peas, blocks, dolls, and so on. What happens when too many objects are piled onto the "boat"?

Playing with blocks. A set of unit blocks (i.e., ones of varying, yet proportionate, size—e.g., a unit, double units, triple units, half-units, etc.), which may be found in most preschool rooms, can help children begin to discover size and number relationships. Often, discussing the need for a particular block or blocks can help children learn these fundamental relationships. (See Chapter 7 for a discussion of mathematical development.)

Discovering hot and cold. In every environment some objects may be warmer than others. Such thermal differences should be demonstrated to children whenever it is safe to do so. Warm and cool pipes can be touched, as can hot and cold water. Children can touch cars of different colors that have been standing in the sun and see whether some colors retain more heat than others. Ice can be melted in a dish and the results observed. An ice cube tray full of water is placed in the freezer compartment of a refrigerator. How long does it take before the water turns to ice? (See Chapter 18 in which further activities with heat are described.)

Explorations with salt, sand, or soil. Sand, soil, and salt are excellent materials for youngsters to work with. These media may be poured from one container to another as in water play, but some interesting new possibilities arise as sand, salt, or soil is mixed with varying amounts of water. What are the effects when a pail full of dry sand is turned over? When a pail full of wet sand is turned over?

Pie tins, scoops, cookie cutters, and small plastic refrigerator containers may be used with wet sand to produce interesting results.

A tray of salt is very interesting for children to run their fingers through or to print their names in. Food coloring may be added to salt to produce colors and to mix colors.

Rocks and shells. Young children can begin to assemble rock and shell collections. Rather than identifying each rock or shell type, it is important for the children to be able to describe similarities and differences in rock or shell color, shape, and texture. The children may devise different ways of storing and displaying their collections. (See Chapter 16 for a full discussion of rocks and minerals.)

Weather observations. Young children should be helped to become attuned to the daily weather and its changes. The weather is often discussed in class. Often teachers find it helpful for children to keep monthly weather calendars. (See pages 12 and 367.) Cloud formations can be observed and discussed. Children may be eager to express particular feelings, emotions, or sensations that they experience with certain kinds of weather.

The change of seasons is an appropriate topic of investigation with young children. The characteristics of each season, in terms of weather, vegetation, animal life and outdoor activities, may be discussed. The gradual changes of seasons may be recorded by studying a specific tree, plot of soil, or schoolyard. Keeping photographs of a tree or schoolyard at various intervals helps children remember how it appeared a few months back. (See Chapter 17 for a discussion of weather and weather-related activities.)

Using magnets. Magnets are particularly intriguing implements for preschoolers. Although such young children cannot understand the theory underlying the operation of a magnet, they can nonetheless try to find magnetic and nonmagnetic substances. Children ought to be provided with a variety of magnet types (e.g., horseshoe magnets, bar magnets, "U" magnets, round and rectangular magnets, as well as natural magnets or lodestones). Young children enjoy being given an assortment of objects that they must test and then place into one of two piles, depending on whether or not each object is attracted to a magnet. (See Chapter 19 for more details on dealing with magnetism and electricity in the classroom.)

These are but a few of the countless activities in which young children can become involved. They all are concrete in nature and involve materials to investigate or firsthand observations that children can make. The concepts developed are simple and direct, yet they can serve as a basis for further experiences.

The scope and nature of science experiences for young children are limited only by the imagination of the teacher and youngsters. Children are often the initiators of investigations; such activities can be a natural outgrowth of their daily questions and concerns.

THE UNIT INVESTIGATION

The unit investigation is a relatively in-depth, long-term set of interrelated experiences leading to broad understanding of a particular topic or theme. (See pages 40–44 for a discussion of science units and unit planning.) Obviously, the topics of units must be selected with care. Within a nursery school year, there is only sufficient time to cover a limited number of units. The topics to be studied should be of great interest to the children if they are to sustain their attention and involvement, as activities related to the topic will occur over a period of days or weeks. Topics for unit studies may include sound and light, investigations of air and weather, animal and plant studies, environmental investigations, self-study, the five senses, seasonal changes, and so forth. Such long-term investigations are cumulative, each day's experiences building on those of the previous day. Young children's thinking is often based in the "here and now," so the prior developments may have to be reviewed briefly as each related experience is introduced. The following pages describe two units in which a class of kindergarteners were involved.

Animal Training

Some basic concepts of behavioral sciences may be introduced by involving children in animal training experiences. A large "T" maze can be constructed with building blocks. The children should be involved in its construction. The maze should be high enough so that a gerbil or rat cannot climb over its sides. A box with a makeshift door or gate is required in order to release the animal into the maze. (See Figure 6–4.)

The following problem can be posed to the children: "What can we do to make sure that every time the animal is released into the maze, it will go only to one side?" Several ideas may be suggested. Try to elicit the response that for the animal to go to one side of the maze, there must be something on that side (e.g., objects, food) that the animal likes or wants. The children in one class suggested that toys, playthings, an exercise wheel, and sunflower seeds should be placed at one end of the maze (the

Figure 6–4 A simple "T" maze constructed by a group of young children.

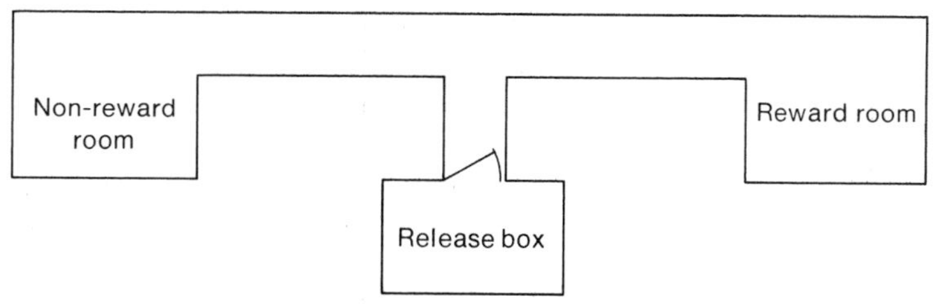

Figure 6–5 A diagram of the major features of the maze constructed by the children.

reward room) and that the other end of the maze (the nonreward room) should be empty. (See Figure 6–5.) Assign jobs to the pupils. One can equip the reward area with cardboard tubes, newspaper, sunflower seeds, and other things that the animal will want. Another child can be the "releaser"; he or she releases the rat or gerbil from the release box and operates the gate. A third child, the "counter," makes a record, using slash marks, to indicate how many times the animal goes to the reward room or the nonreward room. A fourth child may be the "room blocker"; he or she blocks the animal's egress from the room that it enters for a specified period. The teacher (or a child who is able to use a watch with a second hand) is the "timer." The "timer" allows the animal to stay in the reward or nonreward room for three minutes once it has entered the room. After three minutes have passed, the animal is removed and placed back in the release box for another trial. After many such trials over a period of days, the animal will learn to go to the reward area, particularly if these trials are done before it is fed each morning.

This unit was rather successful with a kindergarten class. After eight days of trials, a gerbil did learn to go to the reward area on approximately 90 percent of the trials. Even more important than the gerbil's learning, though, was the understanding the children gained. They continued to devise different "learning" tasks for the gerbil and apparently enjoyed the experience. Some began building other mazes; others were interested in the likes and dislikes of gerbils. Some children suggested different rewards and tried them.

This unit was not attempted until after the children were familiar with the gerbils as classroom animals. Previously they had been involved in caring for the gerbils and discussing their size, color, texture, habits, and growth. The youngsters would delight in observing a gerbil chew through a cardboard tube or "work out" in an exercise wheel. The children had developed a sense of responsibility and kindness toward the animals, and this particular unit was an outgrowth of a few youngsters' having asked whether there were any "tricks" the gerbils could learn. It was explained that teaching an animal "tricks" is often more amusing to humans than beneficial to the animal, but that most animals could learn to do something. At first, this unit seemed to deal with concepts that are too difficult for young children to comprehend; however, as the children became more and more involved, it was apparent that their interest in the matter of teaching the gerbils helped them deal effectively with the concept of animal conditioning. Their suggestions for variations on the training tasks evidenced their ability to apply what they had learned.

Studying Plant Communities

A careful look at a small-scale ecological community is an appropriate unit investigation for young children. Teachers can move beyond "bulletin board ecology" or an occasional cleanup of a schoolyard or park by providing children with more substantive activities related to fundamental environmental phenomena. One way of accomplishing this is by studying two plant communities: a wooded area and a vacant lot.

On the first day of such a study, take the children to a wooded area.

Figure 6–6 Children explore plant samples in a wooded area.

Use the walk to the area to point out trees, shrubs, and animal habitats along the way. At the site, ask the children to pay attention only to plants that are below waist height. (See Figure 6–6.) The various identifiable parts of a plant (roots, stem, leaves) ought to be demonstrated and discussed. Try to dentify some leaves and small plants and collect a few samples in a bag. A day or so after this excursion, display the samples in the classroom. Have the children group the samples according to similarity of physical characteristics.

Divide the children into subgroups of four and give each group two sheets of waxed paper and two sheets of typing paper. Have them arrange one sample of each plant type between the two sheets of waxed paper and then place a sheet of typing paper over each of the sheets of waxed paper. Then iron the "sandwich" lightly until the wax from the waxed paper seals the plant specimen. The result, when mounted on paper or cardboard, serves as a simple, nonwritten key to help the children identify some plants in the field.

A few days later revisit the wooded area and have each group of children take its "keys" and try to match the specimens in the keys with actual plants or leaves on the ground. Some children may be able to identify some of the samples by name.

The topic of community helpers (firemen, grocers, policemen, doctors, teachers, etc.) is commonly covered in preschool programs. The

Figure 6–7 The children look for plant samples in a "vacant" lot.

above activities can provide an analogy: Just as people live together in a defined area and provide services for one another, so too do plants live together in a plant community. (Some plants provide nutrients for others; some provide shade for others.)

Repeat the entire procedure just outlined, but this time take the children to a "vacant" lot. Such sites usually have a wealth of little plants growing out of concrete cracks, lot borders, and so on. (See Figure 6–7.)

Review the plants found in both communities. Keys from each area should be distributed to each group of youngsters. Have them identify similarities and differences among the plant forms in each area. For example, there are overt differences between a blade of grass and a small clover plant. Plants that survive in a vacant lot also tend to be broad and flat, being able to endure constant squashing by people walking in the area.

After a review of these experiences, the keys may be cut up into individual plant samples that the children can sort according to various criteria, identifying the community from which each sample came.

Conclusion

In the units just described, a common feature is the slow, steady development of concepts in the simplest terms. Through intensive involvement the pupils gain a number of experiences that help them grasp concepts that they might not otherwise be able to master. The activities also provide a basic foundation on which further study and learning can be built.

The activities discussed in this chapter, as well as those in the preceding two chapters, provide a point of departure for developing science programs for young children. Continual provision of concrete materials to manipulate, with time for children to explore freely, is a vital consideration. Designing activities in response to children's questions and curiosity is a way of capitalizing on their interest and enthusiasm. Also, asking children questions to stimulate further investigation fosters conceptual development and helps them discover new relationships and interests.

REFERENCES

ALMY, MILLIE C. *Young Children's Thinking.* New York: Teachers College Press, 1966. A classic study in which much of Piaget's basic research was replicated with American children. Provides many insights into intellectual development.

CROFT, DOREEN J. AND ROBERT D. HESS. *An Activities Handbook for Teachers of*

Young Children. Boston: Houghton Mifflin, 1972. Contains a wealth of teaching ideas, with a section entitled "Experiences in Pre-science."

GOLDBERG, LAZAR. *Children and Science.* New York: Scribner's, 1970. A book that can help teachers inspire interest in science through children's curiosity and natural instincts. A professional guide that conveys human values and respect for children.

ROBISON, HELEN F. AND BERNARD SPODEK. *New Directions in the Kindergarten.* New York: Teachers College Press, 1965. An important guide for kindergarten teachers, with content suggestions in a variety of areas.

WEBER, EVELYN. *The Kindergarten—Its Encounter with Educational Thought in America.* New York: Teachers College Press, 1969. The history and evolution of the kindergarten. Includes discussion of major trends and curricular issues.

Science and Mathematical Development

CHAPTER 7

Mathematics is both the queen and the servant of the sciences.[1] Mathematicians often think of mathematics as the purest of the sciences. Certainly, some events and phenomena in the universe and within our bodies can be investigated by mathematicians using the methods and processes of mathematics. In fact, some of the most profound discoveries in science, those made by people such as Newton and Einstein, have resulted largely from mathematical investigations. Also, mathematics is used in almost all of the sciences. The use of mathematics leads to greater precision and helps scientists see relationships that they otherwise might not be able to see. With mathematics we can gain a better understanding of the degree of certainty with which we know what we know, and we can make better predictions based on the data that we have gathered. Modern science would be impossible without mathematics.

Developing an understanding of basic mathematics and the ability to use it is a widely accepted goal of elementary education. As with reading, writing, and other skills, there is value to teaching and learning within the context of the areas in which these skills are needed and used.

[1]In the development of this chapter, we have benefited from discussions with Elizabeth Meng.

Mathematics is needed and used in the sciences. Why not develop mathematical skills through science experiences?

In this chapter we discuss some of the skills and understandings that are important in mathematics and in the sciences and describe science experiences through which these understandings and skills can be developed. Teachers may wish to emphasize mathematical operations as they guide children through these activities. The experiences described in this chapter are only a sample of many science experiences that have significant mathematical dimensions. Many of the science activities described throughout this book can also be used for the development of mathematical understandings and skills.

MEASUREMENT AND UNITS OF MEASURE

Measurement is comparison. We may compare one pencil with another to see which is longer. This direct comparison is called *superposition* and is perhaps the simplest form of measurement. Or we may compare the volume of two containers by seeing which will contain the most liquid. In more sophisticated measurement we use units of measurement to compare objects. For example, we may compare the lengths of pencils by seeing how many paper clips long each of them is. The areas of two table tops can be compared by seeing how many paper squares (or styrofoam trays) of the same dimensions are needed to cover the tops. Or we may use a tumbler to see how many tumblers of liquid are needed to fill each of two containers.

To communicate the results of our measurements, we use standard units of measurement. The meter is a standard unit of measurement for length and the liter for volume. Standard units of measurement are operationally defined so that scientists and mathematicians can check the units being used. The standard meter, for example, is based on the wavelength of the orange–red light emitted by a krypton 86 lamp. Practically, it is defined as the length of the meter sticks used in schools.

Young children should have experience with direct comparisons. The heights of two children can be compared by having them stand back to back. The weights of objects can be compared by "hefting" them. When they are comparing objects of about the same weight, children can be asked to shift the objects from hand to hand. "Does it make a difference which hand is holding the object?" For objects that are of almost the same weight, an equal-arms balance can be used in making the comparisons. Direct comparisons of volumes of liquids can be made by pouring the liquids into similar containers such as paper milk cartons and comparing the heights of the liquids. (See page 25.)

Units of measurement are arbitrary, and children should have the experience of suggesting and using their own. A convenient unit for measuring length is the paper clip. However, a longer or shorter unit may be more convenient for some measurements. Usually we have a variety of standard units that can be used for measurement.

How long (wide) is a book, desk, or table? Have the children measure how many paper clips long and wide a book is. If they have grasped the concept of fractions, they may be able to determine the dimensions to an estimated fraction of a paper clip. Make the same measurements of a desk.

To measure the dimensions of a table, it might be convenient to have a longer standard unit. A nail whose length is a multiple of that of a paper clip might be especially useful. Then the dimensions could be expressed in such terms as "It is 12 nails and 2 paper clips long."

Children should be encouraged to suggest other units that might be used and to discuss the advantages and disadvantages of different units. For example, variation in the sizes of units (e.g., in the lengths of paper clips) would limit their usefulness.

There are great advantages in having standard units of measurement that can be used and understood by everyone. The metric system of standard units is almost universally used in the sciences. It is also the official system of units used in most nations. Since it is a decimal system and we have a numbering system with a base of ten, it is easier to use than most other systems. The metric system is used throughout this book and in most elementary school science programs. The children will find it useful to learn the following prefixes in decimal numeration:

milli one-thousandth
centi one-hundredth
deci one-tenth
kilo 1,000

The following are some of the common metric units and their relationships:

Length
1 kilometer = 1,000 meters
1 meter = 10 decimeters = 100 centimeters = 1,000 millimeters

Mass
1 kilogram = 1,000 grams
1 gram = 100 centigrams = 1,000 milligrams

Capacity (based on the liter, which is the volume of one kilogram of pure water under standard conditions).

1 liter = 100 centiliters = 1,000 milliliters
1 kiloliter = 1,000 liters

Making a milk carton graduate[2]. In science laboratories graduates of various sizes are used to measure volumes of liquids and granulated solids. Children can gain a clearer understanding of volume and the devices used to measure volume by making their own liter graduate. A usable graduate can be made from a paper milk carton.

Have the children measure the two sides of the bottom of a milk carton. Then have them multiply the lengths of the two sides to find the area of the base. The usual quart milk carton has linear dimensions of 7.1 cm (2.8 in) on each side. This makes the area of the base about 50 cm^2. (7.8 in^2)

Since a liter is equivalent to 1,000 cubic centimeters, have the children divide 1,000 by the figure for the area of the base of the milk carton to find the height at which they should mark 1 liter on their milk carton graduate. Then, using a ruler, they can mark off ten equal spaces between the base and the 1 liter mark. If the area of the base is 50 cm^2, the 1 liter mark on the milk carton will be 20 cm from the base ($1{,}000\ cm^3 \div 50\ cm^2 = 20\ cm$).

The height of a liquid or a granulated solid in the milk carton graduate can easily be seen on the outside of the carton by shining a light (e.g., a desk lamp or flashlight) down onto the material in the graduate.

PRECISION OF MEASUREMENT AND SIGNIFICANT DIGITS

In science it is usually important to be as precise as possible in measurement, but we are usually limited by the tools of measurement we use. Children can become aware of such limitations when using a meter stick to measure the length of a book.

How long is your book? Figure 7-1 shows part of a meter stick, with one segment magnified. In this illustration it is clear that the book is more than 23 cm long. We can estimate that the length is probably about 8 mm beyond 23 cm. This would make the length of the book 23.8 cm, but that is as precise as we can be with this particular meter stick.

How long is your book? What is the limit to your precision?

The precision of data should be considered when we handle data, and the results of our calculations should not suggest a greater degree of

[2]We are indebted to the late John Sugarbaker for this idea.

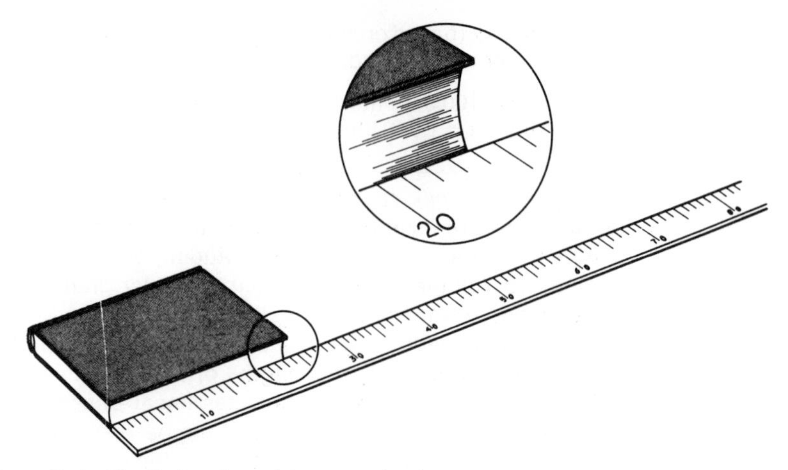

Figure 7–1 The limits of precision.

precision than the data actually have. Usually the answers to our calculations should not have more significant digits than the smallest number of significant digits in any of the numbers we have used in our calculations. For example:

$$\text{Area} = \text{width} \times \text{length} = 19.3 \text{ cm} \times 23.8 \text{ cm}$$

Three is the smallest number of significant digits in the numbers used in this calculation. The area of this particular book cover should *not* be written 459.34 cm². Such a number implies a precision of measurement that is not there. Instead, the answer should be 459 cm².

What is the area of the table? Measure the length and width of a table. What is the limit to the precision of your measurements?

Multiply the length times the width to find the area. How many significant digits are there in your answer?

COUNTING; ADDITION AND SUBTRACTION; NUMBERS AND NUMBER LINES

A key to the development of mathematical skills and concepts is "meaningful use." Science provides many opportunities for meaningful use of basic mathematical ideas and operations. The science experiences described in this section are intended to be suggestive of the wide range of mathematics-related science activities that can be developed.

Sets

A *set* is a collection, often of objects, in which the objects have something in common. In our everyday language we refer to sets of dishes and chemistry sets. In the process of *classification,* we describe the properties of objects and separate objects on the basis of whether or not they have a certain property. Children can develop their own classification schemes as they classify such objects as leaves or peanuts. (See pages 9–10 and 86–88 for descriptions of activities involving classification.) Children practice deciding whether objects belong in particular sets and making distinctions between sets.

Counting

There are many opportunities for young children to practice counting in science. For example, a group of first-graders were studying the trees that grew in their schoolyard. They collected one leaf from each of the trees. First they counted the leaves to find the total number. Then they classified the leaves on the basis of color and placed all the leaves of the same color in the same set. They counted the number of sets and the number of leaves in each set. They classified on the basis of other properties and again counted the sets and the number of members of each set.

Systems and Subsystems

A group of objects that may be considered to have some relationship to each other is called a *system.* For example, the stomach, large intestine, and small intestine have a relationship to each other and are parts of the digestive system. Systems that are parts of larger systems are called *subsystems.* For example, the villi, muscles, and some digestive fluids are a subsystem within the digestive system.

In counting and other forms of data collection and manipulation, children have to make judgments as to whether or not objects are parts of systems or subsystems. There are many opportunities for practice in delineating systems and subsystems. For example, many such judgments have to be made in constructing a *population profile.* A population profile is a graph showing the number of individuals of different ages in a group. In the following exercise children have to make decisions as to what individuals should be included in a system.

What is the population profile of the members of the families of children in a classroom? Have the children list the ages of all the members of their families on a slip of paper and indicate whether each individual is male

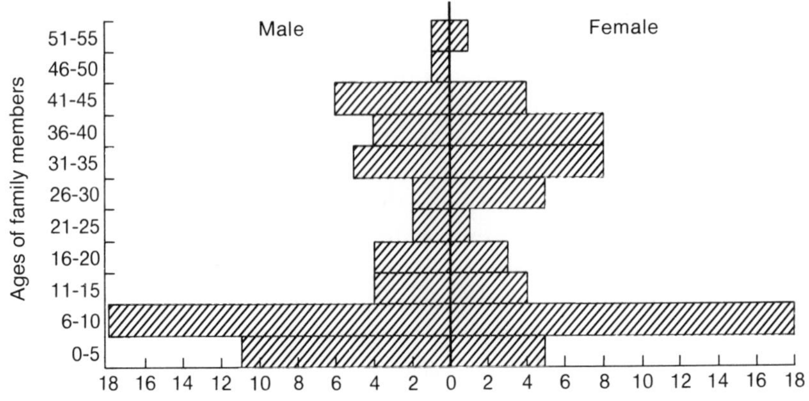

Figure 7–2 The population profile of members of families of a fourth grade class.

or female. Then have a committee make a horizontal histogram showing the number of individuals of each sex in each group. Figure 7–2 is a population profile made by a fourth-grade class.[3]

In this science experience the children are asked to identify the system. Usually they decide that all the members of the class and their families are the system. What are the subsystems? Each of the families might be considered a subsystem, but the children have to decide who will be included in the subsystem. Similar kinds of decisions have to be made in surveys, record keeping, and other science experiences.

Number Lines; Positive and Negative Numbers

A *number line* is a line on which numbers or units are indicated. Number lines are useful in teaching addition and subtraction and the

−6 −5 −4 −3 −2 −1 0 1 2 3 4 5 6

concepts of negative and positive numbers.

The temperature scale on a thermometer is a number line that is used widely in science. Children can practice reading thermometers and dealing with such questions as "How much colder (or warmer) is it today than at the same time yesterday?" Simulated thermometer scales can be sketched on cardboard and a red strip of paper moved up and down to give children additional experiences in dealing with numbers and num-

[3]We are indebted to Marilyn Cohen and her fourth-grade class at the Benjamin Franklin School in Yorktown Heights, New York, for this population profile.

ber lines. A wide range of experiences in reading, adding, and subtracting positive and negative numbers can be developed using the simulated thermometer.

Addition and Subtraction

Addition is related to counting. On a number line, if we want to add $2 + 3$ we can locate the number 2 and count 3 intervals in the positive direction. Counting is very time-consuming, and children learn to add numbers instead. However, it is occasionally helpful to refer back to a number line.

Subtraction is the opposite of addition. On a number line, if we wish to solve $3 - 2$ we locate the number 3 and count 2 intervals in the negative direction. For both addition and subtraction, the thermometer can be a number line, and work with temperatures can provide practice in these arithmetic operations within a context in which the operations are needed.

The "countdown" provides a unique opportunity for counting "backwards" that is essentially subtraction. Before releasing a balloon rocket or allowing a pendulum to swing or starting the timing of some operation, have the children count down: "10–9–8–7–. . ." There may be times when the children should be asked, "How many intervals have we counted." "How many are left before we start?"

Another approach to subtraction is to use the "texture box," in which objects are placed in a bag or an old sock and children are asked to describe how the objects "feel." (See pages 81–84.) The children can be told the total number of objects in the box. When they have removed a certain number of objects, how many are left?

THE EQUAL-ARMS BALANCE

The equal-arms balance is used in many science experiences to compare masses of objects. Figure 7–3 shows a simple equal-arms balance that can be constructed with a beam (e.g., a half-meter stick), a clamp, and some kind of support. A small wire rider is sometimes placed on the beam to balance the two sides.

Objects of about the same weight, such as paper clips or washers, can be used to carry out exercises in addition and subtraction. The teacher asks questions like the following and then carries out the corresponding operation on the equal-arms balance.

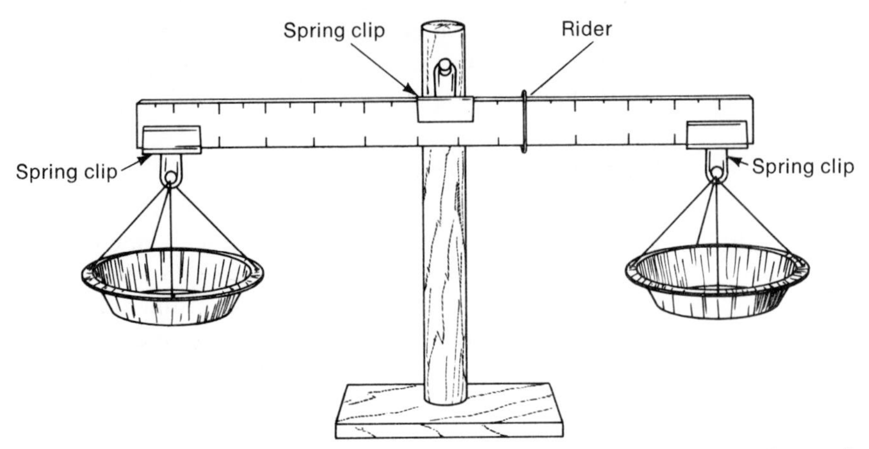

Figure 7–3 Equal arm balance made with a half meter stick, three spring clips, and wire rider.

"If we have 10 paper clips on one side, how many paper clips will we need on the other side to balance them?"

"If we have 12 paper clips on one side and 7 paper clips on the other side, how many paper clips will we have to add to the other side to balance them?"

"If we have 17 paper clips on one side and 9 paper clips on the other side, how many paper clips will we have to take off the first side to have the sides balance?"

Eventually, the equal-arms balance can be used as a physical model of the equation: To maintain balance, operations carried out on one side have to be matched by similar operations on the other side.

DENSITIES

Finding the density of an object is an exercise that gives practice in the use of various arithmetic operations.

$$\text{Density} = \frac{\text{mass}}{\text{volume}}$$

What is the density of a rock? Obtain a smooth rock and weigh it with a spring scale. (Actually, weight and mass are not the same. But for this investigation and for most operations in everyday life they are represented by essentially the same number.)

To measure the volume of the rock, fill an overflow can or beaker with water. The can or beaker should be set up so that all the water that will overflow when the rock is submerged in the water can be captured. Lower the stone into the water and capture all of the overflow water. Measure the volume of this water; its volume is equal to the volume of the stone.

Divide the volume of the stone into its weight (mass) to find the density.

FRACTIONS

A *fraction* is one or more equal parts of a whole unit; a one-third slice of an apple is one of the three parts that make up the whole apple. In the study and use of fractions, the emphasis should be on their meanings. In science experiences there are many opportunities to develop the meanings of fractions in terms of concrete materials and the mathematical operations needed in science investigations.

How much water will a particular kind of soil hold? In an investigation of how much water a particular kind of soil will hold (see pages 331–32), the results can be stated as a fraction. The soil is weighed and then put in a hollow cylinder or lamp chimney that has a piece of cloth fastened across the bottom. Then the entire system is weighed. Water is slowly poured onto the soil until it begins to drip out the bottom. The system is weighed again and the original weight of the system is subtracted from this weight. What fraction of the weight of the wet soil is water?

A fraction may be one or more equal parts of a set. If there are five guppies in an aquarium and two are removed, two-fifths of the guppies were removed and three-fifths remain. Children should recognize that the members of a set can differ. The set might be composed of 15 fish; 5 of these fish might be guppies and 10 might be goldfish. If the 5 goldfish are removed, one-third of the fish will have been removed and two-thirds will remain.

What fraction of the objects are metallic? magnetic? In sorting objects according to various properties (see pages 86–88), different sets can be defined and the fractions of those sets that have some other property can be determined. In a third-grade classroom the children were asked to separate all the objects in a box that were metallic. What fraction of the objects in the box were metallic? Using a magnet, the children identified all the metallic objects that were attracted to the magnet. What fraction of the metallic objects were attracted to the magnet? What fraction of all the objects were attracted to a magnet?

A fraction can be a ratio that compares one number with another. The use of fractions in comparisons can be more meaningful than the raw data. For example, children may find that two objects out of five that they tested were attracted to a magnet. This can be stated as "Two-fifths of the metallic objects tested were attracted to magnets."

The eye color of children in a class. Some children were investigating the eye color of the members of their class. They found that 15 of the 30 members of the class had brown eyes. This was stated in a fraction as "One-half of the members of our class have brown eyes." The children then raised questions like the following:

What fraction of us have the same eye color as our mothers? our fathers?

What fraction of us have mothers and fathers who have the same eye color as we have?

What fraction of those of us who have blue eyes have a mother who has blue eyes? a father? both parents?

Fractions also represent one number divided by another number. The fraction 3/5 indicates $3 \div 5$. Often in science a ratio such as 3/5 may be given but there is a need to find out how many members of a set this ratio indicates.

How many will have blue eyes? Children may have found that half of the members of their class have blue eyes, but they want to predict how many children in the adjacent classroom, where there are 26 children, might have blue eyes. They carry out the following arithmetic operation:

$$\frac{1 \times 26 \text{ children}}{2} = 13 \text{ children}$$

Science activities provide many such opportunities for practice in the use of fractions.

DECIMALS

Decimals are essentially fractions with a denominator of 10 or some power of 10 such as 100 or 1,000. Since the metric system of measures has a base of 10 or powers of 10 and is used so widely in science, this area of the curriculum provides additional opportunities for practice in the use of decimals.

Children can use meter sticks and tapes to measure lengths, graduates to measure volumes, protractors to measure angles, and stopwatches to measure time. The measurements can be to tenths or hundredths of such units as meter, centimeter, liter, and second.

The basic units in some linear measuring devices may be divided only into halves. Such measuring devices can be constructed with string or tape. When using such measuring devices, children can be asked to estimate lengths to tenths of units.

Meter sticks or tapes can be used as number lines on which children can add and subtract decimal fractions. For example, if you want to add 3.8 centimeters to the length of a pendulum that is 40.2 cm long, this addition operation can be carried out on a meter stick number line. Similarly, decimal subtraction operations can be carried out in this fashion.

Most odometers on bicycles and automobiles register tenths of miles. Children can practice adding, subtracting, multiplying, and dividing readings from odometers on their bicycles or family cars.

Again, science provides many opportunities for practice in various arithmetic manipulations with decimals. When mixing solutions, for example, children may calculate how much salt they should add to a given amount of water so that this solution will have .1 as much salt as another solution. Or they may be asked to add .2 more sand to a mixture of sand and iron filings. Or, in an experiment involving the growth of plants, children may want to calculate the effect of adding .3 more fertilizer to one set of plants than to the other; then they also have the experience of using decimals in weighing the fertilizer.

Decimal units can be simulated with objects such as paper clips. Each paper clip is a unit. Chains of 10 paper clips can be assembled to simulate the next-higher unit. These units can be used on one pan of an equal-arms balance (see page 126) to balance an object placed on the other pan. For example, a group of children balanced a quantity of sand with 3 chains of 10 paper clips and 6 separate paper clips. They were then asked for a quantity of sand that was .5 as large. They calculated the number of paper clips that would be balanced by multiplying 3.6 by .5 as follows: 3.6 paper clip chains $\times$ .5 = 1.8 paper clip chains.

SCIENTIFIC NOTATION

In science very large and very small numbers are sometimes used. The speed of light is approximately 30,000,000,000 centimeters per second. To handle such numbers, scientific notation is used.

In scientific notation such numbers are expressed decimally as a product of two factors. One number is a base number and the other a power of 10 expressed as an exponent. The number 50,000 is written in scientific notation as 5×10^4, which is $5 \times (10 \times 10 \times 10 \times 10)$. The number 3,200,000 is written as 3.2×10^6, which is $3.2 \times (10 \times 10 \times 10 \times 10 \times 10 \times 10)$. The exponent of 10 is equal to the number of places to the right of the decimal point.

Children can practice writing large numbers, such as astronomical distances, in scientific notation. The speed of light, for example, can be written 3×10^{10} cm/sec.

As children progress in science and mathematics, they will learn how to express small numbers in scientific notation. For example, .0032 can be written 3.2×10^{-3}. They will also learn how to multiply numbers expressed in scientific notation by adding the exponents of 10, and to divide by subtracting the exponents.

In scientific notation the power of 10 indicates the position of the decimal point. In the notation 3×10^{10} the power of 10, "10^{10}," indicates that the decimal place is 10 places to the right of 3.

The first part of the number in scientific notation can indicate the precision of measurement. In the notation 3.2×10^6 the "3.2" indicates that there are two significant digits in the number.

GRAPHING AND THE HANDLING AND INTERPRETATION OF DATA

Data are collected in many science investigations; graphs are helpful in interpreting and presenting those data. Graphs are particularly valuable in showing the relationships between two or more variables. For example, a graph can show the relationship between the length of a pendulum and the number of swings per unit of time. (See pages 143–46). Children can use a variety of graphs as they present the results of their investigations to other children. (See Chapter 20.)

Histograms and Bar Graphs

Perhaps the simplest bar graph is made of strips of paper or materials that directly represent a variable. For example, each day some children tore strips of paper equal to the height of their plant and pasted the strips on a piece of cardboard. (See pages 13–14.)

Histograms can be made by making an "X" or "✔" each time an event occurs.

How many peas in a pod? Some children were given a bowl full of pea pods. They counted the peas in each pod and for each pod pasted a square in the appropriate space to indicate the number of peas. The result was a histogram that resembled a bar graph. (See Figure 7–4.)

Histograms are very helpful for showing the distribution of counts.

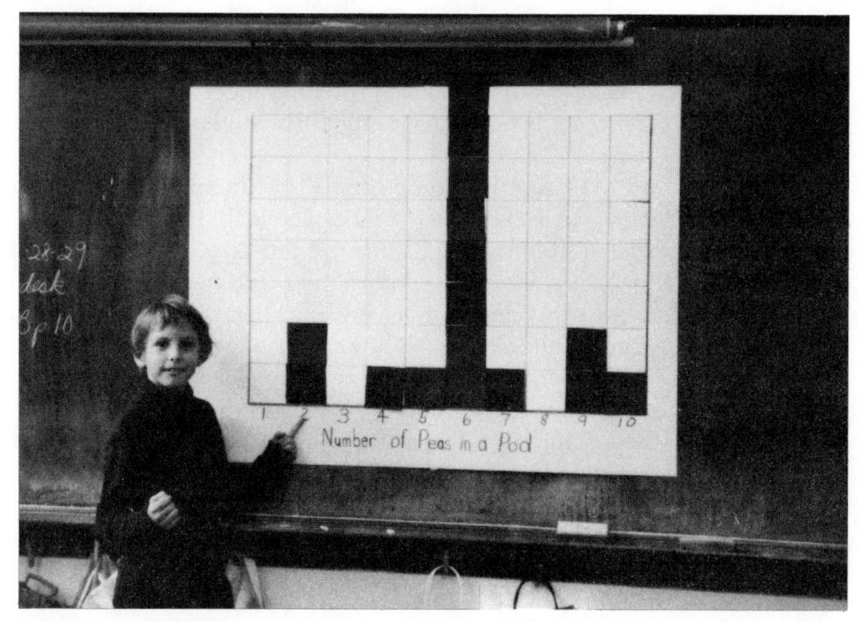

Figure 7–4 A fourth grader's histogram of the number of peas in pea pods.

In *bar graphs* a unit of length represents a certain quantity. For example, a centimeter may represent one pea in a pod. The length of the bars will then be proportional to the number of peas. The bars, of course, can be arranged on the basis of some other variable such as length. Bar graphs are very useful for studying relationships between two factors.

Line Graphs

Line graphs have many of the same attributes as bar graphs. They are especially useful for keeping an ongoing record. For example, a third-grade class checked the outdoor air temperature at 9 A.M. and recorded the readings on a line graph. (See page 299.) The graph helped them detect changes and trends.

One of the advantages of arranging data on graphs is that it can be helpful in projecting or predicting. Children who have made a bar graph of the changes in the height of a plant over time will be better able to project what the height of the plant may be tomorrow or a week from now. If they have made a histogram of the number of peas that they have found in various pods, they will be in a better position to predict the number of peas to be found in the next pod to be opened. Usually it is desirable for children to practice using graphed data to project or predict.

ANGLES

An *angle* is formed when two lines meet at a point. In many science experiences it is necessary to measure angles or to construct angles of a certain size.

To develop some concept of the nature and size of angles, a circle can be compared to the face of a clock. In the classroom a reference line, such as a line going down the center of the room, can be agreed upon. From a particular point on the reference line, children can practice giving the directions of objects, for example, "The aquarium is in the direction of four o'clock." They can also be asked to name "the big object that is at 9 o'clock."

The size of angles is measured with a protractor. After learning how to measure the sizes of angles, the children can practice by measuring the size of each of the angles of a triangle and finding the sum of these measurements. Are the sums of the angles of a triangle always equal? Another science experience involving the measurement of angles is the study of light, in which the angles of incidence and reflection are measured and compared. (See pages 403–4.)

In some science experiences it is necessary to use a protractor to construct an angle of a given size. Astronomers do this when they measure the distance to nearby stars using the method of parallax. Children construct angles of a determined size when they find the distance to an object, using a similar method of parallax. (See page 319.)

Perhaps the most direct way to find the latitude of a place in the Northern Hemisphere is to sight on the north star (Polaris) and measure the angle this line of sight makes with a vertical line. The direction of the vertical is the direction of a string to which a weight has been attached when it has come to a position of rest. The direction of the sightings can be determined using a protractor with an attached drinking straw, as described on pages 459–60.

MEASURING TIME

Time is sometimes referred to as the "fourth dimension." Through science experiences children can develop more refined concepts of various periods of time. Early in life they become aware of days and of periods shorter than the day. As their experience increases, they become aware of seasons and years. As they mature, they will develop clearer concepts of the human life span and the periods within the life span. Although many adults may not have these concepts, through such teaching devices as the "geological clock analogy" (see page 344) they may even gain some concept of the great expanses of geological time.

Early in life children learn how to use watches and clocks to measure time. Some children may need practice in the use of these devices. After they have become proficient, they may be asked to plan their day without the use of watches or clocks. Some teachers have covered the face of the classroom clock and asked the children to devise other ways of "telling time."

Sundials

Some object, such as a pencil, stuck in a piece of clay, can be placed upright in the sunlight. The passage of time is indicated by the movement of the object's shadow. With the help of a watch, the dial can be calibrated so that the position of the shadow will indicate the time of day. Sundials are not effective in measuring short periods and, of course, cannot be used when the sun is not shining.

Sand Timers and Water Clocks

Both of these timing devices are based on the steady flow of sand or water from a container. Sand timers are available commercially to time such operations as the boiling of eggs. A sand timer can be made by punching a small hole in the tip of a filter paper and placing the filter paper in a funnel. Fine sand can be poured into the filter and the sand allowed to trickle into a narrow, straight-sided tumbler. The tumbler can be calibrated so that the height of the sand will indicate the amount of time that has elapsed. Water clocks are constructed so that the amount of elapsed time will be indicated by the amount of water that has dripped into a container. An empty frozen juice can or styrofoam coffee cup with a small hole punched in the bottom with a thumb tack will make an excellent water clock. Sand timers and water clocks are appropriate for measuring shorter periods than can be effectively measured with a sundial.

Rhythmic Counting: the Human Pulse Clock; the Metronome

A number of simple procedures are available for measuring relatively short periods. One of the simplest is to count "one thousand one, one thousand two, etc.," with the time needed to state each number approximately 1 sec. Although the human pulse rate increases with exercise, it does beat regularly and can be used as a timing device. The unit of time is "one pulse." Some children might be interested to learn that Galileo used his pulse to time the swings of the chandelier in a cathedral that led to some of his important studies of swinging pendulums. The metronome is a timing device used in music that can also be used for timing events in science.

The Seconds Pendulum

The swinging pendulum is the basic movement in many clocks and other timing devices. A pendulum that makes a swing in one direction (one-half of a complete back-and-forth swing) in one second is called a "seconds pendulum."

Attach a weight to a string and hang it so that the weight is free to swing and the length of the pendulum can be adjusted. Pull the weight back, release it, and count the number of swings in 10 seconds. Now adjust the length of the pendulum until there are 10 swings in either direction in 10 seconds. A seconds pendulum is about 99 cm (39 in) long. Children can practice using the seconds pendulum by measuring the length of time it takes to bring a flask of water to a boil, dissolve a given quantity of salt in water, or walk up a flight of stairs.

MEASURES OF CENTRAL TENDENCIES—MEAN, MEDIAN, MODE

Undoubtedly, some statistical literacy will be important for our future citizens—the children in our classrooms. Among the concepts that are important are various measures of central tendencies. Many science investigations provide a setting for developing these concepts.

The following are three measures of central tendencies, with an example of a science experience in which this concept can be developed.

What is the mean? median? mode? The *mean* is the average of a set of numbers. Children can find the average of the temperature readings that they have taken over several days.

The *median* is the middle number in a sequence of numbers. Children can measure the lengths of the mealworms in their colony, arrange them in order of length, and note the length of the one in the middle.

The *mode* is the most frequent measure that occurs among a group of numbers. Again, children can measure the lengths of their mealworms; the most frequent length is the mode.

PROBABILITY

Many possible events cannot be predicted with complete certainty. For example, in weather forecasts we often hear such statements as "There is an 80 percent chance of rain tomorrow." This is a statement of the probability of rain. The statement does not state that there will be rain; rather, it is a statement of the probability of rain. Similarly, if we toss a coin the probability of "heads" or "tails" is 50 percent. In many science investigations statements of probability make it possible to evaluate the degree of precision or reliability of measurements and to have some knowledge of the degree of dependability of predictions.

Probability is usually stated as a fraction:

$$\text{Probability} = \frac{\text{number of ways an event can occur}}{\text{number of possible outcomes}}$$

In the tossing of a coin, heads (or tails) can occur in only one way, but the number of possible outcomes is two. Therefore, when tossing a coin the probability of heads or tails is

$$P = \frac{1}{2}$$

Such a probability can easily be converted into a decimal or a percent, as it often is in the statement of the probability of certain kinds of weather.

The probability of heads or tails when a coin is tossed is 1/2. This is a *theoretical probability*. However, when you toss a coin a number of times a head or a tail may not actually be up half the time. Instead, you will get a figure that may approximate 1/2 but will not be exactly that fraction. The fraction that you actually get when you carry out an operation such as tossing a coin is called *empirical probability* (P).

$$P = \frac{\text{frequency of an outcome } (f)}{\text{total number of observations } (n)}$$

This is the probability that is derived from direct experience.

In many cases we do not know all the possible outcomes, and we cannot know all the factors that might affect the outcomes. Then we try to make estimates of probability from actual experience—empirical probability.

What is the empirical probability of heads when a coin is tossed? Under usual conditions the theoretical probability of heads when a coin is tossed is 1/2. How often does a head turn up when a coin is tossed?

Usually it is desirable for the children to work in groups of two. One child does the coin tossing while the other records the results. Have them toss a coin 100 times and calculate the empirical probability using the following formula:

$$P = \frac{(f) \text{ number of heads}}{(n) \text{ total number of tosses}}$$

To what extent does their empirical probability differ from the theoretical probability? What might be an explanation for any discrepancy found?

What are the theoretical and empirical probabilities when two coins are tossed? To determine the theoretical probabilities, all possible outcomes can be listed:

First coin	*Second coin*
heads	heads
tails	heads
heads	tails
tails	tails

In this case there are four possible outcomes. The theoretical probability of both coins turning up heads or tails is

$$P \text{ (both heads)} = \frac{1}{4}$$

The theoretical probability of a head and a tail turning up is

$$P \text{ (head and tail)} = \frac{2}{4} = \frac{1}{2}$$

Now toss three coins 100 times and keep a record of the results. What is the empirical probability of having two heads turn up? two tails? How closely do the empirical probabilities correspond to the theoretical probabilities? How might one explain any discrepancies noted?

Some children may wish to determine the theoretical and empirical probabilities for the results when four or more coins are tossed.

REPRESENTATION AND SCALING

In many activities in everyday life, it is necessary to make representations of objects, and often these representations must be on a different scale than the object that is being represented. Maps, for example, have definite scales so that a relatively small map can represent a very large area. Direct experiences with representation and scaling are very important in helping children learn that these symbolic representations have concrete referents. In a variety of science activities, children gain experience in representing objects and arrangements in two-dimensional sketches and, less frequently, as three-dimensional models.

Perhaps the simplest representation is the *outline trace* of familiar objects. Very young children can trace the outlines of their hands, scissors, pencils, books, and other objects on paper.

Schematic sketches can be made to depict and better understand various arrangements for experiments and to convey the meanings of various terms. A group of children wished to wire their dollhouse so that it could be illuminated with flashlight bulbs. They made a schematic sketch of how the wires would be connected to the dry cells and the bulbs. They were able to check their sketches to make certain that they had complete circuits. Other children made sketches of the thermostat control in their aquarium heater and labeled the various parts. (See page 185.) Many children have dissected flowers and made sketches of the flower, labeled the parts, and used these sketches to explain how flowers are pollinated.

Children should have experiences in enlarging and reducing representations such as sketches. A simple way of enlarging a representation of an object is to put the object on an overhead projector or an opaque projector. The image can be projected onto a sheet of paper taped to the wall. If a sketch is first made on a grid paper, a sketch of either larger or smaller scale can be made by representing one grid in the original sketch by more or fewer grids in the new sketch. Some children may wish to practice using the *pantograph* in making sketches of different scales. However, this should not be merely mechanical representation; the children should learn how the pantograph works.

Maps drawn to scale are used in several areas of science. (See pages 465–66.) An interesting exercise in which children can learn the meaning of scale is to make a map of their desk and later a map of their classroom. In mapping a classroom a scale of 1 cm to 1 m is often used. As in many maps, this scale can be represented by line segments that represent various lengths. After making an outline map of the classroom, the children can try to represent as accurately as possible the positions of various objects in the room on the map. An important use of mapping in science is to indicate the locations of various plantings in a garden. A third-grade class, for example, mapped a flower garden to indicate the locations of the various seeds. A somewhat different application is to make a map of a square meter of schoolyard and indicate the locations of various plants. (See page 193.)

Children can also have experience in using maps to locate positions or objects. Large-scale topographical quadrangles of the local community are especially useful. (See page 154 for information about topographical quadrangles.) A group of fifth-graders, for example, used the topographical map of their locality to locate a bench mark that had become concealed in underbrush.

Children can also be given experience in locating objects on a map by using a grid system. One way of doing this is to make a transparency from grid paper. The transparency can be laid on top of a map. The vertical and horizontal lines on the grid can be numbered, and these numbers can be used to give the locations of positions on the map. Many local maps have a grid system.

REFERENCES

American Association for the Advancement of Science. *Science—A Process Approach.* Lexington, Mass.: Xerox, 1974. Many of the activities in this science program feature the development of mathematical concepts and skills. The processes entitled "Using Numbers," "Measurement," and "Space/Time Relations" provide many opportunities for mathematical development.

Elementary Science Study. *Mapping,* 1971; *Match and Measure,* 1971; *Pattern Blocks,* 1960. New York: McGraw-Hill. Three modules that can contribute to the development of mathematical skills.

Jacobson, Willard J., Edward Victor, Maurice A. Bullett, Richard O. Konicek, Matthew F. Vessel and Herbert H. Wong. *Measurement.* New York: American Book, 1972. A unit in the *Investigating in Science* series that emphasizes approaches to measurement and ways of measuring length, area, volume, mass, weight, and time.

Katagiri, George, Doris Trojcak and Douglas Brown. *Skill Builders.* Morristown, N.J.: Silver Burdett, 1976. One of the *Self-Paced Investigations for Elementary Science* that features the development of various skills within the context of science.

Swartz, Clifford E. and Roy A. Gallant. *Measure and Find Out.* Glenview, Ill.: Scott, Foresman, 1969. This is a series of three books in which science activities that feature a quantitative approach are described.

University of Illinois Astronomy Project. *Charting the Universe.* New York: Harper & Row, 1969. This first of a series of books on astronomy provides experiences with lines, angles, and measurement that are related to the study of astronomy.

Science and the Development of Logical, Formal Thought and Concepts

CHAPTER 8

Think about the friends that you have. We hope there are many who love to play with ideas and enjoy speculating about the possible consequences of ideas. Some are good at defining problems and suggesting a number of reasons why some problem occurred or something went wrong. They are able to separate reasonable causes, and they are imaginative in suggesting possible solutions. They also try to think through what might happen if one idea or another were tried. Sometimes they suggest ways in which ideas can be tested. They always seem to learn from experience and make fewer mistakes next time. These friends are operating with logical, formal thought in at least some dimensions of their lives. One of our most important goals in education is to make it possible for children to reach this level of intellectual operations. We now know that they are unlikely to reach it without certain kinds of educational experiences. The nature and examples of some of these experiences are discussed in this chapter.[1]

Thinking has been defined as using what we know to deal with prob-

[1]In the preparation of this chapter, we have profited greatly from materials like those listed as References. Many of these have been written by colleagues in the field of science education. Discussions with Doris Withers, who has a profound understanding of the nature of logical and formal thought and the kinds of experiences that contribute to its development, have been especially helpful.

lems. Thus, in order to think, a person has to know something and know how to use it to deal with problems. What one knows is derived from the wide variety of experiences that every individual has. But it is possible to learn how to learn from experience. (Most of us can recount cases in which people seem to make the same mistakes over and over again.) It is also possible to learn how to make better use of the knowledge that we have to deal with problems. (See Chapter 3 for ways in which knowledge can be used in the formulation of hypotheses.) Even more important, children should be encouraged and helped to reflect on the procedures they have used in dealing with problems.

Science concepts are the broad, general ideas we have of the nature of ourselves and the world in which we live. In elementary school science inquiry and concept development have often been separated, and one has been emphasized to the detriment of the other. Obviously, both are important. We believe that both should receive attention, and this can be done *by engaging children in science investigations from which they learn how to deal with problems and also develop significant science concepts.* Central to this approach are science investigations, and throughout the science content chapters of this book there are suggested investigations in which children can become engaged. Chapter 20 contains additional investigations. Experience in such investigations should lead to concept development as well as greater skill in dealing with problems.

Investigations into the nature of human intellectual development support the thesis that most individuals pass through several stages of intellectual development.[2] (See Chapter 2 for a detailed discussion of these stages.) Passage through the sensory–motor and preoperational stages apparently takes place even without formal education. For example, children who have been deprived of formal schooling have been tested with various conservation tasks and found to be capable of concrete operations (i.e., they are able to reason with concrete materials before them). However, passage through concrete operations and the attainment of formal, logical operations require education, and this education apparently has to be of a very special kind. An education that emphasizes rote memorization, for example, may actually inhibit the development of concepts and logical, formal thought. Since formal, logical operations provide the individual with much greater intellectual power, it is of critical importance that children have these kinds of experiences in the elementary school. And sci-

[2]These investigations are usually linked to Jean Piaget, who throughout a long and distinguished career has investigated the nature of human intellectual development. Because science education can make such important contributions to the development of formal, logical thought, a number of science education researchers have turned their attention to the study of intellectual development. Among these investigators are Gary Bates, Anton Lawson, Robert Karplus, Ronald Raven, John Renner, and Doris Withers.

ence offers experiences that are especially appropriate for the development of formal, logical thought.

In this chapter we discuss some of the characteristics of formal, logical thought and the nature of the educational experiences that can contribute to the development of these intellectual powers. For each characteristic we provide an example of a science activity. There are many other such activities in the content chapters of this book. In the latter part of the chapter, we discuss concepts and how children's science concepts can be developed.

Science provides an almost ideal context for the development and use of concepts and of logical, formal thought. Almost all researchers in this area agree that children should have a wide range of experiences with concrete materials; from these experiences ideas develop that can provide the basis for thought. Also, it is important to have experience with activities in which the relevant factors can be perceived and controlled. For example, it is easier to grasp the relevant factors in a pendulum experiment than in some complex social problem. The order and regularities in the physical environment help children suggest hypotheses and possible tests of the hypotheses. Also, many experiments can be repeated so that children can go back and check their observations or try a different set of operations.

SEPARATING VARIABLES

When several factors may be influential, it is important to be able to separate variables so that their effects can be studied. In science and throughout life, it is often important to find out what factor or factors affect a situation. If we can separate and identify these factors, it may be possible to do something about them. Science activities such as the one that follows give children practice in separating variables. Then it is possible to design experiments to study the possible influence of these factors.

What factors affect the germination of seeds? In this investigation two factors, sunlight and moisture, will be varied. The children are asked to determine which of the factors affects the germination of seeds.

Fill four paper milk carton bottoms or small flower pots with the same kind of dry soil. You may have to heat the soil in an oven to make certain that it is completely dry. Plant four seeds of the same kind (corn or lima bean seeds work well) in each of the cartons of soil. The seeds should be as nearly alike as possible and should be planted at the same depth and in the same way.

Place two of the cartons in a place where they are exposed to sunlight and the other two in a place where they receive no light. Add water to one of the cartons in the sunlight and one that is in the dark. The soil in the other two cartons should be kept dry. The arrangement of the investigation is diagramed in Figure 8–1.

The children should keep a journal describing how the investigation was set up, when they added water and how much, and any observations they make.

How many seeds germinated in each carton? What variable seems to affect the germination of these seeds? Which factor seems to have no affect? The children may not get exactly the results that they expected. For example, most, but not all, of the seeds may be viable, and it is difficult to control for this.

If the children succeed in identifying the influential variable, you may wish to set up investigations in which there are three or more variables, for example, different kinds of seeds and different temperatures.

Out of investigations like these, children can refine their concepts of the conditions that are necessary for seeds to germinate and for plants to thrive. A "green thumb" is really a colloquial expression that implies the ability to use concepts of what plants need to live and grow.

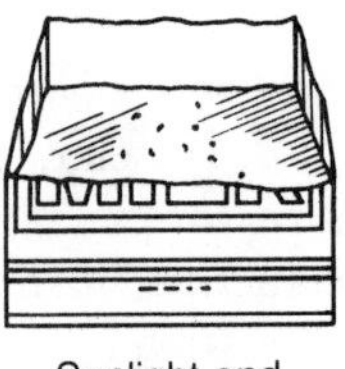

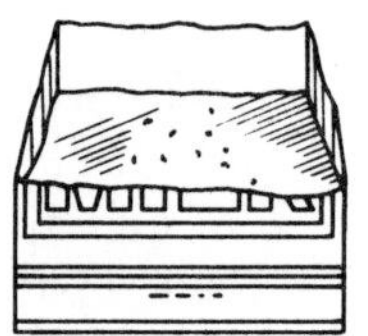

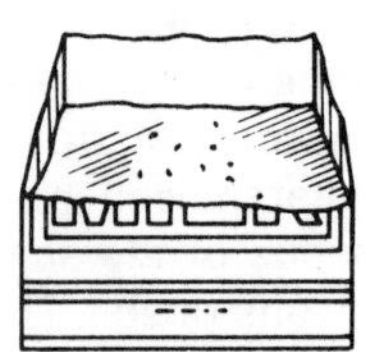

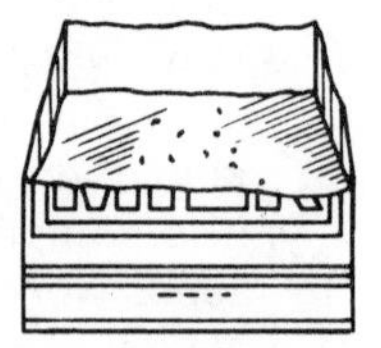

Figure 8–1 Which factor, sunlight or moisture, affects the germination of seeds?

CONTROLLING VARIABLES

Experiments may be designed so that the effect of varying one factor can be studied. Some fourth-grade children had set out two dishes of water. The water in one dish disappeared while there was considerable water in the other dish. Why? One child suggested that "there was more water to begin in one dish." Another said, "I believe that it was because one dish was placed in the sunlight and the other was not." A third child said, "I believe that it was because air was blown across one dish and not the other." Still another child said, "It is because the dishes have different shapes." In order to find out which of these factors led to the water in one dish disappearing while water remained in the other, various experiments could be designed in which all factors except the hypothesized one would be controlled. But if two or more factors are varied at the same time, as children are often wont to do, they cannot learn which of the factors led to the disappearance of the water. In this case the children hypothesized that water in one dish disappeared because the dish was placed in a warmer place. They controlled for all other factors and found that their hypothesis was correct.

The following experiments with a pendulum are designed to give children experience in designing experiments, collecting and organizing data, and interpreting information. Simple and easily obtainable materials are used. We suggest that children work at these experiments in pairs.

What factors affect the number of times a pendulum will swing in a given period? These experiments can be done with a string, washers or other weights of equal size, a measuring stick, and a clock or watch with a second hand. (Many classroom clocks have second hands.) The pendulum should be hung from a support so that it is free to swing without striking an obstruction. Some children have found it convenient to make a hook out of a paper clip and fasten it to the end of the string so that weights

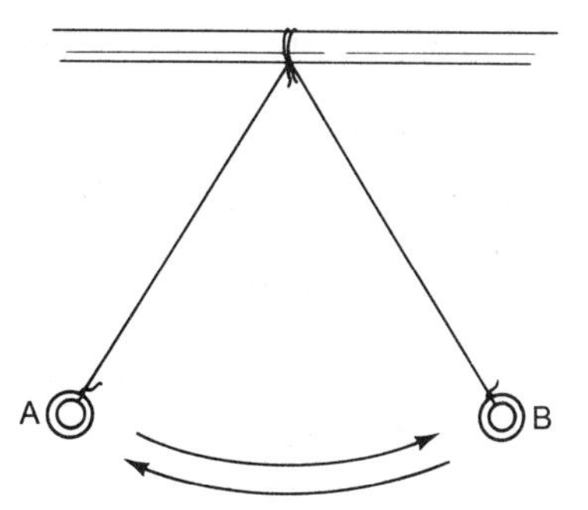

Figure 8–2 A complete swing of the pendulum is from A to B and back to A.

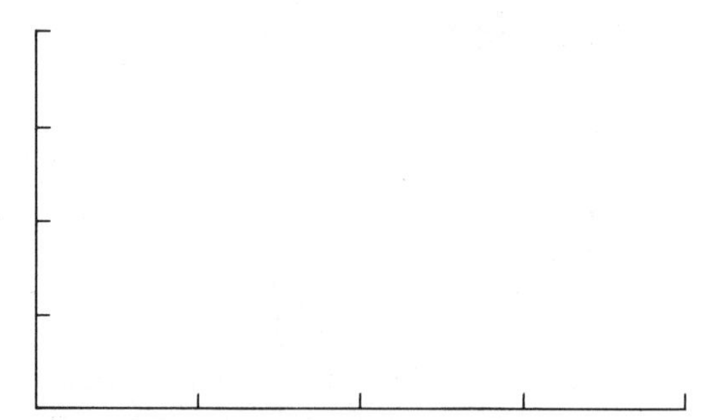

Figure 8–3 Graphs are helpful in interpreting data. Children should have practice in selecting the most useful coordinates, plotting data, and making interpretations of data.

can be added easily. Two operational definitions are useful: (1) A *complete swing* is from A to B and back to A. (See Figure 8–2.) You may wish to discuss with the children the precision with which complete swings can be counted. For example, "Is it possible to count half-swings of the pendulum?" (2) The *amplitude* is the greatest distance that the pendulum bob moves from its rest position.

After playing around with their pendulums, the children may notice that under some conditions pendulums will swing faster than under other conditions. "What factors affect the rate at which a pendulum will swing?" Among the factors that may be suggested are amplitude, weight of the bob, and length of the pendulum. In each of the following experiments, the children should try to control for all factors except the one that is being tested.

In these kinds of experiments, it is usually desirable to make several observations. Then the results can be averaged. You may wish to call attention to the importance of *significant digits.* You should have no more digits in your average than the least number that there are in the numbers you are averaging. For example, if you average the numbers 31.6, 31, 25.6, 32, and 63.9, you should not have more than two digits in your average because the least number of digits among those averaged (31) is two.

To help interpret these data and to generalize, it is often useful to graph the data that are collected. You may wish to use this opportunity to teach the children how to set up a graph and select the most useful coordinates. (See Figure 8–3.)

Amplitude. Some children may hypothesize that if you pull the pendulum bob farther back there will be more swings. To test this they will have to keep constant such factors as length and weight of the bob. It is suggested that a pendulum one meter long be used. The weight of the bob should not be changed.

It is suggested that the number of complete swings there are in 15 seconds be counted when the pendulum bob is pulled back a distance of 10 cm from the rest position. Make three observations, average, and graph the results. Repeat with amplitudes of 20 cm, 30 cm, 40 cm, and 50 cm. Does changing the amplitude affect the number of swings completed in 15 seconds?

Weight of the bob. It can be hypothesized that if you increase the weight of the pendulum bob there will be more swings. The weight of the bob can be progressively increased by adding washers or other equal-sized weights to the paper clip hook at the end of the pendulum string. To test this hypothesis such factors as length and amplitude will have to be held constant. Some children have used a pendulum length of 1 m and an amplitude of 20 cm.

Figure 8–4 What affects the number of times a pendulum will swing in fifteen seconds?

Have the children make three observations of the number of swings when one weight (washer) has been added. Find the average and plot the result on a graph. Repeat with two, three, and four weights. Connect the dots on the graph. Does changing the weight of the bob affect the number of swings completed in 15 seconds?

Length. It can be hypothesized that if you shorten the length of the pendulum the number of complete swings in a given period of time will increase. In this experiment amplitude and weight will have to be held constant. Some children have used an amplitude of 20 cm and a pendulum weight of one washer.

Count the number of complete swings when the pendulum length is one meter. Make three observations. Find the average of the observations and plot it on a graph. Repeat for pendulum lengths of 80 cm, 60 cm, 40 cm, and 20 cm. Does changing the length of the pendulum affect the number of swings completed in 15 seconds?

After the children have completed these three experiments, they should be encouraged to generalize from their experiences. "From your experiments, what factors do and do not affect the number of times a pendulum will swing in a given period?"

HYPOTHETICO-DEDUCTIVE REASONING

Suggesting hypotheses and making logical deductions from those hypotheses are important aspects of logical thought. Hypotheses are suggested answers or solutions to questions or problems. We use all that we know about a question or problem to suggest a possible hypothesis. Thus, the process of suggesting possible hypotheses is closely related to thinking. If we are concerned with educating for thought and thinking, children should have many opportunities to use what they already know to suggest hypotheses. (See Chapter 3 for a discussion of hypotheses and the nature of science.)

One of the important byproducts of encouraging children to use what they know to suggest hypotheses is that they utilize many ideas that they have learned and discover that "much knowledge is of some use." Too often children learn a fact or concept and then never again have an occasion to use it. As a result some children seem to learn that "you never use what you learn in school." The imaginative teacher provides many settings in which children can use much of what they have learned as they struggle to come up with hypotheses. If we can help children think more effectively—and we believe we can—then encouraging children to use what they know to suggest possible answers or solutions is a key element in this kind of education.

Once hypotheses have been suggested, logical deductions from these hypotheses should be made and tested. For example, when a new school was opened some children set up an aquarium, complete with goldfish, in their new classroom. They set up the aquarium exactly as they had in their old school. They used water that had been allowed to stand for several days, a clean aquarium and sand, and healthy goldfish. They used an aquarium heater to maintain the recommended water temperature. But the fish died! The children tried again, and they were even more careful to set up the aquarium as they had before. Again the fish died. Both the children and their teacher were puzzled. One of the children remembered having heard that even a very small amount of copper in aquarium water can kill fish. They hypothesized that the water in their new classroom had flowed through copper pipes and a very small amount of copper might have been dissolved in the water. They partially checked this hypothesis by finding out from a school custodian that copper tubing was used in the plumbing in the new building. They deduced, "If we set up an aquarium using water from a building that does not have copper plumbing, then the fish should live." Two children helped the teacher bring water from his home. In this water the fish lived. In this series of experiences, the children and the teacher encountered a problem. They used an idea that one child had learned to suggest a hypothesis. Then they devised an investigation by which they could test this hypothesis.

What happens to a streak of water on a chalkboard? Dip a finger into some water and make a wet streak on the chalkboard. Ask the children to observe carefully what happens. Urge them to describe their observations as carefully as possible. If appropriate, probe for the meanings of such terms as *evaporation.* Then ask, "What happened to the streak of water on the chalkboard?"

There may be several hypotheses. (See page 63 for some of the hypothesis that children have suggested.) However, in our experience children tend to hypothesize that "the water evaporated" "the water went into the air."

Ask, "How can we show that the water went into the air?" The children may be genuinely perplexed; be sure to give them time to think. Too often, teachers ask difficult questions that should require considerable thought but then allow no time for thought. (When we did this in a fifth-grade class, one youngster gave an ingenious suggestion: "Place some cobalt chloride paper, which changes color when the relative humidity changes, inside a glass tumbler and place the glass tumbler over the streak of water. If the water goes into the air, the color of the cobalt chloride paper should change.") Some children have suggested that "if the water goes into the air, then if we move more air across the streak it should dry faster."

Ask, "How can you test this idea?" Several groups of children with whom we have tried this have suggested that we put two streaks that are exactly alike on the chalkboard and fan one and not the other. They have usually managed to make two streaks that seem to be very much alike. The fanning is usually vigorous and the results are what they expected.

Regularities in the physical universe provide a basis for suggesting hypotheses and deducing the logical consequences from these hypotheses. This is one of the reasons that science is such a fertile field for the development of logical, formal operations. For example, an important physical regularity is that a mixture of water and ice will remain at 0°C until all of the ice has melted. A group of 10-year-old children checked the temperature of a water–ice mixture in their classroom and found that it was indeed 0°. However, when they were asked what the temperature of a water–ice mixture would be tomorrow, they replied, "You can't tell ahead of time; it depends on how hot it will be tomorrow." Apparently, they had not recognized that this regularity will hold for tomorrow and, of course, has held in the past. When asked what the temperature of a water–ice mixture would be in Alaska, one youngster replied, "Alaska is very cold. Therefore, the mixture would be much colder there." This youngster had yet to recognize that this is a regularity that holds everywhere on earth and probably throughout the universe. When youngsters understand the universality of such regularities, they have a better basis for suggesting hypotheses, deducing consequences of hypotheses, and suggesting investigations to check on the deduced consequences.

PREDICTION

What will happen if we try it? One of the powerful features of logical, formal thought is the disposition and ability to predict possible consequences of ideas. To develop this capacity it is important for children to have a variety of experiences in which they predict the consequences of operations with concrete materials. Children at the concrete-operations level can do this. After gaining this skill they may be able to use symbols to represent the operations and to predict the possible consequences of the application of ideas.

"What will happen if we try it?" Children are usually eager to try it and find out. But at least some of the time the teacher should press the children to predict what will happen when they try it. Again, they will need to use what they already know.

Mixing equal amounts of water at different temperatures. The children should measure out equal amounts of cool water and warm water and place the samples in separate polyfoam cups. These cups, which are often used for coffee, are used to reduce the amount of heat lost by the liquids. The cool water should be at about 20°C (64°F) and the warm water at 30°C (86°F). However, the children should measure and record the temperature of the water in each of the cups.

Have the children predict and record the temperature that they think will result when the water from one cup is poured into the second cup.

Have the children pour the water from one cup into the second cup and take the temperature. How close is the resulting temperature to their prediction? Why did they predict as they did? What may have caused any difference between the actual temperature found and that predicted?

Mixing unequal amounts of water at different temperatures. This is a somewhat more complicated investigation of mixtures in which the children are asked to predict the temperature of the resulting mixture when unequal volumes are mixed.

Have the children put three measures of warm water (40°C) (104°F) into one polyfoam cup and one measure of cool water (20°C) (64°F) into another polyfoam cup. Have the children measure and record the temperature of the water in each of the cups.

Ask them to record their predictions of the temperature that will result when the cool water is poured into the warm water.

Have the children pour the cool water into the warm water and stir with the thermometer. How close is the resulting temperature to their prediction? Ask them to explain why they made the prediction that they did and what may have caused any discrepancy between the actual and predicted temperatures.

Many of the investigations in elementary school science that are designed to develop skills and intellectual power also contribute to the further development of important concepts. The investigations involving the mixing of liquids can contribute to more sophisticated concepts of the conservation of heat energy. The amount of heat energy in water depends on both the temperature and the amount of water. There is more heat energy in a volume of water at 40°C (104°F) than in an equal volume of water at 20°C (64°F). Similarly, there is more heat energy in three volumes of water at 40°C (104°F) than in one volume of water at 40°C (104°F). Although a small amount of heat energy may be lost to the surrounding environment during an experiment, the heat energy is substantially conserved. This concept of the conservation of heat energy has wide application and can be used to make predictions when known amounts of water at different temperatures are mixed.

DEALING WITH MANY POSSIBLE SOLUTIONS

When dealing with many possible solutions, it is important to suggest ways in which the most likely solution can be identified. In many life situations there is a variety of possible solutions and combinations of solutions to problems. At the formal, logical level of operations, individuals are able to keep this variety of possible solutions in mind and to manipulate them mentally. Sometimes some of the possible solutions are contradictory, so that if one is proven to be possible the contradictory ones are impossible. Usually, some solutions are more likely than others; children should be encouraged to use their imaginations to suggest possible solutions. However, as other possible solutions are suggested, some of the "wilder" suggestions tend to fall by the wayside. Sometimes investigations can be planned to test possible solutions. For other suggested solutions it may not be possible to suggest empirical tests; these ideas can often be analyzed philosophically for their possible consequences.

In the following science activity, students are asked to suggest as many possible solutions as they can imagine. Later they are asked to suggest possible tests for these ideas. We have listed some of the hypotheses and tests for these hypotheses that have been suggested by several groups of children with whom we have worked.

What could have happened to the water on the chalkboard? Dip your finger into some water and make a streak on the chalkboard. Ask the children to suggest as many explanations as possible for the phenomena that they observe. Encourage them to be imaginative. Then ask them to suggest possible tests for the various hypotheses suggested.

The following are some of the hypotheses that have been suggested by groups of children with whom we have worked, along with the tests of the hypotheses that they have proposed.

1. "The water went into the air."
   Test: "If the water went into the air, then it should disappear faster if we fan it."
2. "The water went into the chalkboard."
   Test: "We should get some more of the material that the chalkboard is made of. If we put water on one side, we should be able to see it come through on the other side."
3. "The water was soaked up in the chalk."
   Test: "We can wipe off all of the chalk from a part of the chalkboard and see if the water still disappears."
4. "The water spread out in a thin film across the board."
   Test: "We can put a streak on a very small piece of chalkboard material and see if the water disappears and if we can detect a thin film."
5. "The water disappears in all of the ways that we have suggested."

Test: "This is hard to prove. But we may be able to show that the water does not disappear in some of the ways that we have suggested."

6. "The water just disappears."
Test: This suggestion almost always has evoked a great deal of animated discussion. "But if you can say that something 'just disappears', then you really don't have science. Maybe you have magic."

This investigation can contribute to a more sophisticated understanding of change of phase or state, particularly the change from a liquid to a gas. To further illustrate such changes, an ice cube can be left in a glass tumbler for several days. What changes take place in the ice cube?

INFERENCE

We can sometimes make reasonable inferences from observations that we make. However, distinctions should be made between observations and inferences. Careful observation and description are very important processes in science. Children can describe objects in terms of color, size, shape, texture, and smell. They can also describe interaction and change in terms of sounds that are made, light that is given off, heat that is generated, changes in position, and other evidences of interaction.

However, it is possible to go beyond direct observation with the senses and make *inferences.* For example, a package is placed on a dry table top. After a while the package wrapping and the table top become wet. We *observe* the wetness, but we can *infer* that there is a liquid in the package. We may also infer that there is a strong likelihood that the liquid's container has been broken or has begun to leak. We may infer, but with much less certainty, that the package has been dropped or handled roughly.

It is important that children learn to make distinctions between observation and inference. For example, we observe directly by seeing, touching, and possibly smelling that the wrapping of the package is wet. We can infer that the package has been dropped, but we know this with much less certainty than when we use direct observation.

As children become more sophisticated, they begin to judge inferences in terms of the degree of certainty or probability that they are true. Children may use such phrases as "more likely," "less likely," or "quite likely." Later, as they continue their studies, they may learn to state probabilities in quantitative terms.

What is in the package? Place an object such as a pencil or a paper clip into a box and tape the cover of the box shut. Then show the children a variety of objects (e.g., a rubber ball, a block, a bottle cap, a coin, and an eraser), but be sure to include among the objects one that is similar to the one you have placed inside the box. Provide a second box that is similar to the one that contains the object.

"How can we find out what kind of object is in the box?" The children may suggest various operations. They may suggest that the box be shaken. After it has been shaken, place each of the other objects in turn into the second box and shake them. The children may wish to suggest other operations that can be carried out in the second box. What kind of object do they infer is in the taped box?

Place some other object with which all the children are familiar into a box and tape the box shut. Ask the children to make as many observations as they can about the object inside the box. Weight? Smell? Does it roll or slide when the box is tipped? What kind of sound is made when the box is shaken? Emphasize that these are observations. Ask them to infer on the basis of these observations what kind of object is inside the box.

The distinction between observation and inference is an important one in the sciences. For example, no one has made a direct observation of the nature of the center of the earth. But observations have been made of how earthquake waves travel through the earth. We also observe that the earth has a magnetic field, and the mass of the earth has been found. On the basis of many kinds of observations, we infer that the earth has a core, that at least some of the core consists of nickel and iron, and that at least the outer part of the core is not solid. Obviously, these inferences are not as soundly based as direct observation and are more likely to be subject to change as we make more and more observations.

SAMPLING AND PROPORTIONAL REASONING

Intellectual power is increased with the development of the ability to state and interpret relationships in mathematical terms. (See Chapter 7 for further discussion of science experiences and development of mathematical abilities.) In both science and mathematics, stating and interpreting relationships is important. We see relationships, for example, among all species in which mothers can nurse their offspring, and we call them mammals. We may compare the lengths of objects directly or indirectly (when we use the metric scale). Sampling and proportional reasoning are ways of stating relationships that provide children with powerful tools for investigating and interpreting phenomena.

Sampling

A *sample* is a set of observations drawn from a larger population of observations. For example, it is very difficult and expensive to study the characteristics of all the people in a nation; in the United States a census is taken only every ten years. However, a great deal can be learned about a population by studying a sample that is believed to be representative of the total population. In the activity that follows, try to learn the size of the total population by studying a sample. The children should be urged to be critical in determining whether the sample is truly representative of the entire population.

How many beans are there in the jar? A contest often included in fairs and bazaars involves guessing the number of beans (or kernels of corn or pieces of candy) there are in a jar. One approach to solving this puzzle is to use sampling.

Fill several jars with beans and distribute them to groups of children. Ask the children to guess how many beans there are in the jar and write the number on a sheet of paper. Then ask them to suggest ways in which we might make more accurate estimates. When we have done this with children, we have found that they almost always suggest some kind of sampling: "Let's count the number of beans in the layer at the bottom and then count the number of layers there are." Suggest that the children try to do this. Of course, they will run into difficulties, but they may eventually learn that some of these difficulties are almost inherent in these kinds of investigation. Make certain that they make a judgment as to whether the beans in the bottom layer are like all the other beans. "Now, how many beans do you think there are in the bottle?" "How close is this estimate to your first estimate?"

Have the children in each group empty the beans onto a table top and divide the beans among them. Each child should count his or her share of the beans, and the group should calculate the total count of beans in the jar. Compare this number with their two estimates. "Which estimate was the more accurate?" "How do you think we could make more accurate estimates?" Also, "Is there a chance that mistakes were made in counting the beans?" When studying large populations, such as the population of the United States, more mistakes are likely to be made in the study of the large population than with a much smaller, but carefully selected, representative sample.

Scales

Scales are ways in which we represent distances on maps. For example, one centimeter on the map may equal 10 kilometers. The scale is usually printed at the bottom of the map. By using maps and scales, children can get a fairly accurate indication of the distances between points.

How far is it? Any map that has a scale can be used. However, the children may find it especially interesting to use the topographical map on which their school is located. Then distances from their school to various points in the area can be found. (Index maps for your state that show the quadrangle for your community can be obtained from the U.S. Geological Survey, Washington, D.C. 20242. The actual quadrangles can be obtained from the same address or from federal map distribution centers or local map dealers.)

To find the distance between two points on a map, the children can stretch a string between those two points and then compare the length of the string with the scale. Since the scale is usually also stated in terms of some unit of length (e.g., one inch equals one mile), the children can also measure the length of the string and convert it into kilometers or miles.

When we stretch a string between two points, we can find the distance "as the crow flies." Children can also find the distance by road by laying a string along the roads that they would follow, measuring the length of the string, and using the map scale to compute the distance that would have to be traveled.

Proportional Reasoning

The ability to use proportional reasoning is a powerful intellectual tool. For example, if we know that the ratio between two numbers is the same as the ratio between a known number and an unknown number, we can find the unknown number.

How many paper clips tall are you? One of the ways of answering this question, of course, is to make a chain of paper clips that dangles from the top of your head to the floor and then to count the number of paper clips in the chain. Ask the children to suggest some of the inaccuracies involved in this method. (One inaccuracy is that there is a loss in length with each connection between separate paper clips.)

Another way to find out how many paper clips tall you are is to measure the length of a paper clip in centimeters and then to use a meter stick to measure your height in centimeters.

Many children will then determine their height in paper clips in much the same way that they use scales. However, some children will be ready for another way of stating the mathematical relationship:

$$\frac{\text{Height in paper clips}}{\text{Height in centimeters}} = \frac{\text{one paper clip}}{\text{length of one paper clip in cm}}$$

$$\text{Height in paper clips} = \frac{\text{height in centimeters} \times 1}{\text{length of one paper clip in cm}}$$

We can learn by reflecting on the experiences we have. Since we learn by reflecting or thinking about what we have done, an essential dimension of teaching is to lead children into reflecting on the experiences that they have had. At first, children may have to be prompted to think about what they have done, to evaluate various operations they have carried out, and to try to learn from their experiences. Eventually, we can hope that they will develop a predilection to reflect—to try to learn from their experiences. Wanting to learn and knowing how to learn from experience can be one of the most important outcomes of education.

The following experiences in a fourth grade illustrates how one child had learned from a science experience and had probably learned a great deal about how to learn from experience.

How can we obtain and evaluate information? A fourth-grade girl had just completed a rather extensive report for her group on the solar system. The report included considerable information about the sun, each of the planets and its satellites, and the orbits of the planets around the sun, and some information about how two of the planets had been discovered. We asked the girl where her group had gotten all of this information.

"We got almost all of our information from books in the library. The librarian helped us a lot. We looked in encyclopedias and in 31 science books. But we really didn't read any book all the way through. We looked up in the books the information we wanted to find. Janet's father, who is a scientist, said that he usually used books that way too—we looked up the information we wanted to find."

We asked them what they did if the books or sources differed.

"We tried to check on the authors and how reliable they might be. We also checked on the publication dates. Usually, if an old book and a new book disagreed, we took the information from the newer book."

A boy in another group thought he had detected an error in the report. "You said that only Mercury, Venus, and Mars are smaller than the Earth. But Pluto is also smaller. So there are four planets smaller than the earth." These children also demonstrated that they were also able to recognize the limits to what we know.

"We checked that in several books. Most of the books said that we really do not know the size of Pluto. Therefore, since we don't actually know the size of Pluto, we thought it better not to say that it was smaller than the Earth."

Certainly, these children had learned a great deal about how to evaluate sources of information and ways of obtaining information. The discussion after their report is an example of how children can reflect on their experiences. Much of our learning is a result of such reflection.

We as teachers have a special responsibility to encourage and lead children to reflect on their experiences; for many children this will not occur without the encouragement and example of a reflective adult. Certainly, there are many adults who seldom reflect on the experiences they have. It is critically important for children to ponder what they have done and how they have done it and to think about what they will do differently next time. If children develop a predisposition to reflect, in a sense they will have "learned how to learn."

DEVELOPING CONCEPTS[3]

Science concepts can be developed through a variety of science experiences. Concepts are broad, general ideas with which we view ourselves and the world in which we live. They are general ways of viewing the world and therefore are usually much more useful than particular facts. We could ask a child to learn that there may be a million different kinds of living animals, but it is probably more useful to have a concept of the variety in the animal kingdom. Children can learn that frogs have a long, sticky tongue, but it is probably more useful to develop a concept of some of the adaptations that organisms have for getting food. Concepts have sometimes been likened to crystals that settle out of solutions. In this metaphor the solutions are all of the experiences that we have and the crystals are the broad, general ideas that crystallize out of these experiences.

Children develop concepts as they become aware of what is happening in their investigations, experiments, and other science experiences and then relate this awareness to their own previous experiences and to the experiences of others. Too often science experiences are humdrum affairs from which little meaning is derived. But it was once said that the ideal education would be to sit on one end of a log and have Mark Hopkins sit on the other. For Mark Hopkins we can substitute the imaginative teacher. Then the log or any other phenomenon becomes much more than something to sit on—it becomes the bark and the heartwood, the lichens and the mosses, the animals burrowing within the log, the interrelationships between the decaying log and the living plants and animals, and eventually the decaying log and soil and the organisms around it. It is thus that the teacher can help children "see where they have only looked before."

Throughout their education children should have opportunities to refine and extend their concepts. We can have concepts of ourselves and of the universe that are of varying levels of sophistication, and one of the

[3]Joseph D. Novak's *A Theory of Education* is one of the best discussions of concept development.

functions of education is to raise the level of sophistication of our concepts. For example, concepts of the conservation of matter and energy can range from the limited and simple to the general and refined, from "You can't get something for nothing" to "You have to balance the equations because you don't get more or less matter just because you have a chemical reaction. Matter can be neither created or destroyed." At scientific meetings learned scholars discuss interactions in a particle accelerator or the interior of a star and assume that matter and energy will be conserved.

Every youngster will develop and refine many broad, general concepts. The nature and extent of that development will depend in large part on the teachers who help children become aware of what is happening and relate these experiences to broader ideas.

What children learn is related to what they already know. We can help children develop concepts by finding out what they know and relating new experiences to what is already known. Children learn as they relate new knowledge to what they already know. The effective teacher tries to find out as much as possible about the experiences that children have had and the ideas that they have derived from those experiences. The professional teacher has considerable knowledge of a general nature about children of a particular age. But of even greater importance is the particular knowledge that the sensitive teacher gains about what the children in a class already know.

A teacher should be alert and sensitive throughout all experiences with children. We can learn in classrooms, in lunchrooms, on playgrounds, and on walks and talks with children anywhere. It is important for children to have exploratory experiences, just "messing around."[4] But these experiences provide the teacher with marvelous opportunities to find out more about what children know. Others will have different approaches, but we have found it especially helpful to sit down with one or a few youngsters and watch and listen as they play and experiment with science materials. Once in a while, but not too often, a question can be interjected as you seek a better understanding of what children know. Actually, a critically important dimension of teaching is finding out as much as we can about what children already know and then planning our teaching accordingly.

Whenever possible we should help children relate their experiences to what they already know. This can be done in many ways, and many suggestions are made in the content chapters of this book. How-

[4]For a discussion of the importance of "messing around," see David Hawkins "Messing About in Science," in *The ESS Reader* (Newton, Mass.: Education Development Center, 1970), pp. 37–44. This article was also published in *Science and Children,* 2, no. 5 (February 1965).

ever, the analogy is one of the most direct approaches. An *analogy* is a stated relationship between the strange and unknown with something that is known and familiar. Analogies can be important approaches to thinking and scientific investigation; Newlands, for example, is supposed to have made an analogy between the musical scale and the periodic appearance of similar characteristics in chemical elements, and Faraday is supposed to have compared magnetic lines of force to stretched rubber bands.

In education analogies are useful devices for explaining the strange and unknown. We have suggested (see page 344) comparing the face of a clock to geological eras as a useful device for developing the concept of the comparative lengths of various spans of geological time. As another example, we have used the following analogy to develop a concept of how the human population has grown.[5]

How has the human population grown? For most of the time that humans have existed on earth, the population has grown very slowly. It is estimated that the human population size has doubled every 1,000 years. Recently, however, it has been increasing very rapidly. It is estimated that now the human population is doubling every 39 years. To develop a concept of this rate of increase, an analogy can be used in which one minute of time equals 1,000 years and one paper clip is the equivalent of 10 million people.

Have the children count out piles of 1, 30, 20, 50, 100, 100, and 100 paper clips. Assign a child to each of these piles, with the responsibility for putting his or her pile of paper clips into the "earth" container at the appropriate time. For this analogy we assume that the first human emerged about 3 million years ago, and our analogy can begin at 09:00. Two days later the first paper clip, representing 10 million people, is added to the earth. The remaining piles are added as shown in Table 8-1.

Table 8.1[a]

Time	Years Ago	Clips Added	Population Size
09:00	the beginning	0	very small
11:00 (2 days later)	3 million	1	10 million
11:10	10,000	30	300 million
11:11:30	1,500	20	500 million
11:11:51	350	50	1 billion
11:11:56	75	100	2 billion
11:11:58	35	100	3 billion
11:11:59	15	100	4 billion

[a]1 paper clip = 10 million people; 1 minute = 1,000 years.

[5]This analogue is from Willard J. Jacobson, *Population Education: A Knowledge Base* (New York: Teachers College Press, 1979) p. 35.

We use analogies whenever we use such phrases as "It is something like . . ." – "Have you ever seen______; it is something like that." Analogies are powerful tools for concept development. As we learn more and more about the children with whom we work, we become better able to use them.

A word of caution is appropriate with regard to analogies. Analogies are powerful tools for explanation, but they should not be used to prove statements. For example, it may be useful to think of the force of gravitation as being something like a rubber band as a planet is held in orbit. But the nature of rubber bands – for example, that there is a limit to how far they can be stretched – cannot be used to suggest that gravitational forces have the same characteristic. It is important that children become aware of this distinction between the usefulness of analogies as tools for explanation and their inappropriateness as approaches to proof.

Concepts are what we think with. Children should be able to practice using their concepts in new situations. Children need opportunities to use concepts in new and different ways. If it becomes obvious over time that the concepts they struggle to acquire are seldom if ever used, children may begin to take their education less seriously. "What is the use of it?" is not always an appropriate question, but certainly children should have a variety of experiences in which they use what they have learned.

Actually, if our science experiences are developed in light of our continuing study of the children we teach and our science goals for those children, many situations will naturally arise in which children use the concepts they have developed. For example, if we provide many opportunities for children to develop a concept of the vastness of the universe, they will also have many occasions for using their concepts of geological time and the reference frames from which we can view our position on earth, in the solar system, and in our galaxy. However, it is also useful to plan for ways in which children can use the concepts that they have developed.

General concepts can be applied to conditions and events in children's everyday lives. For example, "habitat" is a general concept, but children can apply it to the plants and animals that they have at school and at home – "What is the nature of the habitat of your pet dog?" One group of children, after having had experiences with the oxygen–carbon dioxide cycle, nitrogen cycle, water cycle, and rock cycle, looked for cycles in their immediate environment. Their study of the cycle of red, green, and yellow signals of the traffic light at the street corner next to their school was a model of how a general concept can be applied to our everyday world.

Science concepts can be applied in other areas. Most elementary school teachers work with children in many curriculum areas. One of the advantages of this practice is that concepts and skills developed in one curriculum area, such as science, can be applied in others. The concept of conservation of matter, for example, underlies many problems in mathematics, and it is useful to have children see this. The concept of reference frames is useful in developing an understanding of perspective in art. The operational definitions used in science can be compared with the logical and literary definitions used in literature. Cause and effect, a useful concept in much science work, may have limitations in other fields. However, it is useful to compare cause and effect with related concepts when they are used.

REFERENCES

ENNEVER, LON AND WYNNE HARLEN. *With Objectives in Mind.* London: Macdonald Educational, 1972. The British Science 5/13 materials were designed to foster children's intellectual development. This book describes the rationale for the project and includes discussions of the development of formal, logical thought.

GINSBERG, HERBERT AND SYLVIA OPPER. *Piaget's Theory of Intellectual Development.* Englewood Cliffs, N.J.: Prentice-Hall, 1969. This is one of the clearest and most practical discussions of Piaget's theory of intellectual development.

KARPLUS, ROBERT AND RITA PETERSON. *Formal Thought.* San Francisco: Davidson Films, 1971. A film that shows some of the characteristics of formal thought. *Classification* and *Conservation* are two films in this series that deal with concrete operations. Available from Davidson Films.

RENNER, JOHN W., ET AL. *Research, Teaching and Learning with the Piaget Model.* Norman: University of Oklahoma Press, 1976. Based on research and development work in Oklahoma schools. The chapters "Formal Operational Thought and Its Identification" and "Curriculum Experiences and Movement from Concrete Operational Thought" are of special interest.

VAIDYA, NARENDERA. *Some Aspects of Piaget's Work and Science Teaching.* New Delhi: S. Chand, 1970. We can learn from education in other cultures. In this book one of India's leading science educators shows ways in which some Piagetian ideas can be applied in science teaching. Chapter V, "The Maintenance of Reflective Atmosphere in Our Classrooms," is especially good.

Science for the Handicapped

CHAPTER 9

Several years ago a group of visitors was touring an elementary school known for its many curricular innovations. The principal, in reviewing the special programs that the visitors would be seeing, also mentioned that in the third grade there were three educable mentally handicapped youngsters. This school was attempting to "mainstream" handicapped children long before the practice was common.[1] The visitors looked in on the reading and math labs, the audio-visual center, the school museum, class projects, and music and art classes. In the third grade the visitors saw the children working with different types of circuits making bulbs light and bells ring. The visitors spent about twenty minutes in the third grade. At lunch the guests chatted with the principal and remarked about how impressed they were with what they had seen. One visitor said, "It's too bad the handicapped children were not in the third grade today, I would have loved to have seen how the teacher deals with them." The principal beamed as she said, "Oh, but they were in the room and I'm so happy that it's working out so well."

[1]*Mainstreaming* is the practice of integrating handicapped or otherwise disabled children into regular classes for all or at least part of the school day. In most quarters mainstreaming is considered preferable to segregating exceptional youngsters because it is most similar to situations in the "outside world." Also, separating handicapped youngsters and placing them in special education classes often stigmatizes those children.

This story is not unique. It has occurred in many schools and in many parts of the country. Successfully mainstreamed youngsters are in many respects undetectable from their nonhandicapped peers. This is a tribute to the success of the important practice of mainstreaming.

SCIENCE IS FOR ALL CHILDREN

If teachers were to select a general, overall goal for instruction, it might be to simply help each youngster develop his or her potential to the fullest. This implies that we must know the children we teach—their strengths and their weaknesses; their talents and their disabilities; their exhilarations and their frustrations. All children have special needs and interests, and once we are aware of these, we can attempt to meet these needs.

The handicapped have a right to benefit from the same educational opportunities as other children. If we see all children as individuals, we can hope to meet the needs of children who are in some way different. But what is meant by "handicapped"? For our purposes handicapped children are those who have a speech, hearing, or visual impairment or are crippled, learning disabled, or mentally or emotionally disturbed. Although some people suggest that gifted children and those who are socially or economically disadvantaged are also special, we will not consider these situations in this chapter.

Participation in science holds much promise for handicapped children. In science children have many opportunities to work with materials. This widens their contacts with the physical world, contacts that have often been neglected. Moreover, children can work independently with their own materials. This gives those who might need to work a little more slowly the time to explore and investigate without the time constraints inherent in full-group activities. In science, too, each child's curiosity and creative potential can be tapped without heavy reliance on reading. Children whose language development has been stunted can begin to attach real meanings to words and experiences.

LEGISLATION REGARDING THE EDUCATION OF THE HANDICAPPED

It has been estimated that one out of every ten school-age children is handicapped in some way, that is, requires some help beyond that which is offered in the regular classroom. Many states now mandate that the public schools provide appropriate educational services to meet special needs in addition to a regular school program for handicapped children. Some states authorize diagnostic and remedial educational programs for children with obvious handicaps at birth. Some states require the provision of such services beyond the regular maximum of 18 years of age and extend schooling to 21 or even 23 years of age.

The trend in recent legislation has been towards the treatment of exceptional children in their local schools. Most notably, the passage of Public Law 94-142, The Education of All Handicapped Children Act of 1975, mandates that handicapped children be placed in "the least restrictive environment," a concept defined by each state. Usually this has come to mean that handicapped children will be educated in the same classroom with their non-handicapped peers, except in cases where special aids or services needed are unavailable. Another provision of P.L. 94-142 requires the development of an Individualized Education Plan (I.E.P.) for each handicapped youngster at the beginning of the child's school career. This plan is to be reviewed and re-evaluated annually. Among some of its other provisions, P.L. 94-142 mandates that teacher education institutions review and revise their programs to assist teachers in accepting, understanding the nature of special needs, planning for, and integrating handicapped children into their regular classrooms.

Whenever possible, handicapped children are educated in regular classrooms, with assistance provided to the classroom teacher where appropriate. Special-education consultants may meet with the classroom teachers as the need arises or on a regularly scheduled basis to help plan activities, demonstrate special materials, or suggest specific approaches to be used when working with children who have special educational needs. Some schools have itinerant specialists who work with handicapped children in or out of their regular classrooms for a few sessions each week. Other schools have "resource rooms" staffed with a special-education teacher where handicapped children can receive help related to their individual needs for a part of their school day, spending the rest of their day mainstreamed in the regular classroom. Children with more severe problems may be assigned to a full-time special-education class in their home or a neighboring school. In still more severe cases, the children may attend a special day school or a residential school; only in the most severe cases are residential hospitals or institutions used.

MAINSTREAMING AND THE REGULAR CLASSROOM TEACHER

Most teachers, unaccustomed to working with special students, are somewhat uneasy about the prospect of having a handicapped child in their classroom. True concern about whether the child's special needs can be met is an initial worry. Certainly, some modifications may have to be made, but once this is done, and it is realized that the mere integration of the handicapped youngster into the regular classroom goes a long way toward meeting some needs, the experience can be among the most rewarding and satisfying ones that teachers can have.

All of the children in the class will be richer as they come to recognize the diversity in human talents, difficulties, and disabilities. Integrating handicapped children into the regular classroom also helps children realize how *similar* we all are. The similarities among humans far outweigh the differences, and normal children can be taught to treat handicapped children with compassion and friendship.

One of the key ingredients of a successful mainstreaming experience is the teacher's attitude toward the individual child and the whole concept of mainstreaming. It seems essential that the relationship between the handicapped child and the teacher and class be a positive one. It has been said that if teachers value diversity in children, they are more likely to achieve success in mainstreaming. As the handicapped child is helped to become a participating, contributing member of the class, the adjustment for all parties involved will be eased. Teachers who have successful experiences with mainstreaming are those who see *all* children as individuals, get to know their strengths and weaknesses, and attempt to meet their individual needs.

There are some basic, general principles that teachers may follow in working with handicapped children:

1. The individual pupil–teacher relationship is of paramount importance. The relationship should be based on the nature of the handicap, mutual trust, praise for positive action, and a great deal of patience.
2. Learning should proceed slowly and gradually, yet deliberately. Concepts may have to be developed over long periods.
3. Science learning should begin with the objects, events, and words that are most familiar to the child and then, if necessary, go on to the unfamiliar.
4. In most handicapping conditions language growth will be needed, since the child has suffered from limited exposure.
5. Time should be taken to get to know how the individual child learns best–the child's learning style, rate of learning, and interests.

Further suggestions are offered as specific handicapping conditions are discussed.

SCIENCE FOR THE DEAF AND THE HEARING IMPAIRED[2]

Deaf children's needs to explore, wonder, and probe their environment are the same as those of all children. The world is full of interesting

[2]We have benefited here from the work of Allan Leitman. The results of some of his research have been published as *Science for Deaf Children* (Washington, D.C.: Alexander Graham Bell Association for the Deaf, 1968).

phenomena and events for deaf children too. Their natural curiosity and enthusiasm are *not* impaired by their hearing loss.

Deafness occurs in several degrees of severity. Some individuals may have a mild or moderate hearing loss, while others may have a severe hearing impairment or a total hearing loss, as in the case of the profoundly deaf. Deaf or hearing-impaired children may be educated in a variety of settings ranging from a regular classroom to a residential school for the deaf, depending on the extent of hearing loss and the availability of local services. A variety of specific teaching methods is used in working with deaf children, depending on the training of the teacher and the degree of hearing loss. This discussion, though, does not advocate a special or specific method but is a *general* treatment of science teaching and deaf children.

Language growth is a critical need for deaf children. Language, of course, in this context goes beyond the mere vocalization of words and suggests that the *meanings* of words, objects, events, and sensations need to be understood. Science experiences can broaden the child's view of his or her environment through extensive use of physical and natural materials. In order to build language, deaf children need countless such encounters with physical objects and events. These direct experiences should both precede and occur simultaneously with labeling and vocabulary building. As this is done, the child should be at eye level and spoken to directly.

As in all handicapping conditions, the relationship of the teacher to the deaf child is critical. This relationship can impede progress or help a child reach new heights, depending on the quality of the interaction. The child—his or her interests, moods, strengths, and weaknesses—must be known well. The provision of a supportive environment with plenty of successful experiences is essential. The deaf child must also be challenged to stretch his or her potential and realize growth in new areas.

When working with deaf children in science, adaptations may have to be made in some activities. For example, when description of the sounds of objects is called for, the deaf child can concentrate instead upon physical characteristics that are apparent visually or on the texture of the objects. Teachers of deaf children should not be too quick to provide answers or solutions to problems. This can convey impatience and help children learn that they do not have to exert themselves—if they wait a few seconds, the answers will be given to them. Instructions for certain activities may have to be provided through pantomime, pictorial representations, setup models, printed words, or in rebus form. Older children can obtain instructions from task cards. Completed activities should be summarized in large experience charts, demonstrations, or written reports.

Repetition is often required when teaching deaf children. Children will take cues from demonstrated actions and models. A model provides a continuous setup or procedure for children to follow and can help avoid repetition. When working with young children, pictures should be used extensively to define words, concepts, or actions. In general, the pace of instruction should be slow; it should proceed in manageable increments and build on past familiar experiences.

Opportunities should be provided for youngsters to freely explore materials before formal directions are given. "Playing" with the materials will be an aid to concept development. As children investigate on their own, they examine, manipulate, and discover new relationships and possibilities. When instructions are given, they should be presented as clearly as possible using any or all of the means described earlier.

Science for deaf children (or, for that matter, all children) should be a "doing" program. The classroom should abound with concrete materials and opportunities to cook and to work with paints, clay, crafts, printing, and other expressive activities. Excursions and trips are particularly valuable for deaf children. Local stores and community services (post office, police station, firehouse, etc.) are good starting points. Factories, offices, nature centers, and museums are also fine places to visit. When taking trips with deaf children, extra precautions will have to be taken to ensure the safety of youngsters, particularly if machinery is involved. Where fully hearing and hearing-impaired children are in the same class, they can be paired for trips and excursions.

When dealing with young deaf children, the science program should not be very different from any good experience-oriented science program. High priority should be given to free exploration with materials. The activities described in Chapters 4, 5, and 6 are all appropriate for young deaf children (except, of course, the activities that deal with sound).

When working with older children, reading becomes more appropriate. The science program should still be experiential, but instructions can be given in the form of task cards, pictorial representations, and models. The observations older children make should be sharper and more discriminating. The children should deal with real problems and develop concepts of natural systems (e.g., food chains, water cycles, and other ecological systems). The parts of a system should be related to the whole so that the children can interpret the whole system. Success in science experiences can help older children develop lifelong interests as they work with plants and animals and are assigned special projects, investigations, and reports. It is not usually difficult to motivate interest in science.

Science work can easily be individualized for deaf children. Once the child's special abilities are known, projects can be assigned; group or committee work can be set into action (with particular responsibilities outlined); and learning centers may be devised. In a class in which hearing-impaired children are integrated with non-hearing-impaired children, pairing of youngsters has been found to be most effective. Each member of the pair can decide what aspects of an activity he or she will pursue.

Existing commercial science programs can also be adapted for use with deaf children. The Science Curriculum Improvement Study (SCIS) has a version that has been adapted for deaf children. With careful advance planning and the provision of primary encounters with materials before concepts are developed, most science activities can be readily adapted for deaf students.

SCIENCE FOR THE BLIND AND THE VISUALLY IMPAIRED[3]

Blindness is a term that is usually used to describe a variety of visual impairments. Legal blindness is defined by a certain degree of uncorrectable vision but does not necessarily mean that the individual has no sight at all. Often, legally blind children can see objects (although perhaps not at all clearly) or at least shadows. In any case a science program has to be adapted to be appropriate for blind children. Most often, totally blind youngsters are educated in special classes or special schools, whereas visually impaired children can be mainstreamed more easily. Blind youngsters often lack experience with concrete materials and, thus, may also be deficient in their language development.

Science instruction can go a long way toward helping to overcome an experiential or language deficit. As mentioned several times in this and other chapters, direct experiences with natural and physical objects provide an essential basis for language growth. Blind children need to examine objects and phenomena using their auditory and tactile skills. This can help them learn new words, apply basic math concepts, exercise manipulative skills, group objects, organize information, and learn to use their intact tools of investigation to best advantage.

Some specific procedures or approaches have been found most

[3]Here we have been influenced by Herbert Thier and his associates, *Science Activities for the Visually Handicapped* (Berkeley: University of California, Lawrence Hall of Science, 1977).

successful when working with blind children in science. These may be summarized as follows:

1. Stress tactile and auditory observations in science work. Children can also uncover information about objects through smell or taste, and these senses should not be overlooked.
2. Whenever possible, blind or visually impaired children should participate in science work along with their sighted peers. Pairing children is often a successful strategy.
3. Sufficient opportunities for free and individual investigations should also be provided. Blind children often need a greater amount of time to explore objects than their sighted peers. As children get to know a variety of objects well, their confidence and independence will grow.
4. Adapt exercises or activities so that blind children can participate in them (e.g., instead of sorting objects from a kit or commercial program according to purely visual characteristics, the blind children can accomplish the goal of sorting if they group objects according to their shape, texture, odor, etc.).
5. Instructions for activities can be provided using tape recorders, Braille labels (if available), or directions read to the blind student by the teacher or another child. When providing instructions either verbally or on tape, it is best to use terms that are already familiar to the child.
6. Language development can be encouraged *while* the child is manipulating an object or performing an operation (e.g., as a child feels a block, he or she should be asked to describe it fully). If new terms are to be provided, they should be done clearly and while the child is experiencing the meaning, sensation, or characteristics associated with the word.
7. Successful experiences should be planned for in order to build independence. While praise should not be given indiscriminately, it should be provided often enough to sustain and build on growing confidence.
8. Materials to be used with blind children should be selected with great care so as to provide maximum sensation while taking precautions not to introduce dangerous (e.g., sharp or pointed) objects.

Commercial science programs have been adapted for use with blind children, and certain organizations have been instituted for this purpose. Adapting Science Materials for the Blind (ASMB) is a project that was started at the University of California, Berkeley in the 1970s. (See the listing of special organizations for the handicapped at the end of this chapter.) ASMB attempts just what its name implies: Its staff analyzes existing science programs and materials and develops necessary adaptations and new materials to make the programs suitable for use with blind children. Specifically, the ASMB staff has produced a version of SCIS fully adapted for the blind. SCIS was selected for the modification because of its ungraded, sequential development of concepts, extensive use of concrete materials, and provision for free exploration before further instruction. (These qualities, of course, also exist in other commercial science programs.) Adapted units include those dealing with material objects, electric circuits, pulley systems, solutions, organisms, and others.

Some of the activities involve novel modifications. For example, in the organisms unit children set up an aquarium and examine the size, shape, movements, and other characteristics of a goldfish. In order to facilitate this examination, the children use two plastic aquariums—one of which has holes in it. The aquarium with holes in it is nested inside the regular watertight aquarium and the fish is placed in the inner aquarium. As the child slowly lifts the inner aquarium, the water drains out of it and the goldfish (a hardy fish indeed) is left at the bottom of the nearly dry aquarium, ready to be examined by the blind student.

The American Printing House for the Blind (APH) publishes literature and produces educational aids for use by blind students. (See the list of organizations at the end of this chapter.) The APH also adapts and develops materials compatible with existing science curricula. For example, members of the APH staff have invented a light-sensing device that emits sounds of varying pitch and volume when exposed to different intensities of light. The device can be a tremendous aid in helping blind students investigate solutions and suspensions, determine relative amounts of pollutants in water samples, and the like. The Braille clock is another innovation that allows blind students to time procedures or experiments. Such devices open new avenues for scientific pursuits for the blind both in schools and in industry.

More recently, a group of educators at the Lawrence Hall of Science (University of California, Berkeley), the prior location of the SCIS adaptation, have developed a new science program called Science Activities for the Visually Impaired, or SAVI. Working in cooperation with the APH, the SAVI staff has devised a set of integrated science activities for visually impaired children between the ages of 9 and 12. The program consists of concrete science experiences using real objects and organisms, and seeks to improve logical development, promote everyday living skills, and improve manipulative abilities. Oral and written language skill development is an important goal of the program. More than five modules have been marketed, each with its own set of activities.

Among some of the interesting activities included in SAVI is the examination of crayfish. These animals are hardy, slow-moving, and not slippery. SAVI crayfish have had their pincers covered so that they cannot hurt the children investigating them. Children thus can study the movements, structure, speed, and general behavior of these animals.

Other SAVI units include studies of mixtures and solutions, sounds and other forms of communication, plants and animals, and several activities in scientific reasoning.

SCIENCE FOR THE EDUCABLE MENTALLY HANDICAPPED[4]

Science is an area that is most suitable for involving the educable mentally handicapped (hereafter referred to as EMH) because of its concrete nature and the wide use of natural and physical materials. There is a tremendous diversity of ability and special skills among EMH children—just as there is in the population at large. There are, however, some characteristics that are typical of EMH children. Usually, the children exhibit developmental lags in general maturation, cognitive ability, and social adjustment. They are usually slower to learn and less able to retain material than other children. Students with such mental handicaps often have difficulty recalling, transferring, conceptualizing, and abstracting material; for example, they may have difficulty making generalizations from several illustrations of a particular concept. The attention span of EMH children is usually somewhat short; they are easily distracted. Often their visual-motor coordination is poor, and their language development may be deficient. As a result of these difficulties, EMH children often have a poor self-image and are easily discouraged from pursuits that may be frustrating. Operating primarily at Piaget's concrete-operations level of cognitive development, EMH individuals rarely reach the stage of formal operations. (See pages 26–28.) Yet despite this seemingly discouraging list of deficits, it is also true that EMH children have the same needs, hopes, dreams, and curiosity as other children. They have the same love of interesting activities and experiences, and in all ways *they are children.* All of these characteristics have important implications for teachers. It is useful to be aware of some limitations that might be encountered in working with EMH children, but it is also important not to lose sight of how very similar all children are.

The Value of Science Activities for the EMH

Science activities are quite successful with and important for EMH children and should not be neglected. For years science was thought to be "too hard" for EMH children because of the often complicated concepts involved, but this is simply not true. The use of concrete materials

[4]Robert Walrich has been very helpful in the development of this section. The pictorial riddles were developed and tested with EMH children by Mr. Walrich.

is an advantage of science. Firsthand experiences are a particular need of EMH children, as it helps them become familiar with the world around them. Other subject areas often rely heavily on reading; science experiences can be accomplished without a great deal of printed matter. Science can help develop the important basic skills of observation, description, comparison, application, and inference, all of which are needed for everyday living. Science can help children discover alternatives for problem solving and new relationships among common phenomena. Science can foster language development by expanding the child's store of mental images through the use of concrete materials. Science can help children function in a technologically oriented society by familiarizing them with object functions. Science can help children develop new manipulative skills as they construct models and perform operations and transformations with materials. Science can help children learn more about themselves as they are involved in units dealing with their bodies and bodily functions. Science can help children develop lifelong interests in particular phenomena and aspects of their world. Science can also improve a child's self-image through the development of competence in handling materials and investigating phenomena as well as increased self-awareness.

Suggestions for Teaching

In general, when working with EMH children the expectations held for them should not be too limited. Maintaining reasonably high standards (while understanding some basic limitations) can serve to stimulate and challenge EMH students. Some broad guidelines regarding *general* teaching conditions are as follows:

1. Children should be provided with experiences in which they achieve some degree of success.
2. Children should receive praise for their accomplishments and be helped through their difficulties.
3. Children should participate fully in all aspects of an activity and be helped to do their best.
4. Children should be provided with activities that have some relevance to their everyday lives.
5. The classroom environment should not be distracting and the children should not be expected to ignore other stimuli in their environment.
6. The teacher–child relationship should be a positive one built on mutual trust and respect.
7. The science program should be flexible enough to follow the natural curiosity of the child, when appropriate and practicable.

The factors just listed deal largely with attitudes toward learning and learners. The pupil–teacher relationship is of critical importance in working toward an effective science program for EMH children. This relationship is more important than specific materials, and a warm positive rapport can go a long way toward helping children reach higher levels of achievement. EMH children are usually sensitive to their failures and are quick to note disdain or disapproval, whether verbal or nonverbal (e.g., facial expressions or gestures).

There are also some specific techniques for teaching EMH children. They learn best when:

1. The activities are relatively uncomplicated in terms of the steps involved, and the presentation of material is clear.
2. The science program includes a great many primary experiences with concrete objects and natural materials.
3. The children are helped to complete an activity that they have started.
4. Instruction begins with familiar concepts and materials.
5. The progression of an activity is slow and sequential, with each new step building on and directly related to the preceding one.
6. Opportunities for review and repetition are built into each lesson or activity.
7. A model or sample of a finished product is provided against which children may check their work.
8. The activities in which the children are engaged are not too dependent on reading skills.
9. The children are given opportunities to apply the knowledge they obtain to new situations and events.
10. The activities provide some opportunities for movement within the classroom or large-muscle involvement.
11. A variety of teaching techniques and learner interaction settings (e.g., small groups, individual projects, paired experiences, etc.) are employed.

With regard to the last point in this list, several teaching strategies can be effective. Learning centers (described in Chapter 2) can be most helpful in that they allow children to work at their own pace and the material may be selected for a particular individual. Learning contracts offer the same benefits, and also provide an opportunity to plan the work to be covered in conjunction with the learner. Individual contracts can be designed to meet a special need or interest in a particular area.

The use of pictorial riddles with EMH children is an example of a technique that has been found effective. The pictorial riddle is a drawing that can be thought provoking, tells a story, has a previously determined degree of detail, and can communicate a variety of concepts. Consider,

Figure 9–1 An example of a pictorial riddle.

for example, the drawing in Figure 9–1. The following concepts may be developed using this drawing:

1. Living safely in a city environment requires some precautions.
2. The principle of momentum can have some dangerous applications.
3. Testing inferences can confirm ideas.

The teacher might begin with the following discussion questions:

1. What is happening in this picture?
2. What do you think might happen to the children in this picture?
3. Why do you think these things can happen?
4. How can we test our predictions of what might happen using materials available in the classroom?
5. What do you think would make this situation a safer one?

Pictorial riddles can serve as motivational devices to capture children's interest and lead them into investigations that will help them check inferences made from interpretations of a picture. Such drawings can be developed around a wide variety of situations and phenomena that are familiar to the children in their everyday lives.

Special Science Programs for the EMH

The Biological Sciences Curriculum Study (BSCS) has developed a life sciences program specifically for EMH students in the upper elementary and junior high school grades. The program has two main parts: *Me Now*, a two-year sequence, and *Me and My Environment*, a three-year sequence. Initially funded by the Bureau of the Handicapped of the United States Office of Education (Department of Health, Education and Welfare), the program is now available commercially. Using a multimedia approach, the program helps children develop their problem-solving skills, achieve successful experiences in science, and refine their manipulative skills. *Me Now* deals with the systems of the human body and their structure and function; children explore the interactions in their own bodies using a variety of perceptual modalities. *Me and My Environment* launches the students into investigations of their own environment and interrelationships within it. Both sequences have built-in overlearning and repetition. The associated reading materials are highly motivating, but with a relatively low reading level. The tasks are uncomplicated and generally familiar to the students. The activities are brief and sequential, and progress in small increments. The students keep records on data sheets, and a certain degree of success is always attainable. (See Figure 9–2.)

Some school systems have been adapting *Science—A Process Approach* for use with EMH students. The deliberate hierarchy of skills and defined sequence make this program (as well as others) suitable for use with EMH children. In the future we can expect to see more science curricula for handicapped children as well as adaptations of existing programs.

Figure 9–2 Children working on units from *Me and My Environment.* Reprinted with permission of Biological Sciences Curriculum Study.

The content areas most commonly dealt with in working with EMH children are plants, animals, human physiology, mental health, weather, seasons, the earth, and machines. Such topics are close to the child and his or her immediate surroundings.

SCIENCE AND THE LEARNING DISABLED

Learning disabilities are encountered more frequently than any of the other conditions described previously. Learning disabilities can be broadly described as difficulties that interfere with an individual's being able to cope with material taught. Usually detected when a broad discrepancy occurs between a child's school achievement and his or her intellectual potential, learning disabilities may appear in a variety of areas. Some disabilities are specifically perceptual; others are more connected with language. Some occur in math work; others are speech problems; and still others may be due to hyperactivity.

Often learning disabilities go undetected for years and the child is thought of as slow or limited in intellectual ability. With proper diagnosis and educational prescription, the regular classroom teacher can usually provide great assistance and achieve successful results with learning-disabled children.

Teachers usually suspect some sort of problem when children experience difficulty in learning to read or in language work. Learning-disabled children are often discouraged and their self-esteem is negatively affected. Frustration may become apparent as children see their classmates "breezing through" material with which they have difficulty.

You may have heard the terms *dyslexia, aphasia, minimal cerebral dysfunction, hyperactivity, visual–motor integration difficulties, auditory processing problems, laterality problems, language disorders,* and the like. All refer to a specific type of learning disability, but all are beyond the scope of this book. We cannot hope to deal with these difficulties in any detail in a section of a chapter, but some general guidelines for working with learning-disabled children can be set forth.

When working with learning-disabled youngsters, it is important for the teacher to know the child well—in terms of personality, strengths and weaknesses, preferred modalities for learning, and so forth. School support specialists, administrators, and of course parents can help teachers pinpoint particular problem areas. Once this has been done, special approaches or materials may be helpful.

Children with learning disabilities may be brilliant in areas other than the one specifically affected. Science is a subject in which children with learning difficulties may shine, particularly in cases in which reading or language achievement is a problem. Here children may become involved and absorbed and can use their manipulative skills. In situations in which the reading of instructions or directions would be inappropriate, instructions may be recorded on tape and played back step by step for the students. The use of learning contracts that take advantage of the child's preferred modality can also be helpful. When the pace of instruction can be altered, this too may have good results. Success may not be immediate, since learning-disabled youngsters often suffer from a lack of self-esteem, but with patience and support success will come in time.

THE CREATIVE CHILD

In some instances creativity is considered a handicap in that the child's thought processes may not lead him or her to the expected or "right" answer but may result in new and novel interpretations, insights, and meanings. The creative child may be "out of tune" with his or her classmates. Creativity, when recognized, should be valued and encouraged. Children who seem particularly creative can be assigned special projects and assignments to maintain stimulation and interest.

A WORD ABOUT GIFTED AND TALENTED CHILDREN

When we speak of "exceptional" rather than handicapped children, the former term usually refers to children with the conditions discussed in this chapter plus children who are gifted or unusually talented. Such children are often a bit out of step with their classmates in that they require additional stimulation and other special considerations. It is sometimes necessary to alter techniques and materials for gifted children and assign them extra projects, supplementary reading materials, and special activities. Investigations that call upon the gifted child's inner resources can be devised and incorporated into the child's regular science program. Some of the activities suggested in Chapter 20 are particularly suitable. Gifted and talented children can be called upon to give classroom demonstrations, devise media and materials, conduct special environmental investigations, or become involved in community service projects. During regular class sessions the intellectual curiosity of gifted children can be stimulated by open-ended questions that lead them into new investigations.

All children have special abilities and disabilities—no two are the same. As we get to know the children we work with and value their diversity, we can begin to meet individual needs. This, along with patience, perseverance, and special assistance, can help children who could never have dreamed of attending their community school find acceptance, self-worth, and satisfying educational experiences.

JOURNALS DEALING WITH THE HANDICAPPED

American Journal of Mental Deficiency, published by the American Association on Mental Deficiency.

Education and Training of the Mentally Retarded, published by the Council for Exceptional Children.

Education of the Visually Handicapped, published by the Association for the Education of the Visually Handicapped.

**Exceptional Children,* published by the Council for Exceptional Children.

**Exceptional Children in the Regular Classroom,* published by the Council for Exceptional Children.

Focus on Exceptional Children, published by Love Publications Company.

**Journal of Learning Disabilities,* published by Professional Press, Inc.

**Journal of Special Education,* published by Grune and Stratton, Inc.

Journal of Visual Impairment and Blindness, published by the American Foundation for the Blind.

**Teaching Exceptional Children,* published by the Council for Exceptional Children.

The Volta Review, published by the Alexander Graham Bell Association for the Deaf.

ORGANIZATIONS DISPENSING INFORMATION AND EDUCATIONAL MATERIALS ON CERTAIN HANDICAPPING CONDITIONS

Adapting Science Materials for the Blind (ASMB)
Lawrence Hall of Science
University of California at Berkeley
Berkeley, California 94720

American Printing House for the Blind (APH)
P.O. Box 6085
Louisville, Kentucky 40206

*Journals that are particularly useful to teachers.

Alexander Graham Bell Association for the Deaf
3417 Volta Place, N.W.
Washington, D.C. 20007

Bureau of Education for the Handicapped
Department of Health, Education and Welfare
Office of Education
Washington, D.C.

Council for Exceptional Children
1920 Association Drive
Reston, Virginia 22091

Institute for Research on Exceptional Children
University of Illinois
Urbana, Illinois

National Center on Educational Media and Materials for the Handicapped
Ohio State University
Columbus, Ohio 43210

Science for the Handicapped
Department of Elementary Education
University of Wisconsin, Eau Claire
Eau Claire, Wisconsin 54701

Society for Crippled Children
2800 Thirteenth Street, N.W.
Washington, D.C. 20009

REFERENCES

BLUHM, DONNA L. *Teaching the Retarded Visually Handicapped: Indeed They Are Children.* Philadelphia: W. B. Saunders, 1968. A somewhat detailed guide to dealing with this special group of children.

COUNCIL FOR EXCEPTIONAL CHILDREN. *Science: A Guide for Teaching the Handicapped,* EC 003 1252, ED 046 168, November 1970. Iowa City: Iowa University, Special Education Curriculum, 1970.

HUMPHREY, JAMES. *Teaching Elementary School Science Through Motor Learning.* Springfield, Ill.: Charles C Thomas, 1975. Motor activities are used to stimulate perceptual and academic skills. Games and techniques for accomplishing this goal are described.

KOLSTOE, OLIVER P. *Teaching Educable Mentally Retarded Children.* New York: Holt, Rinehart and Winston, 1970. A general treatment of practices and methods in teaching the educable mentally handicapped.

LEITMAN, ALLAN. *Science for Deaf Children.* Washington, D.C.: Alexander Graham Bell Association for the Deaf, 1968. A guide to science teaching for deaf children based on Dr. Leitman's classic study. Many specific activities are included.

MAYER, WILLIAM V., ED. *Planning Curriculum Development: With Examples from Projects for the Mentally Retarded.* Boulder, Colo.: Biological Sciences Curriculum Study, 1975. Papers concerning various special-education curriculum projects.

NORTHCOTT, WINIFRED H. *The Hearing Impaired Child in a Regular Classroom.* Washington, D.C.: Alexander Graham Bell Association for the Deaf, 1975. A guide to successful integration of hearing-impaired children in the regular classroom.

ROWE, MARY BUDD. *Teaching Science as Continuous Inquiry.* New York: McGraw-Hill, 1973. Contains a chapter on science for handicapped students.

Science
and
Children
PART TWO

Science is the investigation and interpretation of events in the natural physical environment and within our bodies. Most, perhaps all, children are born curious about the natural physical environment in which they live. Science experiences should nourish and stimulate this curiosity. The scientific enterprise is humankind's endeavor to investigate and interpret events in our environment and in our bodies. Children have a right to benefit from the results of this cooperative endeavor, to which scientists throughout the world have contributed.

In our work with children in science, we should strive to develop a child's concepts of the big ideas of science. There is a danger of becoming trapped in facts and minutiae, but it is the big ideas that should be stressed. Learning the distances to planets and the dimensions of our galaxy may be of some value, but it is much more important to develop a concept of the vastness of the universe. Focusing on the disparate characteristics of thousands of different kinds of living organisms can be a deadening experience. It is much more important to develop a concept of the great variety of living organisms. In turn, to work effectively with children in science it is important for teachers to continue to develop their concepts of the big ideas of science.

In science experiences children should use many of the processes that are characteristic of the scientific endeavor. They ask questions, suggest hypotheses or possible answers, collect and interpret data, communicate and criticize ideas, and check their results with the results of others. Science is a "doing" subject, but it is not activity for activity's sake. Science activities should involve children with many of the processes of science.

Science in Our Lives—The Study of Science in Schools and Schoolgrounds

CHAPTER 10

All of our lives have been greatly affected by developments in science and technology. You might ask children, "How do our lives differ from those of people who lived 100 years ago?" Here are some differences:

Perhaps half of children in the classroom would not be there; they would have died before the age of 5. The most important factor leading to greater longevity has been the sharp reduction in the death rate among young children.

The children would not have access to telephones, radio, or television. These modern communication devices are based on basic scientific developments resulting from the study of electricity and electronics.

The children would not be able to ride in automobiles or school busses. The harnessing of the internal-combustion engine to transport ourselves and our goods has revolutionized the way we live.

We would not have any of the objects in our homes and schools that are made of plastics, aluminum, synthetic textiles, rubber, and other materials that have been developed within the past century.

We would not have electricity and would have to do by hand many of the tasks that are now done by electricity.

Imaginative children may add many items to the list, but the central idea is that our lives and the way we live have been greatly influenced by developments in science and technology.

We can study some very important principles of science and technology in our schools and schoolgrounds. Usually, everything within our schools and schoolgrounds is readily accessible and the logistical arrangements for field trips can be kept to a minimum. Because it is relatively easy to arrange to study what is near at hand, the children will be able to study whatever changes take place throughout the school year. The various people who help operate a school—from the people who run the lunchroom to those who maintain the buildings and equipment—should become involved in some of these experiences. An important by-product of these experiences is that the children become more aware of the contributions these people make to the operation of the school. Of special importance are opportunities to see that the applications of science and technology are not always distant and far removed; they are being used to make it possible for us to live, study, work, play, and enjoy in our schools and schoolgrounds.

As with all field experiences, it is important to plan carefully. (For a more extensive discussion of planning field experiences, see pages 46–48.) Often it is helpful to discuss the objectives of an experience and to identify questions that will help children see more than they ordinarily would. It is very important to make prior arrangements with those who operate and care for the school. The first time, it may be desirable for you to make many of the observations first and discuss with other school personnel how the children can gain most from the experience. For example, for observations where there may be noisy machinery it is usually desirable to arrange for discussions in a quieter environment both before and after the observations.

In the sections that follow are some of the basic principles of science that can be studied in schools and schoolgrounds. Questions that can help children see more are listed after each principle. Some activities that can be carried out in the classroom in order to gain a better understanding of these principles are described.

TEMPERATURE

Humans can exist only within a certain temperature range. The temperature range within which they are comfortable is quite narrow. The human body must maintain a temperature of about 37°C (98.6°F). A variation of just a few degrees can mean death. Our bodies have such adaptations as changes in the rate of metabolism of food, shivering, and perspiration, which help maintain this constant body temperature. In addition, we use clothes and shelter to help maintain temperatures within our tolerances.

With the help of proper clothing and shelter, we have been able to live in Antarctica, on the tops of mountains, and in other places where great extremes of temperatures are found.

Our school building is one of the shelters that helps us maintain temperatures within the range in which we are comfortable. In your school,

1. What kinds of fuel are used to generate heat?
   a. Where does it come from?
   b. How much is used?
   c. How much does it cost? How have the costs changed?
   d. Why is this kind of fuel used rather than some other kind?
   e. Where and how is it stored?
2. Where and how is the fuel burned?
3. How is the heating system controlled?
4. What are the temperature ranges that are maintained in the classrooms and offices?

How does a thermostat work? A thermostat is a device for keeping temperatures within a certain range. A *compound blade* can be used to build a simple thermostat. A compound blade is made up of two different metals bonded together that expand at different rates when heated. As a result the blade will bend and can be adjusted to make and break an electrical circuit.

The construction of a simple thermostat is shown in Figure 10.1. When the circuit is closed, the flashlight bulb will light. Instead of a flashlight bulb, in an actual heating or cooling system a furnace or an air conditioner might be turned on or off as the thermostat closes or opens the circuit.

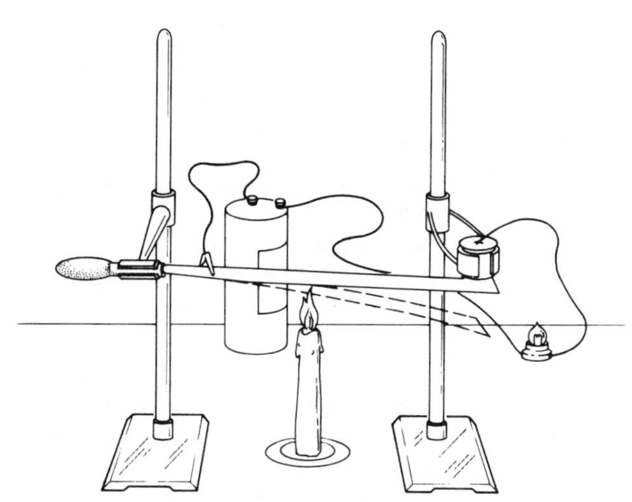

Figure 10–1 When the compound blade is heated, it bends and the circuit is broken. When it cools, the circuit is closed again.

Energy can be changed from one form to another. Children may have experienced energy in the forms of light, sound, and heat, as mechanical energy, and as chemical energy in fuels. Energy may be found in many of these forms in a school. There will also be examples of energy being changed from one form to another. In your school,

1. How is energy brought into the school?
2. How is energy in a chemical form (fuel) changed into other forms?
3. How is light generated?
4. How is sound generated?
5. How is heat generated?
6. How is mechanical energy generated?

How can different forms of energy be generated? Have the children try to generate energy in the forms of light, sound, heat, and mechanical energy. In each case have them try to analyze what energy transformations are taking place.

Heat Energy

Heat energy tends to flow from regions of high temperature to regions of low temperature. This is a very important application of the second law of thermodynamics. In many school heating systems, a fuel is burned at a high temperature in a furnace. The heat energy is transferred by conduction, convection, or radiation to classrooms or offices that are at lower temperatures. (See pages 396–98 for further discussion of heat transfer.) Eventually the energy is dissipated into the surrounding environment. In your school,

1. At what temperatures is the heat energy generated in the furnace or boiler?
2. How is heat energy transferred throughout the building?
3. At what temperatures is heat used?
4. What eventually happens to the heat?

How does heat travel along a metal rod? For this experiment a metal rod such as a curtain rod or a thin pipe is needed. Fasten tacks at intervals along the sides of the rod by dripping wax from a burning candle onto the rod and then holding the tacks in the wax until the wax hardens. Support the rod

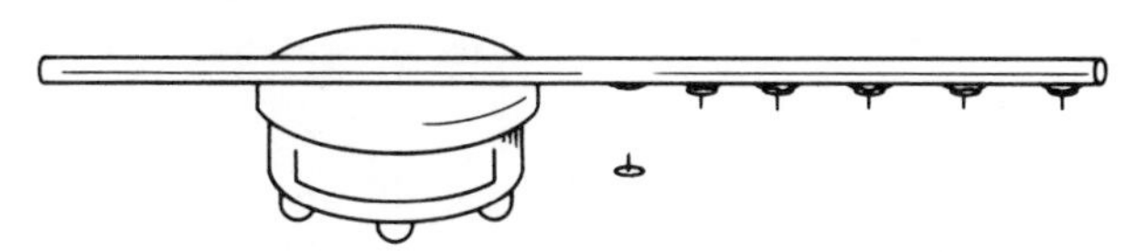

Figure 10–2 The tacks are fastened to the metal bar with candle wax. What happens when one end of the bar is heated?

with a clamp or on top of two metal cans. (See Figure 10–2.) Begin to heat the rod at one end. What happens to the tacks? Is there a pattern in what happens? How can these observations be explained?

Electricity

Electricity is one of our most convenient forms of energy. Electricity can be conducted long distances from where it is generated. Within the school it can be conducted to the classrooms and offices where it is to be used. It can be turned on and off by the mere flicking of a switch. Electrical energy can be converted into light, heat, sound, and mechanical energy. In your school,

1. Where does the electricity come from?
   a. How is it generated?
   b. What kind of fuel is used to generate the electricity?
2. How is electricity conducted to the school building? At what voltage?
   a. How is the voltage changed within the school building?
   b. Try to get a sample of the cable that is used in wiring buildings. How many wires are there in the cable?
3. What are the various ways in which electricity is used within the school? What voltage is used for different purposes? Are there special requirements for certain kinds of equipment?
4. How is a record kept of the flow of electricity? How does the electric utility check on the use of electricity?
5. What safety features are used to help prevent accidents and fires?
6. What is the cost of the electricity? How has it changed?

How can electricity be conducted from one place to another? First connect a wire from one pole of a dry cell to one connection of a flashlight bulb socket. The bulb doesn't light! A *complete circuit* is needed. Now connect the other connection of the flashlight bulb socket to the second pole of the dry cell. Does the bulb light?

Most electrical circuits in any building are *parallel circuits.* By connecting flashlight bulbs as shown in Figure 10–3, we can compare a parallel circuit with a series circuit.

Unscrew one of the bulbs in each of the circuits. What happens?

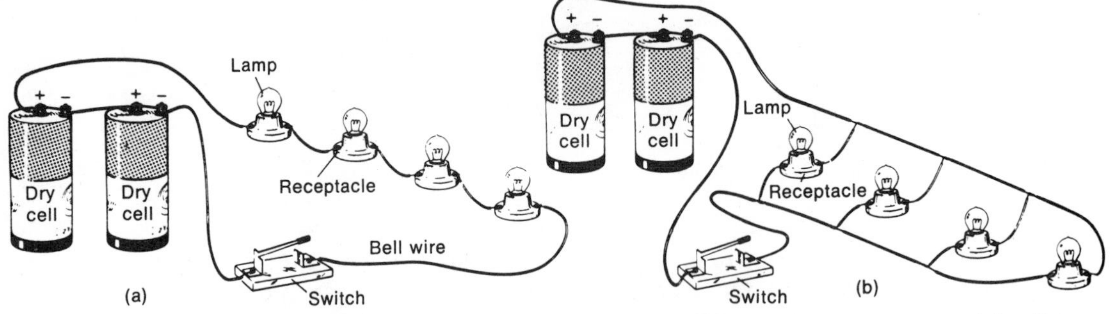

Figure 10–3 What happens when bulbs are added in (a) a series circuit? (b) Parallel circuit? What happens when a bulb is unscrewed?

Add one more bulb to each of the circuits. What happens to the brightness of the lights?

Which circuit, series or parallel, seems to be better for wiring schools and homes?

WATER

Water is one of the substances that are essential for life. Water must be provided wherever people live for any length of time. Help your children find out where the water in the school comes from, where it is stored, and how it is distributed throughout the school. In your school,

1. Where does the water come from?
2. If it is stored in the school, where and how is it stored?
3. How is the water treated? Why?
4. How is the quality of the water checked?
5. How is the water for different uses kept separate and conducted throughout the building?
6. In what ways is water used in the school?

Characteristics of Water

What are some of the physical characteristics of water? The importance of water is due partly to its physical characteristics. Have your children obtain a sample of water from their drinking fountain in a clear glass container. What is its color? its smell? its taste?

Place some water in a metal container and heat it on a hot plate. Using a thermometer, try to find the temperature at which water boils.

Place a metal container containing some of the drinking water in a mixture of ice and salt. Using a thermometer, try to find the temperature at which water freezes.

Water is a relatively heavy liquid. Place equal amounts of water and a liquid such as kerosene into two separate test tubes or olive bottles. Now float two similar wooden pencils in each of the liquids. In which liquid does the pencil float higher? In which liquid would a boat or ship be most likely to float? You may wish to have the children speculate as to how conditions would be different if some other liquid than water filled our streams, lakes, and oceans.

The Movement of Water

Water flows downhill. Energy has to be supplied to move it uphill. All materials near the earth, including water, are acted upon by the force of gravity. Water from a standpipe or a reservoir at an elevation higher than that of the community using the water will flow downhill and can be distributed by pipes throughout the community. In some schools water is pumped into a tank on top of the building from which it can flow throughout the building. In your school,

1. Where does the water come from?
   a. Who supplies the water?
   b. How much does it cost?
2. Is the water stored at the school? Where? How?
   a. How is it transferred to the storage area?
   b. How does it flow from the storage tank?
3. How is water distributed throughout the building?

What is the level of the water? Connect two pieces of glass tubing with rubber tubing. Then pour water into one of the pieces of glass tubing until the surface of the water is about halfway up the tubing. (See Figure 10–4.) What is the level of the water in the other piece of tubing? Raise one piece

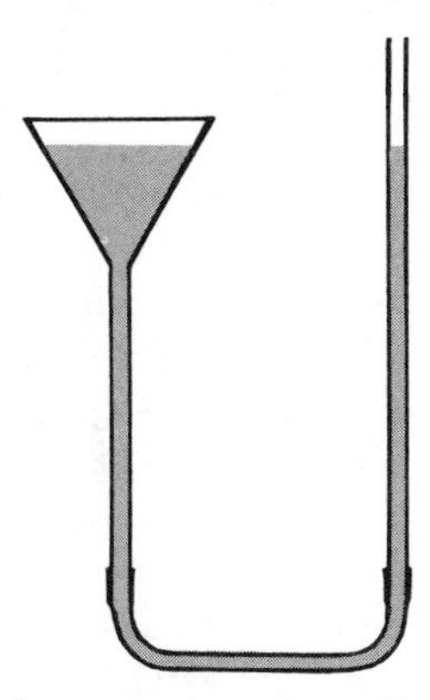

Figure 10–4 What happens when you raise or lower the funnel?

of glass tubing. What happens to the level of the water in it? What about the water in the other piece of glass tubing? Try moving the pieces of glass tubing in various ways to see whether you can change the level of the water in the tubing. Have the children try to explain their observations.

If there is trapped air above the water in one tube, the water may not rise. Have the children hold a finger over the opening in one glass tube. Now will the water rise in that tube?

HEATING AND COOLING

Cooling takes place as liquids are changed into gases. Many schools have air conditioners, and in school kitchens there will probably be refrigerators and freezers. Most of these devices work on the principle that heat energy is needed to change a liquid, the refrigerant, into a gas. This heat energy may come from a room or from the space within a refrigerator or freezer. Since heat energy is withdrawn, that region is cooled. However, energy, usually in the form of electrical energy, is used to bring about this cooling. In your school,

1. What cooling devices are used?
2. With a thermometer, measure the outside temperature and the temperature of the cooled space. What is the difference between the two temperatures?
3. What kind of energy is used to bring about this cooling in the school?

What happens when a liquid changes into a gas (evaporates)? When a liquid changes into a gas, it absorbs heat from the region around it. This can be demonstrated to children by putting a drop of some liquid that evaporates quickly, such as rubbing alcohol, on the back of their hands. What do they feel as the liquid evaporates?

Air is heated when it comes into contact with warmer materials. It is cooled when it comes into contact with cooler materials. Schools and homes are often heated and cooled by having air come into contact with hot or cold materials. For example, air that comes into contact with hot radiators will be heated, and air that contacts the cooling coils of an air conditioner or refrigerator will be cooled. Have the children observe how air is heated and cooled in their school. In your school,

1. How is air cooled?
2. How is air heated?
3. How is air moved throughout the building?

How is air heated? Hold the open end of a paper or plastic sack above a hot plate. (Take care that the sack does not touch the hot plate.) After the bag has been filled, carry it to another part of the room and have the children feel the air in the bag. Is it hotter than the surrounding air? If so, how was it heated?

SENSING DEVICES

Sensing devices can detect changes in the environment. Just as the body has eyes and ears that help us sense changes in our environment, there are sensing devices in our schools that can detect changes in that environment. In some schools there are smoke detectors that can set off fire alarms and heat detectors that will activate fire-extinguishing systems. Thermostats will detect changes in temperature. Photoelectric systems may detect when someone steps into an elevator or an intruder comes into the school at night. In your school,

1. What kinds of sensing devices are used? What changes in the environment do they detect?
2. Where are the messages from the sensing devices sent? Do they control automatic systems?
3. What is supposed to happen when a sensing device detects a change in the environment?

How does a sensing device work? The photoelectric cell is an example of a sensing device. When light strikes the cell, a small electric current is generated that can move a needle, be amplified to activate a garage door opener, and so forth. Show the children a photoelectric cell such as the light meter used in photography. Have the children place the light meter in a dark place. Then gradually add light. How much light must be added to make the needle move?

EXTINGUISHING FIRE

Fires can be extinguished by removing the fuel, blocking access to oxygen, or lowering the temperature of the fuel below the kindling temperature. In most schools there are fire control and extinguishing systems ranging from hand-operated fire extinguishers to smoke-sensing devices and automatic sprinkling systems. The fire-extinguishing systems usually operate by eliminating one or more of the requirements for burning. In your school,

1. How are possible fires detected?
2. What kinds of automatic fire control devices are there in the building?
3. What kinds of fire extinguishers are available and how can they be used?
4. On what principle are the fire control systems and extinguishers based?
5. How are the occupants of the building warned of the possibility of fire?
6. How is the fire department warned that there may be a fire in the building?
7. What provisions are there for evacuating people from the building?

How can a flame be extinguished? Place a candle upright in an aluminum cake or pie container. Ask your children what is needed for a fire. The three usual requirements are fuel, oxygen, and a kindling temperature. Light a match and ignite the candle.

"How can this candle flame be extinguished?" In general, a fire can be extinguished by removing one or more of the requirements for burning. "How can this be done?" One way of removing the fuel is to take pliers or scissors and squeeze the wick just below the flame; this prevents the melted wax from rising up the wick. The flame can be deprived of access to sufficient oxygen by placing a large glass jar over it; the children should watch the flame as it burns under the jar. The kindling temperature can be lowered by pouring a small amount of water over the flame. (It may be that this also denies the flame access to oxygen.) Which of the requirements for burning do the fire extinguishers in your school remove?

LIVING ORGANISMS

Usually, a variety of organisms live together in an environment. Almost always there are many kinds of plants and animals living in schools and schoolgrounds. Have the children find and name some of the plants and animals that live in the school. Certainly, there will be a variety of plants; usually, there will be a variety of insects. In your school,

1. What organisms live in the school and schoolgrounds?
2. What conditions do these organisms require to live?
3. What organisms do we help grow in the school and schoolgrounds? What kind of care do they require? Why?
4. What organisms do we try to eliminate or control in the school and schoolgrounds? How do we try to eliminate or control them? How do these controls work?

What organisms live within a square meter of the schoolground? Have small groups of children each measure out a square meter of surface on the schoolground and enclose it with string. (An ideal place to do this is on a lawn, but it can even be done on a surface that appears to be barren.) Have the children search for and list as many different kinds of plants and animals as they can find within their square meter. It is not always necessary to identify the various organisms specifically, although sources such as the *Golden Nature Guides* will aid children who wish to be more specific in their identifications. Have the groups compare lists. If the groups had different kinds of surfaces, they may have found quite different kinds of organisms.

PEOPLE IN THE SCHOOL

The people who plan, operate, and care for our school facilities do different kinds of work and have different backgrounds and skills. Usually, many people are involved in operating our schools—custodians, school bus drivers, lunchroom personnel, engineers, maintenance personnel, and grounds-keepers. It is important for children to become more aware of the nature of the work that they do and the background and skills that are required. In your school,

1. What are the different kinds of people who help operate the school?
2. What background and skills are required to plan, operate, and care for our school facilities?
3. How can these skills and backgrounds be acquired?

What kinds of work do different people in your school do? What kinds of backgrounds and skills do they have? Whenever the children go to study some aspect of the school, help them find out more about the work done by the people they meet. What kind of work do they do every day? What are some of the problems they encounter? As the children study science, they can also learn more about the people who help maintain and operate *their* school.

RELATED ACTIVITIES

What are the adaptations of some of the plants in environments in your locality?

How many types of insects can be found locally?

What birds are to be seen in our neighborhood?

Which animals live in the local community? p. 244.

Studying plant communities, p. 114.

What adaptations do the trees that live in our cities have? p. 205.

What kinds of trees seem best able to survive under city conditions? p. 205.

What are the adaptations of plants that live in cracks in sidewalks and roads? p. 205.

Keeping a tree diary, p. 13.

How much water can lichens absorb? p. 221.

How can electrical energy be converted into heat? p. 442.

How can electrical energy be converted into light? p. 442.

How can electricity be generated chemically? p. 432.

How does the dry cell generate electricity? p. 433.

How can we detect an electric current? p. 435.

What materials will conduct an electric current? p. 439.

How can you make a bulb light? p. 55.

What are the characteristics of series and parallel circuits? p. 441.

How is electricity conducted in a flashlight? p. 441.

How can a small house be wired? p. 441.

Building an electric motor. p. 445.

What are some of the characteristics of water? p. 345.

How do ocean currents flow? p. 348.

What can we find in a drop of pond water? p. 244.

How much water will different soils hold? p. 331.

How fast does water travel up through different soils? p. 332.

What kinds of materials are there in our rivers and lakes? p. 200.

Demonstrating that air has weight, p. 363.

How can children be helped to realize that air is all around them? p. 363.

Demonstrating the presence of water vapor in air, p. 362.

How can local seasonal variations be studied? p. 250.

The case of the candle and the jar. p. 74.

How can we demonstrate that air is composed of more than one gas? p. 361.

How can a housefly carry wastes? p. 202.

What sounds can be heard in the immediate environment? p. 409

What kinds of materials pollute the atmosphere? p. 200.

How does organic material decay in soil? p. 201.

How does your drinking water compare with water from a nearby river, lake, or pond? p. 198.

Using triangulation to measure the length of the classroom. p. 320.

Mapping the classroom, p. 137.

How far is it? p. 154.

How does our city dispose of its wastesP p. 200.

How are wastes treated before they are disposed of? p. 200.

What evidence of geological processes can be found near the school? p. 353.

What changes can we see taking place in a mud puddle? p. 349.

What evidence of erosion can we find? p. 331.

What are some of the things we can learn from a rock? p. 337.

What minerals are found in rocks in the local area? p. 340.

What is the nature of soil? p. 330.

How can we test soil? p. 219.

What are some of the resources for teaching science in cities? p. 206.

REFERENCES

BLOUGH, GLENN O. *Who Lives in This House?* New York: McGraw-Hill, 1957. People are never alone in a house. Almost always there are other animals inhabiting our homes.

BORLAND, HAL. *Beyond Your Doorstep.* New York: Knopf, 1962. A fascinating book that describes many observations that can be made in the country. You can suggest many of these observations to children.

COOPER, ELIZABETH K. *Science in Your Own Back Yard.* New York: Harcourt Brace Jovanovich, 1958. Contains such enticing chapters as "Exploring the Yard on Your Stomach" and "Exploring the Yard on Your Back." There are many suggestions for activities that can be carried out by children on schoolgrounds and near their homes.

JACOBSON, WILLARD J., ET AL. *Ecology: Field Research in Science.* New York: American Book, 1969. A discussion of some of the biological concepts that can be studied in the local community. The book can be used directly by older children. Teachers can use the ideas in the book as they help younger children derive meaning from the observations they make outdoors.

SCHNEIDER, HERMAN. *Everyday Machines and How They Work.* New York: McGraw-Hill, 1950. A book for children that explains how many of the machines they will encounter work.

SCHNEIDER, HERMAN AND NINA SCHNEIDER. *Let's Look Inside Your House.* An early book by the Schneiders that suggests experiences with electricity, water, and heat.

Science for Children in Cities

CHAPTER 11

The city! A product of science and technology. It may be that the first cities were made possible by the development of agriculture, which freed a few people to do things other than grow the food and fiber that everyone needs to live. Recently, the mechanization of agriculture has made it possible for many more people to leave the farm and go to the cities. In the United States today only 5 percent of the population is engaged in the agricultural tasks of growing food and fiber. This has made large cities possible.

Although most people now live in urban areas that have remarkable resources for science education, we have not made optimum use of the many opportunities for unique science experiences for children in cities. Often elementary school science may seem to be best taught amid the fields and forests of rural areas. True, these areas can provide settings for fine science experiences, but equally excellent experiences can be developed in urban areas. In the pages that follow are examples of high-quality science experiences that can be developed with children in cities. (See Chapter 10 for science experiences that can be developed in schools and schoolgrounds.)

We can wonder about the city, and one of our primary aims in elementary school science is to encourage, stimulate, and fertilize this sense of wonder. The very young child may wonder about where the water that he or she drinks comes from, and for a while the child may justifiably think it comes from the wall where the faucet is attached. An older child may wonder how pressing a button can cause an electric light to go on or an elevator to come to a particular floor, and finds out more as he or she experiments with electrical switches and circuits. But every day the multitudes who live in the city must have food, water, and other essentials for life. Where does it come from? How does it get there? For all the water and other material that comes into the city, about an equal amount has to be disposed of. What is done with it? Where does it go? Communication, transportation, housing, education, health, and sanitation all become more complex in the city, and science and technology are used in carrying out these functions. We should encourage children to become curious about these functions and guide them in their inquiries.

Often cities have remarkable scientific and cultural resources. There may be museums, zoos, botanical gardens, aquariums, industries, utilities, and most important, talented and knowledgeable people. Actually, some of these resources are almost unique to cities; a certain population size is needed to support specialized museums and other cultural resources. Certainly, they should be utilized in developing the best possible science activities for children in cities.

The following section describes some of the ways in which we can study how science and technology are used to provide what people in cities need and want. This is followed by a checklist that suggests cultural and scientific resources that may be available to you and your children in cities.

SCIENCE IN CITIES

Science and technology make cities possible. We can enhance the science education of our children by studying science and technology as they appear in our urban environments. In this section several basic principles are stated and discussed briefly. We also describe science experiences that can be developed to give children a deeper understanding of these principles.

Food, Water, Shelter

People in cities need food, water, shelter, and other necessities of life. People living in cities have about the same requirements as those living anywhere else, but because there are so many people concentrated in a city, the systems for supplying these requirements often require the movement of large amounts of materials through complex supply networks.

Have the children compare the ways in which they get their necessities with the ways in which farmers living 100 years ago—with a garden, cows, a woodlot, and a well—got their necessities. Who is more dependent on others for the basic things all humans need? Who can get the greatest variety of food? What changes have had to be made to make it possible for large numbers of people to live in cities?

Where does our food come from? Visit a supermarket or grocery store and try to find out where the fresh fruits and vegetables come from. Some of this information can be obtained from the boxes and bags in which the food is shipped; clerks and store managers usually can provide additional information. How do the sources of our foods change with the seasons? How is the food transported to the city and to the store?

Have the children examine various food containers in their homes to try to find out where these foods come from. From how far away do we get foods?

How are foods preserved and stored? Ask the children to note the next time they are in a grocery store or supermarket the different ways foods are preserved and stored (canning, drying, bottling, pickling, using chemical preservatives, chilling, and freezing are some of the ways that are often used).

Obtain two or three different kinds of fruits and vegetables. What methods should we use to preserve and store these foods so that we could eat them later? If possible, have the students actually preserve and store the foods in the ways they suggest. Later, check to see how these methods have worked.

Most cities get their drinking water from reservoirs that catch the water that runs off the surrounding hillsides or from nearby rivers and lakes. Philadelphia and Cincinnati get their water from nearby rivers; Chicago and Milwaukee draw their drinking water from Lake Michigan; New York City gets most of its drinking water from reservoirs in the Catskill Mountains; and Los Angeles gets some of its water from the distant Colorado River. Usually the water is treated and brought to the city in large pipes or aqueducts. Often your local water supply department can give you information and materials that describe your city's water system. In some cases it may be possible to take the children to see reservoirs, aqueducts, pumping stations, and water treatment plants.

How does your drinking water compare with water from a nearby river, lake, or pond? Fill glass jars with water from a drinking fountain or faucet and from a nearby river, lake, or pond. Compare them in the following ways:

1. *Smell.* The drinking water will often have a slight smell of chlorine. Badly polluted river or lake water may smell like chemicals or decaying matter.
2. *Clearness.* When you look through the samples of water in the two glass containers, do they appear equally clear?
3. *Sediment.* Allow the water to stand in the glass jars. Does any material settle to the bottoms of the jars? Is there any material floating at the top of the water?
4. *Mineral Matter.* Put equal amounts from each sample of water into small containers in which the water can be evaporated. Have the water in the small containers evaporate by heating it on a hot plate or placing it in the sun. When the water has evaporated, do you see any grayish mineral matter on the bottoms of the containers?
5. *Hardness. Hardness* of water refers to the amount of mineral matter in the water and the relative difficulty of having soap form suds. Put equal amounts of drinking water and river or lake water into two glass containers. Add liquid soap to each of the samples a drop at a time, stirring after each drop. After how many drops do suds begin to form? Which sample of water requires the most soap for suds to form?

Filter the soapy water through paper toweling and examine the residue. This residue was formed as the soap reacted with the minerals in the water. What are the residues like?

Waste Disposal

Cities discharge their wastes into the surrounding environment. How much waste do cities have? Your children may be surprised to learn that cities have about the same quantity of waste as the quantity of food, water, bottles, cans, and all other materials that is shipped into the city. For example, sanitation engineers have to plan to treat and dispose of as much sewage as there is water brought into the city. About as much garbage has to be disposed of as there is solid material hauled in. A basic principle of science is the law of conservation: "Under ordinary circumstances matter can be neither created or destroyed." The city can be considered a system in which this basic conservation law holds. The water and solid materials that are brought into the city may be changed in form, but they are not destroyed and must be disposed of in some way.

The wastes of cities are disposed of in the surrounding environment. The gases from the chimneys of factories and homes and the exhaust from automobiles are discharged into the atmosphere. Sewage and other liquids are discharged, ideally after treatment, into nearby rivers and lakes, or, when there are relatively small amounts, they may be filtered back into the ground water from holding ponds or settling tanks. Solid wastes, or garbage, are often covered with soil in landfills, where much of it will eventually decay. In some cities it may be dumped into the ocean or nearby lakes. Increasingly, cities burn some of the solid waste and use the energy for heating or generating electricity.

How does our city dispose of its wastes? One of the first steps in the study of waste disposal is to find out how the city disposes of its wastes. Ask the garbage collector! The people who help us dispose of our wastes can help us find out how our solid wastes are processed and disposed of, and children should learn more about the work of the people who do this job. Look across the city from a hill or the top of a building and try to see evidence of wastes entering the atmosphere. Try to find out where and how liquid wastes are disposed of.

How are wastes treated before they are disposed of? Help the children find out how smoke and other gases are treated before they are discharged into the atmosphere. A good place to start is to find out how smoke from the school's heating plant is controlled and treated. Through a visit to a sewage treatment plant, children can find out how the sewage of the city is treated before it is released into a river or stream. A visit to a landfill site and a talk with those who operate it will show how solid wastes are disposed of and people are protected from diseases that might be spread by wastes.

What kinds of materials pollute the atmosphere? Tack a piece of cloth or old window shade onto a wooden board, smear vaseline or grease over the cloth, and place it on top of the school or outside the classroom window so that it will collect whatever materials settle out of the atmosphere. (See Figure 11–1.) After about a week examine the materials with the aid of a magnifying glass. Can you identify any of the materials? What may be the source of most of the materials? Are any of the materials rounded, as if they have been melted? (Some people believe that such melted materials may actually be the remains of very small meteors that have melted as they passed through the atmosphere.) If such collections of atmospheric pollution can be made over time and compared, it is possible to make inferences as to changes in air pollution. For example, how does the amount of material collected in the winter compare with the amount collected in the summer?

What kinds of materials are there in our rivers and lakes? Collect bottles of water from a nearby river or lake. Examine it. Does the water look clear? Or is there a noticeable cloudiness (turbidity) to the water? If so, what seems to be causing the turbidity? Is there any material floating at the top? What kind of material does this seem to be? Smell the water. Is there an odor of decay-

Figure 11–1 A board with a piece of cloth tacked over it, on which vaseline has been smeared.

ing material? If you have a microscope, use it to examine a drop of water. Are there any living organisms in the water? Allow the water to stand for a day or more. Is there any material that settles to the bottom?

How does organic material decay in soil? One of the most common ways of disposing of solid wastes is to cover it with soil in landfills in which the organic matter eventually decays.

Bury some apple cores, potato peelings, or other wastes in a box of soil. Have the children note the condition of the organic matter before it is buried. After several months, uncover the organic matter. What has happened to the organic matter?

Organic matter can be used as fertilizer for a garden or flowerpots by composting it. Build a compost pile by placing waste organic matter into a pile and occasionally covering it with soil. For best results the compost pile should be turned over periodically with a spade or fork. After the compost has the appearance of dark soil, mix it with the soil in a garden or for flowerpots. Does the compost seem to help plant growth?

How can a housefly carry wastes? One of the insects that seems to be present almost everywhere and can carry contaminants onto our foods is the common housefly. Swat a housefly and then examine it with the aid of a magnifying glass. What are some of the characteristics of the legs of houseflies? Would materials cling to them? If you have a microscope, examine a housefly leg under higher magnification.

Animals

A variety of animals live in cities. Some animals seem to thrive in close association with humans. Ants, cockroaches, houseflies, and mosquitoes are insects that can be observed, captured, and studied. Pigeons, sparrows, and starlings are among the birds that seem to do well in cities, and a wide variety of birds will fly by, with occasional stopovers, as they migrate between their winter and summer homes. Mice and rats are other animals that live in cities, and if we know where to look we may catch a fleeting glance of them. Occasionally other animals, even pheasants and deer, may wander into the city, and the alert observer may be able to catch sight of them in city parks.

Some of the following animals can be observed and studied in or near schools.

Cockroaches. Two common cockroaches found in North America are the German and American cockroaches. The American cockroach is the larger variety. Cockroaches have adaptations that help them survive in cities. For example, they have flat bodies that enable them to squeeze into narrow cracks. The female cockroach lays a small bean-shaped egg pouch. If this egg pouch is deposited in a grocery container or wrapping, it can be carried into the most immaculate home. Roaches like dark, moist areas and tend to be most active at night. While they do not transmit any specific disease, they probably can carry germs onto any

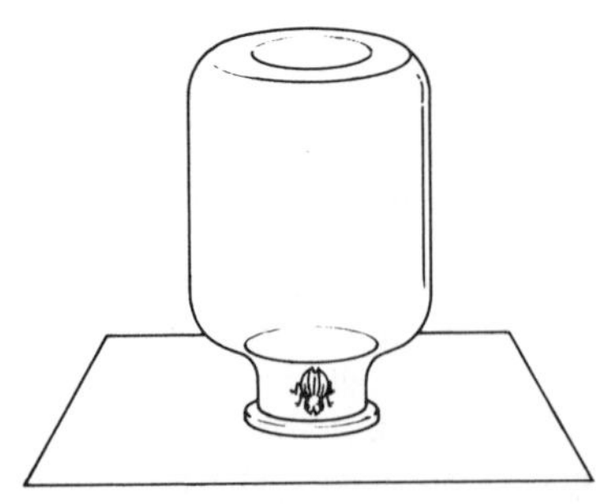

Figure 11–2 To capture a cockroach, place a wide-mouthed bottle over it. Slip a card underneath it and then quickly turn the bottle over.

food they touch. They eat many of the same kinds of foods that humans eat. They will also eat glue from postage stamps and the backs of books.

Like crickets and grasshoppers, cockroaches undergo *incomplete metamorphosis* (i.e., after hatching the insect will always have the appearance of the adult). The eggs hatch into very small cockroaches. The roaches shed their outer coverings (molt) several times before they reach maturity. Adult cockroaches have wings, and some can fly short distances. Cockroaches can often be found in dark, wet places. They can be captured by placing a wide-mouthed bottle over them and slipping a card or a piece of cardboard under them and then quickly turning the bottle over. (See Figure 11–2.) Moist paper should be placed in the bottle so that the cockroach will be able to retreat into dark areas within the folds of the paper.

Have the children observe the cockroach, the paper, and the jar over time. What changes take place? Cockroach droppings look like small black balls. Young cockroaches hatch out of small egg pouches and are very, very small at birth.

Cockroaches use their antennae, the stringlike feelers attached to their heads, to detect food and other objects in their environment. How does the captured cockroach use its antennae?

Observe the wings; try to see whether the cockroach is carrying an egg case at the end of its abdomen.

Fossils of cockroaches have been found that are several hundred million years old. Yet this very old animal survives very well in a relatively new environment—the city.

Mosquitoes. Mosquitoes live part of their life cycle in water and part as the adult mosquito that flies through the air. The mosquito is a common example of an insect that undergoes *complete metamorphosis;* it passes through the egg, larva, pupa, and adult stages (i.e., the insect appears quite different in each stage of its life cycle).

Mosquitoes can be obtained for study by placing a pan of water outside in a place where adult mosquitoes can have access to it. The female mosquito may lay eggs on the surface of the water. After a while you may see what appear to be wriggling worms in the water. These are mosquito larvae. Pour some of the water and the larvae into a gallon jar. With string or a rubber band, fasten a piece of cloth over the mouth of the jar so that the mosquitoes will not be able to escape. Bring the jar into the classroom for further observation and study.

The various stages of the complete metamorphosis of an insect like the mosquito can be observed in the jar. (See Figure 11–3.) The mosquito eggs float in a case on the surface of the water. The larvae move up and down in the water, but they must return to the surface to get air. The pupae float near the surface and get their air through tubes. They even-

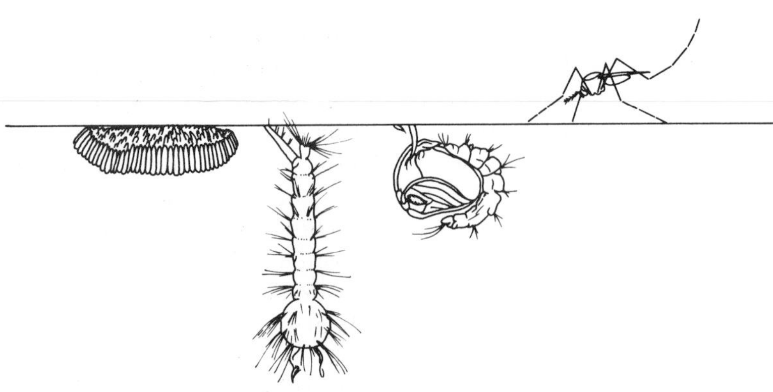

Figure 11–3 Metamorphosis of the adult mosquito. The eggs float on the surface of the water. The larvae move up and down in the water. The pupae remain near the surface. The adult flies through the air.

tually change into adult mosquitoes. The female adult mosquito can lay eggs on the surface of the water and begin the cycle again.

One of the most common ways of preventing the spread of mosquitoes is by preventing the larvae and the pupae from getting air at the surface of the water. Place a few larvae and some water in a small glass jar. Put a drop or two of oil on top of the water. What happens to the larvae and pupae?

If a child has been bitten by a mosquito, he or she may have an itchy red blotch on his or her skin. After the skin is pierced, a small drop of poison is released that expands the blood vessels, prevents the blood from congealing, and causes a small inflammation of the skin. It is the female mosquito that bites. She uses the blood to nurture the eggs that she will lay.

Birds. Pigeons, various kinds of sparrows, and starlings are among the birds that are common in most cities. But there may be many others that make the city their home, and even more may pass through the city on their semiannual migrations. Bird watching is an interesting avocation to which children can be introduced in the elementary school.

Children should be encouraged to observe birds in parks and around their homes. They may begin a "life list" and add to it as they continue to see more birds. A great variety of bird feeders are available for purchase or can be made. A feeder set up within view of the classroom window will increase the number of birds that can be seen. (See p. 262.)

Binoculars make it possible to see much more detail, and they are essential for those who wish to pursue bird watching as a hobby. For the identification of birds and other animals and plants, field guides like the *Golden Nature Guides* can be used.

Special adaptations make it possible for some plants to survive in cities. The city environment is a difficult one for many plants.

The following are some of the difficulties that trees face in a city environment. Have your children examine the environments of several street trees in your city. What difficulties do trees that live in our cities face?

1. *Lack of soil.* Very often most of the ground is covered with concrete or macadam and there is little room for trees.
2. *Lack of water.* When most of the land surface is covered with concrete or macadam, most of the rainwater runs off and does not sink down to the roots of trees.
3. *Limited sunlight.* High buildings may shade a tree for most of the day. Smog will reduce the amount of sunlight it receives.
4. *Salt.* Salt is often spread on icy streets, and some of it sinks down to the roots of the trees.
5. *Dog urine.* Dog urine contains salts and may raise the salt content of the soil above the tree's tolerance level.
6. *Air pollution.* Sulfur dioxide from furnaces and ozone from the breakdown of automobile exhausts are toxic to most plants. Particulate matter, such as coal dust, can cover leaves and make it more difficult for them to get the oxygen and carbon dioxide they need.
7. *Wear and tear by humans, automobiles, bicycles, and animals.* Roots and trunks can be severely damaged by wear and tear.

What kinds of trees seem best able to survive under city conditions? Using a guide such as the *Golden Nature Guide to the Trees,* have the children identify the trees that grow in the city environment. Among the trees that seem to be surviving and even thriving in the city are maples, oaks, ailanthus, ginkgo, and the London plane tree, which very much resembles the sycamore. The American elm and the American chestnut once lined many city streets but have almost completely succumbed to disease.

Some plants are able to survive under the most difficult conditions. For example, plants can be found growing in cracks in sidewalks and streets where there is little soil and water, great extremes in temperature, and much wear and tear from pedestrians and vehicles.

What are the adaptations of plants that live in cracks in sidewalks and roads? Have your children find one or more plants that grow in a crack in a sidewalk or road. Which of the following adaptations does it have?

1. Thick, tough, waxy leaves. The leaves must be able to withstand a great deal of wear and tear. A waxy covering reduces the amount of water lost by the plant.

2. A tendency to cling to the surface of the road or sidewalk. This adaptation will help prevent the plant from being broken and torn.
3. A tough stem. Try pulling a section of the stem apart. With some of the hardiest plants, this is almost impossible to do. This adaptation is essential for the plant to withstand the buffeting it takes at the surface of a road or sidewalk.
4. Well-anchored roots. Pull up on the plant to see how firmly it is anchored in the ground.
5. Ability to withstand extremes of temperature. Take the temperature on a very hot, sunny day. Then take the temperature on a very cold day. What is the range of temperatures that the plant is able to withstand?

RESOURCES FOR SCIENCE IN CITIES

Everywhere there are unique resources for working with children in science. In rural areas there are farms and forests; in urban areas there are zoological parks and botanical gardens. In some places there are great national or state parks and magnificent geological and geographical features; in cities there are museums and a wide variety of exhibits and displays. Everywhere there are important human resources to be tapped; most schools in cities have access to a very rich variety of human resources.

Make the most of what you have! In the city this means utilizing the rich cultural resources of museums, zoos, aquariums, botanical gardens, parks, and exhibits of all kinds. Some school systems have compiled guides to the science resources in a city. These can be a great help in the planning of effective science experiences. Fellow teachers and administrators can be a source of further information about opportunities for science experiences; make the most of your opportunities to chat with them about their plans and yours. Most large museums, zoos, and other such resource institutions have education departments that can suggest possible experiences in their institution and ways in which you and your children can make the best use of these resources.

The following checklist can be suggestive of resources for working with children in science. Broad categories of resources are suggested. The exact nature of these resources will differ from city to city, and under each of the headings it will be helpful to gather and list further information and possibilities for their use. As you use the various resources in the city, a short appraisal can be made for future reference. In some schools teachers have combined their appraisals in resource files.

The checklist can be used in planning science experiences for children. You may wish to scan the major headings of the checklist and identify those that you can or wish to use. Next, review each of the resources under the headings that you believe you might use and the specific contribution the resource might make. Remember, the checklist is suggestive and not inclusive, and you and your colleagues may wish to add items to the list. Once you have identified several possibilities, seek additional information about the resource. Talk to other teachers who have used the resource. Refer to any records of past visits or contributions that may be available. By all means talk to the resource person and visit the resource institution yourself before taking your children. Use your knowledge of your children to determine the most effective way of utilizing the resource. If it is an institution that has an education department, consult it for advice and alert it of your plans.

A CHECKLIST FOR SCIENCE IN CITIES

I. Human resources—list people and the specific contribution they might make.
   A. Children in the school
   B. School personnel
   C. Parents
   D. People in the community
   E. Other human resources

II. Museums (Exhibits and Possible Uses)
   A. Natural History
   B. Science and Industry
   C. Other Museums

III. Botanical Gardens (Exhibits and Possible Uses)

IV. Horticultural Gardens (Exhibits and Possible Uses)

V. Zoos and Zoological Parks (Exhibits and Possible Uses)

VI. Industrial and Commerical Exhibits
   Stores, office buildings, and manufacturing plants often have useful exhibits that have been developed to inform the public.

VII. Shops and Supermarkets (Possible Observations)
   Among other possibilities, these are places where we can learn more about where the things we eat come from and how they are processed.

VIII. Parks
   Almost all parks have features that can be used for science instruction.
   A. Accessible geological features
   B. Trees
   C. Flowers
   D. Soils and examples of erosion
   E. Animal life
   F. Opportunities for astronomical observations
   G. Seeds, fruits, and leaves

H. Opportunities for weather observations
I. Gardening opportunities
J. Bird watching opportunities
K. Rock and mineral study opportunities
L. Insect study and collection opportunities
M. Bodies of water and aquatic features
N. Beaches and beach phenomena

IX. Utilities
Energy and other essentials are provided by public utilities.
A. Production sites
B. Places where important processes can be observed
C. Special materials and exhibits

X. Manufacturing Plants and Other Industries
Most cities have an industrial base that involves science and technology.
A. Industrial sites that can be visited
B. Processes that can be observed
C. Special materials and exhibits

RELATED ACTIVITIES

What are some of the characteristics of water? p. 345.

What is the level of the water? p. 189.

Which animals live in the local community? p. 244.

Maintaining a bird feeder, p. 262.

How great are the differences among various plants in your community? p. 220

What are the adaptations of some of the plants that live in environments in your locality? p. 221.

What plants will begin to grow on open soil? p. 228.

How can terrariums be set up and used as tools for learning? p. 225.

What sounds can be heard in the immediate environment? p. 409.

How can local seasonal variations be studied? p. 250.

What changes can we see taking place in a mud puddle? p. 349.

What minerals are found in rocks in the local area? p. 340.

What are some of the things we can learn from a rock? p. 337.

What kinds of work do different people in your school do? What kinds of backgrounds and skills do they have? p. 193

REFERENCES

ADRIAN, MARY. *Secret Neighbors—Wildlife in a City Lot.* New York: Hastings House, 1972. City wildlife is described, and a story about a field mouse conveys how city animals live through changing seasons.

BUSCH, PHYLLIS S. *City Lots: Living Things in Vacant Spots.* Cleveland: World, 1970. City environments and some of the living things that may be found in them.

COLE, JOANNA. *Cockroaches.* New York: William Morrow, 1971. Basic information about cockroaches that could inspire investigation and activity.

GALLOB, EDWARD. *City Leaves, City Trees.* New York: Charles Scribner's Sons, 1972. A well-illustrated book about trees commonly found in cities.

HAWKINSON, LUCY AND JOHN. *City Birds.* Albert Whitman, 1957. A guide to birds that frequently inhabit cities.

NATIONAL AUDUBON SOCIETY. *A Place to Live.* New York: National Audubon Society, 1970. An elementary school workbook (with accompanying teacher's manual) that includes many interesting investigations for the study of the city as a place to live. Includes walk and city field trip outlines.

RALIN, JOAN ELMA. *Nature in the City: Plants.* Chicago: Children's Press, 1976. Suitable for all elementary grades. Wildflowers commonly found in cities are described.

RUBLOWSKY, JOHN. *Nature in the City.* New York: Basic Books, 1967. A guide to nature studies in cities.

RUSSELL, HELEN ROSS. *City Critters.* New York: Meredith Press, 1969. There are more "critters" in the city than you might think. This delightful book for children describes some of the animal life to be found in city environments.

SCHNEIDER, HERMAN. *Let's Look Under the City.* New York: Scott Publications, 1954. How basic utilities are brought to city buildings through underground pipes, wiring, etc.

WEEKS, MORRIS. *Inside the Zoo.* New York: Simon and Schuster, 1970. A descriptive book about the zoo and its operation.

Children and Plants

CHAPTER 12

Stroll down a park lane and you may see green, cushiony mosses nestled up against a lichen-covered rock; a tall tree providing a shadow for the path you tread; wild flowers with petals of yellow, pink, or lavender; and grasses that make a lush green lawn. These are all living organisms, and they are in the kingdom of living organisms that we call *plants*. The biologist will tell you that the members of the plant kingdom differ from animals in that plant cells have cell walls. But in most cases children can readily tell you which organisms in their environment they call plants.

Through the study of plants, children can learn a great deal about living organisms in general as well as plants in particular. All children should have a chance to grow plants, care for them, and study the changes that take place in them. They should learn more about what living organisms require to live, and through their care and concern for plants they should begin to develop a sense of responsibility for other organisms in their environment.

The study and care of plants can be an ongoing activity for children in the elementary school. Early in the school year, each child can start a plant from seeds, cuttings, or tubers. The plants will require only a few minutes of care each day, but the children should come to realize that this care is essential. They should become sensitive to the seasonal changes that take place in trees and flowering plants. They may see and become concerned about the misfortunes of some lawns, shrubs, or forests, or the trees that line city streets. The study of plants can be an important part of the flexible dimension of elementary science programs in that fresh observations and new questions will continue to arise. The unit on plants will never be finished – it can go on for a lifetime.

CHARACTERISTICS OF PLANTS

A combination of characteristics distinguish living organisms from nonliving material. If you ask a group of children to name the objects in a classroom that are living, they will respond by naming the plants and animals that they can see in the room. (To our surprise, we once had a child seriously suggest that a radio was living.) If we ask "What are the characteristics of living organisms?" the following are often given. However, it is possible to find nonliving material with these characteristics.

Movement. Many living organisms, especially animals, are able to move. Many plants are also able to turn their leaves in the direction of a light source, and plants such as the mimosa and the Venus flytrap may respond quickly to touch. However, a seed or spore may lie dormant for a long time without moving, yet it would be classified as living. While all animals can move during some stage of their life cycle, many animals (e.g., the oyster) do not move during much of their life span. However, "self-propelled" movement is a characteristic of many living things.

Response to stimuli. Most living organisms can make some kind of response to changes in the environment. The leaves of the corn plant will curl when there is a lack of water, and the microscopic ameba will react to chemicals dropped into the water. Higher animals have well-developed nervous systems that are sensitive to stimuli and can detect complicated responses. But again, many plants and animals in the egg stage of development do not respond in obvious ways to stimuli from the environment.

Use of food. All living organisms use food of some kind as a source of energy and raw materials for growth and repair. This food is manufactured by green plants through the process of photosynthesis, and it is used by both plants and animals. However, a number of nonliving objects such as internal-combustion engines use fuel in a way that is similar to the use of food by living organisms.

Respiration. Living organisms obtain energy through the oxidation of food in the living cell. Oxygen is taken into the organism and carbon dioxide is discharged as a waste product. Again, however, internal-combustion engines also use oxygen and discharge carbon dioxide.

Growth and repair. Living organisms are able to use food for growth and repair. In 20 years a 6-pound baby can become a 200-pound fullback. When a finger is cut, the cut will eventually be repaired. Of course, almost all organisms reach a point at which there is no more growth, and if an injury to the organism is serious enough it will not be repaired. But a snowball that rolls down a hillside will grow, and some machines can automatically replace a part when it is worn out.

Reproduction. Many living organisms are able to reproduce their kind, and many students believe that this is the key characteristic of living organisms. But of course many individual organisms do not reproduce. Among humans, for example, young children and old people do not reproduce, yet they certainly are very much alive. A seed crystal dropped into a saturated solution may cause many other crystals to form. Is this reproduction?

Distinguishing living from nonliving matter. There is no clear, unequivocal test that will differentiate between the living and the nonliving. The differentiation has to be made in terms of a number of characteristics like those just listed.

The question is further complicated if organisms that were once alive but are now dead are considered. While dead organisms do not have many of the characteristics of the living, the material is composed of cells and the "stuff of life"—protoplasm. Also, there is the difficult question of when an organism is dead. Human hair, for example, will continue to grow even after the body has been legally declared dead. Is it "really" dead?

Some forms of matter, such as some viruses, are especially difficult to classify as living or nonliving. At times the viruses have the characteristics of living organisms, including the ability to reproduce. But at other times they appear to be more like crystals. The viruses apparently are on the borderline between the living and the nonliving and sometimes are living and at other times not.

Many living organisms contain materials that are nonliving. The cork in trees is nonliving. In fact, much of the material in human bones is nonliving. However, even though organisms may contain considerable nonliving material, the entire organism is considered to be alive.

Is a candle flame alive? Light a candle and place it in a small, shallow dish. Ask the children, "Is the candle flame alive?" Usually, most children will agree that the candle flame is nonliving. Then investigate the candle flame in terms of each of the characteristics of living organisms. (See Figure 12–1.)

Movement. The candle flame moves. If you blow gently, it will flicker a great deal.

Response to Stimuli. Clap your hands near it. It will react almost instantaneously.

Use of Food. The candle is the food for the candle flame, and if it were to burn for a considerable time, the candle would become noticeably shorter.

Respiration. The candle uses oxygen. If a large glass jar is placed over the candle, the flame will eventually go out. In the process of burning or oxidation, the candle flame, like living organisms, produces carbon dioxide. If a little limewater is swished in the jar, it will turn cloudy. This is a test for the presence of carbon dioxide.

Growth and Repair. If the flame is given more fuel, such as a small piece of paper, it will grow. If the wick just below the flame is squeezed, the flame may almost go out. When you stop squeezing, the flame will burn brightly again.

Reproduction. If the wick of another candle is placed in the flame, another candle flame will be produced.

Figure 12–1 Is a candle flame alive?

New plants can be grown from parts of plants. Throughout the living world there is a continuity of life. Humans have children. Animals have offspring. New plants can be grown from various parts of parent plants. It is important to point out that organisms always beget organisms of the same kind. Often there are distinct resemblances between the offspring and the parent organisms.

It is highly desirable that each child have the experience of starting new plants in each of the ways described. Flowerpots are ideal for starting plants, but a styrofoam drinking cup or the lower half of a milk carton can be substituted. Potting soil that can be obtained at any garden store is recommended, but soil from a garden or field can be used. Vermiculite or other porous material is ideal for starting many cuttings, but sand can be used.

How do seeds sprout? Put some paper toweling into a plastic or glass jar. Then place some seeds, such as lima bean or corn seeds, between the paper and the sides of the jar. (See Figure 12–2.) Put a little water in the bottom of the jar. The water will move up the toweling by capillary action and moisten the seeds. Check the seeds each day. In what direction do the roots grow? The shoots?

Figure 12–2 These children are examining some seeds that have sprouted when placed next to moist paper toweling.

How can plants be started from seeds? All annual flowers and vegetables and many other plants can be started from seeds. Corn, lima beans, peas, marigolds, zinnias, pansies, petunias, and lawn grass are among the plants that can be grown from seed in the classroom.

A mixture of two-thirds potting soil and one-third vermiculite (ordinary garden or field soil can be used) should be placed in some container such as a plant tray or an aluminum cake tin. The seeds are sprinkled on top of the soil and then covered with soil. The soil should be sprinkled with water. To keep the soil moist, the container can be covered with clear plastic. After one or two pairs of leaves have appeared on the plants (see Figure 12–3), the small plants should be transplanted into flowerpots or other large containers.

How do potatoes grow? Insert toothpicks into a potato or sweet potato so that half the potato is immersed in water. (See Figure 12–4.) Watch the roots and shoots grow. Then transplant into soil.

How can plants be grown from cuttings? A *cutting* is a small part of a plant that is removed and treated in such a way that roots will form and a new plant will grow. Geraniums, cacti, coleus, hibiscus, and petunias are among the many flowering plants that can be propagated by cuttings. Cuttings are sometimes called slips.

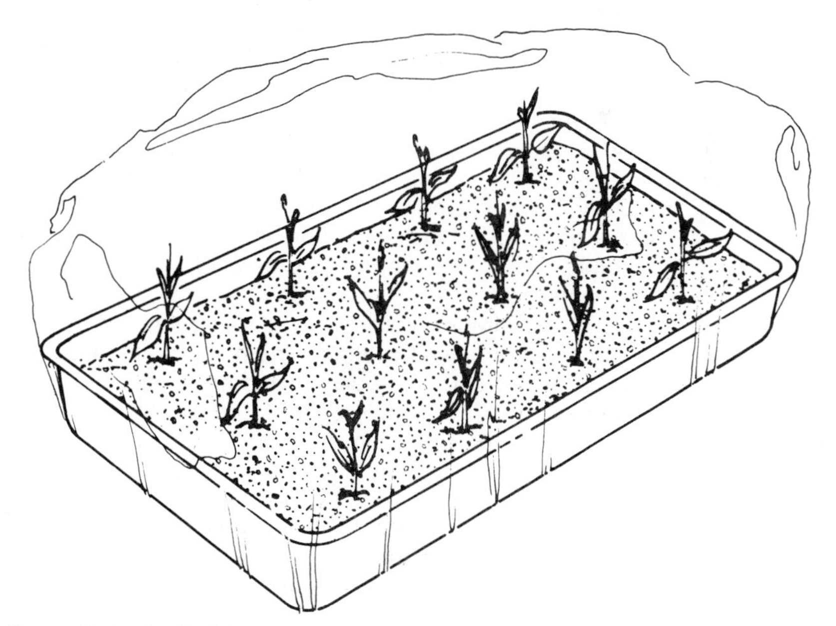

Figure 12–3 Seeds being sprouted in a mixture of potting soil and vermiculite under plastic.

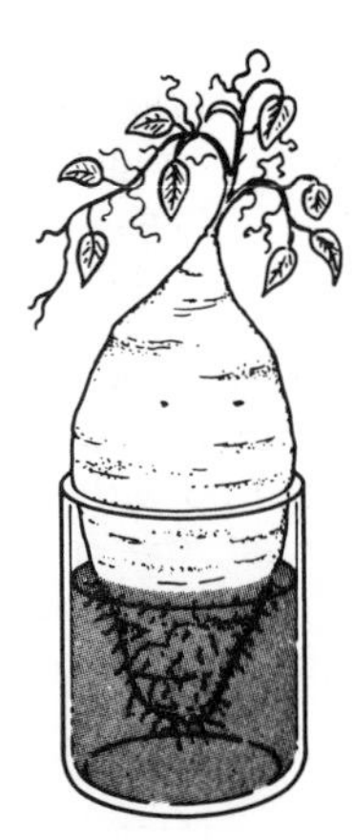

Figure 12–4 When partially immersed in water, a potato or sweet potato will begin to grow.

Cuttings are usually 2 to 4 inches long and are from shoots or new growth on the plant. The cutting should be made with a sharp knife or razor blade. The stem should be cut just below a node. It should be a slanting cut so that more of the growing part of the stem is exposed. (See Figure 12–5.)

All but 2 or 3 of the top leaves should be removed and the cutting sprinkled with water. After a half-hour or so, the cut end of the stem should be inserted into damp sand or vermiculite. The sand or vermiculite should be pressed firmly around the base of the cutting. To retain moisture, the cutting can be covered with plastic. However, the plastic should be supported so that none of it touches the cutting. The cutting should be kept out of direct sunlight for the first few days, and the sand or vermiculite should be kept damp.

When some new growth appears on the cutting, usually after 3 or 4 weeks, the new plant is ready for potting. A mixture of two-thirds potting soil and one-third vermiculite or sand makes an excellent potting medium. The soil should be placed in a flowerpot or some other container and the roots and a small part of the stem of the cutting inserted into a hole in the potting medium. The potting medium is pressed firmly around the roots and the soil kept moist.

Plants with thick and fleshy leaves, such as begonia and gloxinia, can be propagated by leaf cuttings. The leaf of a begonia, for example, can be laid flat on the surface of damp sand or vermiculite. Small pebbles may be placed on top of the leaf. Roots will form at various places along the bottom of the leaf. These can then be potted.

Plants, like other organisms, can produce many more plants. There is an intrinsic capacity for growth in all populations of living organisms. Find a fluffy dandelion and try to count the number of seeds produced by one dandelion plant. Or, observe a tree whose seeds are still suspended from it. How many seeds does each tree produce each year? Examine any flowering plant while its seeds are being formed. While many of the

a. Cut stem just below a node

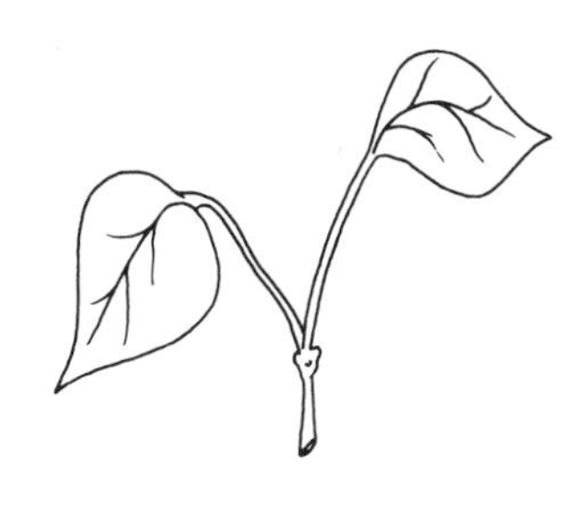

b. Remove all but 2 or 3 leaves

c. Insert cut stem into sand or vermiculite

d. Add water and cover with plastic

Figure 12–5 Steps in propagating plants by cuttings.

seeds will never reach conditions under which new plants can be formed, usually enough new plants will be formed to ensure the survival of the population. The potential for a great increase in numbers is also there.

For the survival and increase of a population, it is desirable that there be mechanisms for the dispersal of seeds. Blow on the fluffy dandelion head and watch as the seeds are wafted great distances before they settle back down to earth. Note the "wings" attached to the seeds of the maple tree which spin like helicopter wings as they fall from high in the trees. The burrs on the burdock and the flat, fuzzy seeds of the sticktight will cling to the clothes and fur of people and animals and receive free rides for considerable distances before they drop or are picked off, possibly to start a new plant if the environment is favorable. Squirrels carry acorns, bury them, and sometimes forget about them. Small seeds may be eaten by birds and remain viable in their droppings. Other seeds are catapulted through the air by a springlike action as the seedpods open. There is a great variety of methods of seed dispersal—almost all of them generally effective.

How are seeds spread? Have the children find seeds that are spread by the wind, by becoming attached to animals, by the helicopter effect as they fall, or by being catapulted away from the plant when the seedpods open. What other methods of seed dispersal can they find?

REQUIREMENTS FOR LIFE

Plants, like other organisms, have certain requirements for life. Children can learn how to care for living organisms. Perhaps the most basic approach to such care is for them to try to think of what all living organisms need to live and thrive. Children can try out their ideas and develop "green thumbs" as they care for their plants with "tender loving care."

Living organisms need food, water, proper temperatures, and general good care. In the list of references at the end of this chapter are a few of the many sources that discuss the general care of plants as well as the specific requirements of different varieties of plants. Children should learn that such references are available and begin to consult them.

Through the process of photosynthesis, green plants manufacture their own food. In fact, all plants and animals are dependent on green plants for their food. To manufacture this food green plants need minerals, water, and light. The most important dimension of proper plant care is that they receive the right amounts of each of these essentials.

In most cases plants get the minerals they need from the soil. The most important minerals are nitrogen, phosphorous, and potassium. When children examine the labels on bags of fertilizer they will usually see three numbers such as 5-10-5. These numbers signify the percentages of the fertilizer that are, respectively, nitrogen, available phosphorus compounds, and potassium. Nitrogen is essential for deep green leaf growth. Plants receiving too little nitrogen will have stunted growth and yellow leaves. Composted materials, manure and such nitrogen-fixing plants as alfalfa are sources of nitrogen for the soil. Phosphorus is necessary for plant growth, root growth, and flower and seed development. Lack of phosphorus will often lead to stunted growth and leaves that have a purplish color. Bone meal and phosphate rock are the major sources of phosphorus fertilizer. Potassium is needed for leaf and root system development. It is especially important for such root crops as potatoes, carrots, and beets. Plants that do not receive sufficient potassium often have leaves that appear dry or scorched at the edges. Wood ashes and seaweed are important sources of potassium fertilizer. Other minerals that are needed are calcium, magnesium, sulfur, and a number of trace elements.

How can we test soil? Soil-testing kits are available from garden centers and department and discount stores. The first soil-testing experience that children should have is that of testing pH, or the acidity or alkalinity of the soil. Kits for testing pH are relatively inexpensive. These kits usually include instructions for soil testing and the pH requirements of some plants.

The pH of a sample is its relative acidity or alkalinity, as shown on the following scale:

Very acid			Neutral			Very alkaline
4	5	6	7	8	9	10

Most plants, but not all, grow best in soils that are neutral (7) or slightly acid (6–7).

When testing soil, children should try to get a sample that is "representative" of the soil in which they are going to grow their plants. If they are going to grow their plants in a flowerpot, then almost any soil they take from the pot will be "representative." But if they are testing soil from a small garden, they should get a little soil from several places.

In most soil pH tests, a small sample of soil is wetted with the solution provided. After the soil particles settle, the color of the liquid is matched with the color guide provided to get the approximate pH reading. Children often find it interesting to compare the results they obtain when testing the same sample of soil. What may have caused any differences in results?

Water is an essential raw material for food production in plants. In photosynthesis water is combined with carbon dioxide from the air to form sugars and starches. Also, all the minerals must be dissolved in water before they become available to plants.

Plants in the classroom must be watered. However, there is often a danger of watering too much so that all the pores in the soil are filled with water, allowing no air to reach the roots. For most plants the soil should be kept moist but not wet. One way of watering house plants is to put the bottom of the flowerpot into a tray and then put some water into the tray. Water will then enter the hole in the bottom of the flowerpot and move up through the soil by capillary action. Desert plants such as cacti should be watered well, and then the soil should be allowed to dry out before watering again.

Like other organisms, plants must be kept clean. Plants need both oxygen and carbon dioxide from the air. Much of this enters through pores on the surfaces of the leaves. The intake of oxygen and carbon dioxide will be impeded if these pores are filled with dirt and grime. In fields and gardens, rainfall and sprinkling will remove much of this dirt. In the home and classroom, it may be desirable to periodically wash the leaves of plants.

Light is essential for green plants to manufacture food. Most plants kept in a classroom should be placed in windows where they will receive the most possible light. Most plants will tend to orient themselves toward the light, a tendency called *phototropism.* When most of the leaves of a plant have been oriented toward the light, have the children turn the plant to see whether the plant will again orient itself toward the light.

VARIETY IN PLANTS

There is great variety among plants. There are extremely small one-celled plants and there are giant redwoods that tower above our heads. There are plants with bright, colorful leaves and flowers, and there are yeasts and mushrooms that have little color. There are plants that have broad, thick leaves and others that have thin, needle-like leaves or no leaves at all. There are plants that live in the desert where there is almost no water, and there are plants that live in the oceans. There are young plants that have just sprouted from seeds, but there are other plants that are the oldest known living organisms.

How great are the differences among various plants in your community? (See page 193 for a description of how to study the organisms that live on one square meter of a schoolyard.) Have the children locate, identify, and list the plants that have the following extreme characteristics:

- largest and smallest
- most colorful and least colorful
- largest leaves and smallest leaves
- largest seeds and smallest seeds
- plants that live in water and those that live where there is practically no water
- plants that live in direct sunlight and those that live where there is practically no light.

How do parts of organisms vary? We can study variety in the size, shape, color, and other physical characteristics of almost any part of any organism. As examples, we can study variety among pea pods and among the children in a group or a classroom.

Obtain a quantity of pea pods and have the children measure and record in a histogram the number of pods of various lengths. Have them count the numbers of peas in each pod and record in a histogram the number of pods containing various numbers of peas. Have the children suggest other ways that pea pods and peas may vary. (See Figure 7–4.)

Have the children make histograms showing the number of children in their classroom of various heights and eye colors. Have the children suggest and study other characteristics by which they vary.

PLANT ADAPTATIONS

Different plants have different adaptations that make it possible for them to live under different conditions. There are desert areas where there appear to be no plants whatsoever; the first rain in years may fall, and the desert is covered with flowering plants. The flowering plants quickly produce seeds that can lie dry and dormant for years until another rainstorm comes.

Climb a rocky mountain where no soil appears to be available. The rocks may be covered with lichens that are able to live even on the surface of barren rock.

How much water can lichens absorb? Find a small rock that is covered with lichens. Measure out a square centimeter of lichen surface and scrape away the other lichens. With a medicine dropper or a piece of glass tubing, add water to the lichens until they can absorb no more water and the water begins to run out onto the rock. How many drops of water can a square centimeter of lichens absorb?

If there are trees in your neighborhood that cast dense shadows, look underneath the trees. What kinds of plants grow there? What kinds of leaves do they have?

The walls of some of the buildings in your community may have ivy or other kinds of plants growing on them. Have the children try to find out how these plants are able to support themselves on vertical walls.

The characteristics of different kinds of plants (and animals) that make it possible for them to live in different kinds of environments are called *adaptations.* The cacti, for example, have deep roots that enable them to tap water deep in the ground, and thick, waxy leaves that limit the amount of water lost by the plant. Thistles have sharp needles that protect them from grazing animals. Some trees, such as the redwoods, have a cork layer in their trunks that helps them survive forest fires. Under natural conditions the plants that grow in an environment have adaptations that make it possible for them to survive in that environment.

What are the adaptations of some of the plants that live in environments in your locality? Suggest to children that they locate plants that live under conditions in which it would seem difficult for plants to exist, and note the adaptations that make it possible for the plants to live there. What are the adaptations that make it possible for plants to live

where there is little light?
in cracks in sidewalks and highways?
on solid rock?
where the plant roots are under water?
where there is very little water?
where there is a great deal of wear and tear from automobile tires?

One of the central ideas in the life sciences is *natural selection*. The individuals in a population that are best fitted to survive and reproduce in an environment will survive and leave the most offspring, and these offspring will tend to have the adaptations that lead to survival. As an example of natural selection, a species of moth in Britain has both light-colored and dark-colored individuals. Many kinds of birds eat this moth. In areas where trees and shrubbery had not been darkened by industrial smoke and fumes, the light-colored moths blended with the environment and were not as easily seen by birds. Here most of the moths were light in color. However, in areas that had been darkened by smoke and fumes, the light-colored moths would be conspicuous against the dark background and they would be easily detected by birds. Dark-colored moths, however, would blend with the background and be more difficult for birds to find, and in these areas there would be many more dark colored moths. In each of these environments, there was natural selection of the individuals in a population that have the adaptations that help them survive. Natural selection over many, many generations is believed to be the mechanism by which evolution has taken place.

INTERRELATIONSHIPS

There are interrelationships between plants and between plants and animals. All of the individuals of a particular species that live in a particular area form a *population*. For example, there is a population of human beings in Zanesville, Ohio, and a population of sequoia trees in Sequoia National Park. However, all the various organisms that live in an area make up a *community* of living organisms. For example, all of the living organisms in a pond or forest form a community. There are interrelationships between many of these living organisms.

Green plants manufacture the food that is used by almost all living organisms. But many other organisms live on green plants or plants that once were living. You may find toadstools growing on a lawn or bracket fungi growing on trees. Molds may be found on bread or oranges. Woodpeckers may be seen and heard pecking at the trunk of a tree, seeking termites that are burrowing in the tree and eating the food once manufactured by the green leaves of the tree. Throughout our environments there are *food chains* and *food webs*. When you see a soaring hawk dive, it may snatch a small snake that has eaten a small mouse that has eaten grain manufactured by a wheat or oat plant. These food chains and webs are based on food manufactured by green plants.

How can we grow bread mold? Obtain a slice or loaf of bread and place it in a pan or aluminum tin. Put a small amount of water in the tin and then cover the bread with plastic to keep it moist. (See Figure 12–6.) Put the bread in a warm, dark place. After a week or ten days, examine the bread. Has any mold grown on the bread? Are there different colors of mold? What is the source of food for the mold?

Lichens are found growing on rocks, tree trunks, and soils and can be brought into and kept in the classroom. Actually, lichens are composed of two plants that have a *symbiotic relationship* (i.e., the two plants live together for their mutual benefit). (See pages 257–58 for a further discussion of symbiotic relationships.) The most conspicuous part of a

Figure 12–6 These children have found that this slice of bread has a rich growth of mold.

lichen is a fungus. This supplies the *alga* with water and minerals. When water is dropped onto lichens, it is absorbed by the fungus. However, the fungus cannot manufacture food. This is done by the green alga. The green alga manufactures food for itself as well as for the fungus. The lichens are an excellent example of an interrelationship between two or more organisms in which each of the organisms depends on the other for survival and growth. There are many such interrelationships in the environment.

There is a very important interrelationship between many flowering plants and insects such as bees. For flowering plants to form seeds, pollen must be transferred from the anthers to the pistil, and usually it is desirable that this transfer be from the anthers of one flower to the pistil of another. For many flowering plants this transfer is carried out by insects. Flowering plants produce nectar that is food for bees and other insects. However, as the insects seek the nectar, powdery grains of pollen become attached to the bodies of the insects. As the insects fly from one flower to another, grains of pollen are deposited on the tip of the pistil; they eventually go down the pistil to fertilize the flower's ovules. While

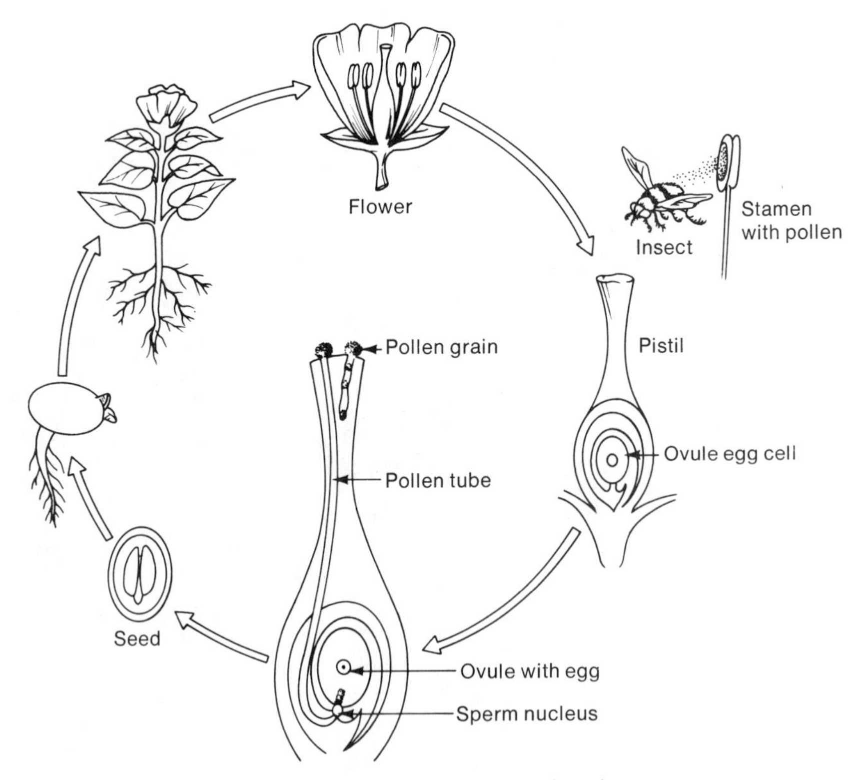

Figure 12–7 Stages in the reproductive cycle of flowering plants.

the flowering plant's nectar provides essential food for insects, many plants are utterly dependent on insects for the transfer of the pollen that is necessary for seed production. In Australia, where there were no bumblebees, clover could not be grown until bumblebees were introduced.

How are flowering plants pollinated? Obtain some flowers that the children can examine and dissect. (Lilies are among the flowers that have easily identifiable parts.) Using the diagram in Figure 12–7, have them identify the various parts of the flower. Have them note the relative positions of the anthers and the pistil. Note that in many flowers the tip of the pistil is above the anthers. This makes it less likely that the pollen from the anthers will fall onto the pistil. Instead, the pollen from some other flower can be brought by an insect and deposited on the pistil.

After locating and identifying the different parts of the flower, have the children describe how the flower might be pollinated. If there are flowering plants outdoors, they may be able to watch bees or other insects move from one flower to another.

Have the children gently dissect the flowers to see whether they can locate ovules in the ovary.

The kinds of plants that live in an environment are determined by such factors as water, temperature, and light. For trees to grow, for example, about 30 inches of rainfall a year are needed. The plains regions of North America usually receive less than this amount of water, and the only trees to be found are along river banks or are watered artificially. In the far north it is too cold for trees to grow, so the characteristic vegetation of the "tundra" is mosses, lichens, some bushes, and grasses. Underneath a dense thicket or on the floor of a forest, only plants that can survive with little light will be found. By setting up various kinds of terrariums we can study some of these interrelationships.

How can terrariums be set up and used as tools for learning? The terrarium is a land habitat for living things, and terrariums can be used to stimulate such environments as desert, bog, grassland, and woodland. If a terrarium is well planned and carefully set up, organisms will thrive in it for long periods. For example, terrariums were used to transport plants from one continent to another in the days of sailing vessels, when weeks or months were required for the long sea voyage.

To set up a terrarium, obtain a container with glass sides. An aquarium, a commercially made terrarium, or a large glass jar can be used. Put a layer of small rocks or pebbles on the bottom and then cover it with a little wood charcoal. The charcoal will tend to absorb any noxious gases that may form. Cover the pebbles and charcoal with soil. (See Figure 12–8.)

Place plants that would grow in the environment to be simulated in the soil. Small cacti are excellent for a desert terrarium, small ferns and tree seedlings for the woodland terrarium (also put in some pieces of decaying

Figure 12–8 These children have just set up a dry land terrarium.

wood); grass seed can be planted in the soil for the grassland terrarium. Water the plants in the terrarium. For all but the desert environment, cover the terrarium with a piece of glass to reduce loss of water.

As children observe organisms growing in the terrarium, ask them such questions as the following:

1. Do plants grow about the same amount on each side (symmetrical growth), or does there tend to be more growth on one side?
2. Do plants grow in the direction of the source of light?
3. What happens when the amount of light changes? The amount of water?
4. Do roots grow in the direction of a source of water? Do they tend to grow downward?
5. What happens when roots encounter a pebble or some other obstacle?

Terrariums can be used to study various interrelationships between organisms and between organisms and their environments. These simple experiments, difficult to duplicate in the natural environment, can be carried out in terrariums:

1. What is the effect of temperature changes on plants and small animals such as insects? Usually a terrarium should be kept at a comfortable room temperature of 20–22°C. (64°–68°F.) A desert terrarium will do better at a higher temperature. Keep a record of the

temperatures. Then place a small source of heat near the terrarium. Watch for changes. Try not to permanently damage the plants in the terrarium. Remove the source of heat. What happens?

2. What is the effect of changes in the amount of light on plants? First expose the terrarium to light for 24 hours a day. What happens? Then cover the terrarium so that it receives no light. What happens when the terrarium is returned to normal light conditions? Again, as in studying the effects of changes in temperature, try not to permanently damage the plants.
3. What is the effect of changes in humidity on plants and animals? Except in the case of a desert terrarium, a glass plate should be kept over the terrarium to prevent the loss of water vapor. Do drops of water form on this glass plate? If so, where do these drops come from? Remove the glass plate from the terrarium for a few days. Do any changes take place? Replace the glass top. (It may be necessary to add a little water after having removed the glass top to replace water that has been lost.)
4. Are there organisms in the terrarium that are harmful to other plants and animals? Such organisms would be considered to be consumers. Are there any signs of damaged organisms? In some cases you may wish to spray with insecticides or other chemicals. What are the effects of the use of these chemicals?
5. What happens when plants grow so that the leaves touch the sides or the top of the terrarium?

As environmental conditions change, the kinds of plants that live in the environment will usually change. It is difficult, yet important, for children to develop a concept of change. Some changes, such as an explosion or the fall of a rock, take place almost instantaneously and can be observed from beginning to end. Other changes, such as the wearing away of mountains or the shifting of continents, take place over extremely long expanses of time and are difficult to conceptualize. (See pages 342–44 for suggestions for developing concepts of long periods of time.) The changes that take place in the communities of plants that live in an environment take place in much less time than many geological changes but still usually require more than a lifetime. Thus, they can be difficult for children to comprehend. However, if children are to become intelligent about their relationship to their environment and the possible consequences of some of their actions, it is important they they begin the process of gaining an understanding of the changes that can take place in their environment.

Environments are constantly changing, and as an environment changes, the kinds of plants (and animals) that live there will change. For example, as a small pond becomes filled with silt, sand, and the remains of vegetation, the waterlilies and cattails that live in the water give way to sedges, grasses, bushes, and perhaps eventually trees. This change in the community of plants that live in a particular environment is called plant succession. Some aspects of plant succession can be found in the neighborhoods of most schools.

At one time much of the land surface of the earth was probably covered with barren rock. Even today there are many prominent rocks on hillsides, mountains, and other exposed areas. Often these rocks have lichens growing on them. (See pages 221–23 for special adaptations of lichens.) Lichens can live on barren rock, and they are the first stage in the succession of plants that may, if conditions permit, eventually lead to a forest.

The lichens and various weathering processes eventually build up enough soil so that mosses can grow. As the mosses grow, die, and decay, more soil is formed and it becomes possible for such plants as the grasses to grow. The soil-building process continues, and when there is enough soil and other conditions permit, bushes and eventually a forest of trees may grow where there once was barren rock. (Figure 12–8 on page 226 shows a terrarium that depicts some stages of succession on dry land.) This plant succession does not take place in distinct stages. Instead, the changes are gradual, and there is usually an intermingling of plant species as one replaces another. However, it is useful to consider every community of plants as being in some stage of succession and to have some conception of what communities have been in the environment in the past and what communities may come in the future.

What plants will begin to grow on open soil? Plant succession will take place in a burned-over area, on an abandoned road, and in a plowed field. Children can see some of the beginnings of succession by turning over a small area of land and removing all the live plants. The succession of plants that begin to grow in this newly "plowed" area should be studied over as long a period as possible. Which are the first plants to grow in the barren area? How did these plants get there? Are they replaced by other plants? Are there significant differences between the vegetation on the "plowed" plot and in the surrounding area?

Eventually the plant succession results in a *climax community* that can continue to exist indefinitely. Where there is sufficient soil and rainfall, as in most of eastern and southern North America, the climax community will be a forest community of beech and maple, oak, pine, or the like. In much of the prairie region, there is not enough rainfall to support forest and the climax communities are various grasses. In the desert regions, where there is even less rainfall, the climax communities may be dominated by sagebrush. The vegetation in all environments that do not have climax vegetation may be considered to be in some stage of succession toward the climax community. When children look at a marsh or rocky hillside, they can try to visualize what the area will be like when it reaches the climax community.

When we cut down trees, plow the land, or cover land with concrete or macadam, we are affecting plant succession. Some actions, such as agriculture, are essential in order to feed and clothe the people who inhabit this planet. Climax communities such as forests and grasslands do not produce a great deal of food and fiber for humans. However, when we do make these changes in the environment it is well to have some concept of the natural succession that we are affecting and to be sensitive to some of the possible consequences of altering the natural succession of plants and animals.

RELATED ACTIVITIES

The Law of Mass Production, p. 73.
Studying plant communities, p. 114.
How many peas in a pod? p. 130.
Caring for and studying plants, p. 106.
How can we study cells in the classroom? p. 270.
Germinating seeds in different ways, p. 66.
What factors affect the germination of seeds? p. 141.
Do green plants need sun? p. 299.
What are the adaptations of plants that live in cracks in sidewalks and roads? p. 205.
What difficulties do the trees that live in our cities face? p. 205.
What kinds of trees seem best able to survive under city conditions? p. 205.
Where did fossilized organisms live? p. 351.
How can youngsters study limiting factors? p. 259.
What organisms live within a square meter of school ground? p. 193.
Sorting leaves, p. 9.
Serial ordering with peanuts, p. 10.
Keeping a tree diary, p. 13.
What colored materials are found in the leaves of plants? p. 466.
How can the aquarium be used as a teaching and learning tool? p. 260.
What is the "black stuff" on the bottom of an aquarium? p. 8.

REFERENCES

For Teaching

CAMP, WENDELL H., VICTOR R. BOSWELL AND JOHN R. MAGNESS. *The World in Your Garden.* Washington, D.C.: National Georgraphic Society, 1957. Where did the plants in the garden come from? How did they get there? A description of the backgrounds of many of our useful plants.

FAUST, JOAN LEE. *Book of Vegetable Gardening.* New York: Quadrangle/New York Times, 1975. A helpful guide to planning, planting, growing, and caring for garden plants.

HILLCOURT, WILLIAM. *Fieldbook of Nature Activities.* New York: G. P. Putnam's Sons, 1950. A classic compilation of nature activities of special value to teachers. Contains many suggested activities involving plants and animals.

The Golden Nature Guides. New York: Western, A series of guides that can be used to identify plants and other organisms in the environment. For the study of plants, the following guides are useful: *Flowers, Trees,* and *Non-Flowering Plants.*

For Young Children

BLACK, IRMA. *Busy Seeds.* New York: Holiday House, 1970. A picture book with poetic text that describes how seeds develop into the kinds of plants from which they came.

GRAY, WILLIAM O. *What We Find When We Look at Molds.* New York: McGraw-Hill, 1970. A well-illustrated book on molds and how they grow.

JACOBSON, WILLARD J., ET AL. *Changes in Animals and Plants.* New York: American Book, 1975. A description of some changes that take place in plants and animals.

LOVOOS, JANICE. *Design Is a Dandelion.* San Carlos, Calif.: Golden Gate Junior Books, 1966. A picture book showing some of the beautiful designs to be found in nature.

LUBELL, WINIFRED AND CECIL LUBELL. *Green Is for Growing.* Chicago: Rand McNally, 1964. A picture book discussing many common green plants.

TRESSELT, ALVIN. *The Dead Tree.* New York: Parents' Magazine Press, 1972. The story of an oak tree that stood for 100 years or more. The tree left a legacy through the acorns it had nurtured.

WONG, HERBERT H. AND MATTHEW F. VESSEL. *My Plant.* Reading, Mass.: Addison-Wesley, 1976. A picture book introduction to planting and growing plants.

For Older Children

BUSCH, PHYLLIS S., *Wildflowers and the Stories Behind Their Names.* New York: Charles Scribner's Sons, 1977. An interesting book about the names of various wildflowers.

FENTON, CARROLL LANE AND HERMIONE B. KITCHEN. *Plants We Live On.* New York: John Day, 1971. A history of many of the common grains and vegetables.

GALLOB, EDWARD. *City Leaves, City Trees.* New York: Charles Scribner's Sons, 1972. Photos and photograms that can help identify many of the trees found in cities in eastern North America.

HAWKINSON, JOHN. *Our Wonderful Wayside.* Chicago: Albert Whitman, 1966. A description of what can be found at different seasons of the year along the little roads that wind through the countryside.

HOKE, JOHN. *Terrariums.* New York: Franklin Watts, 1972. Many ideas for studies that can be carried out in terrariums.

KURTZ, EDWIN B. AND ALLEN CHRIS. *Adventures in Living Plants.* Tucson: University of Arizona Press, 1965. Imaginary adventures with and inside plants. The book has unusual illustrations and many suggestions for things to do and think about. Excellent for elementary school children in the upper grades.

SILVERSTEIN, ALVIN AND VIRGINIA SILVERSTEIN. *Carl Linneaus, the Man Who Put the World of Life in Order.* New York: John Day, 1969. A biography of the man who was largely responsible for our modern systems of classification.

SIMON, SEYMOUR. *Projects with Plants.* New York: Franklin Watts, 1973. Growing plants without soil, plants in a maze, seeds in motion, and other plant projects.

Few events in the elementary school engender as much excitement and enthusiasm as the arrival of a classroom animal. Children quickly volunteer for various duties in the care of such pets—even cleaning cages. The squeals of joy heard as a class visits a farm or zoo are familiar to many an elementary school teacher. What motivates this tremendous affinity to animals? Is it because animals live, move, and remind us of ourselves? Could it be because animals possess an intelligence that intrigues us as we try to communicate with them? Or perhaps the interest stems from the implication of the term pet—an animal that needs care, attention, and love.

ANIMALS IN THE CLASSROOM

In working with children and animals, there are certain goals that we can attempt to achieve. Caring for animals in a classroom makes it possible for children to study firsthand some of the major concepts and principles of animal life as we know it. These major concepts or "big ideas" can help children understand their own place in the natural environment. Classroom pets of all kinds provide many opportunities for children to observe animals, learn their ways, watch them in action, touch them, and be responsible for their general well-being. As the needs of classroom animals are studied and identified, children may volunteer or be selected to perform certain routine duties related to the day-to-day care of the animals. Along with this come responsibilities that most children will be happy to assume. It should become apparent to the youngsters, as they list the various chores involved in caring for an animal, that all of them are important and that an animal's health may be jeopardized if they are not carried out.

Classroom animals will usually generate much enthusiasm and interest, but children should be cautioned not to handle the animals too much or too harshly. Particularly as pets first arrive in a classroom and the children's enthusiasm is high, it is important not to overwhelm the new creature. On the other hand, children who are afraid of animals may, through caring for a classroom pet, rid themselves of some of their fears.

The arrival of a classroom animal should be planned for by the entire class. An adequate cage or environment should be acquired or built. Animals should have sufficient room to move about. Keeping an animal in a tiny, confining cage may affect its behavior and also sets a poor example of how pets should be cared for. Children ought to gather materials that, when placed together, replicate as much as possible the animal's natural environment. It will probably require some pupil research in order to find out just what conditions exist in the animal's natural habitat. The food and water needs of the animal should be determined and sources of supply secured. (Handbooks on the care and feeding of animals are generally available at pet shops and libraries, or through scientific supply houses.) Care for the animals during school vacations must also be arranged for.

The list of animals that can be kept in a classroom is long. Some require more care than others, and in most cases whole teaching units can be built around observations of and changes exhibited by the animal. Some of the animals that are easy to care for are goldfish, birds, ants, mealworms, earthworms, gerbils, hamsters, guinea pigs, rats, mice, salamanders, lizards, and chameleons. Animals that require somewhat more care because of food, cage, temperature, or other environmental requirements are caterpillars, bees, cockroaches, tropical fish, rabbits, crickets, crayfish, snails, snakes, and tadpoles.

Once the pet is well established in the classroom, children can begin to conduct observations to learn more about the animal and its behavior. Activities can be developed that involve the entire class and require each child or a committee of children to study a distinct aspect of the animal's life. Among the questions that may be considered are the following:

How does the animal move? Describe the range of its movements from slowest to fastest. Under what circumstances does it move fastest?

How does the animal get its food? How is the food brought to the animal's mouth? What part of the animal's body moves as it eats?

How does the animal breathe? What parts of its body move as it breathes?

Can you tell when the animal is sleeping? How long does it sleep during the day? Does the animal have active or busy times and quiet or restful times? Are these times predictable during the day?

Describe the animal's skin or covering. Does it change in any way during the year? How is the animal's covering essential to its life?

How much does the animal weigh? Does its weight change over weeks or months? What are the effects on the animal of changes in its diet?

How does the animal reproduce? How do the young enter the world? How many young are born at one time? How does the mother care for its young? Does the male of the species have any role in caring for the young? How long does it take for the young to mature?

How does the animal react to environmental changes? What is its response to changes in heat, light, sound, etc.? Does the animal tend to avoid certain environmental conditions? How is the avoidance displayed?

Can the animal be taught to do something? Can the animal learn to go to one side of a T-maze if certain conditions exist at one arm of the "T" and not at the other? (See pages 112–16.) Is the animal conditioned to associate certain actions with feeding time? (The opening of a can, the rustling of a feed bag, the tapping of aquarium glass often signal to animals that they will soon be fed.) (See Figure 13–1.)

Another important goal in working with children and animals is less academic but no less important, namely, to generate enthusiasm for, interest in, and love for animals. Teachers can help youngsters develop these positive attitudes. Usually, as they observe the gentle handling of animals by the teacher and other children, children can learn to be gentle and caring toward animals. A visit from a representative of your local Society for the Prevention of Cruelty to Animals may be a particularly interesting event for a class and can help foster positive attitudes toward animal care. Raising animals is an interest that is usually developed in childhood but can provide lifelong fulfillment. Many adults can be seen caring for a collection of tropical fish with patience, love, and enthusiasm.

A trip to a local animal shelter may help a class develop a sense of responsibility for pets in a community. Some classes become involved in community service projects by publicizing the needs of a local animal shelter and collecting recyclable materials that can be turned into cash for such an organization.

A trip to a zoo can be an educational venture from many points of view. (See pages 46–48 for field trip guidelines.) Not only can pupils

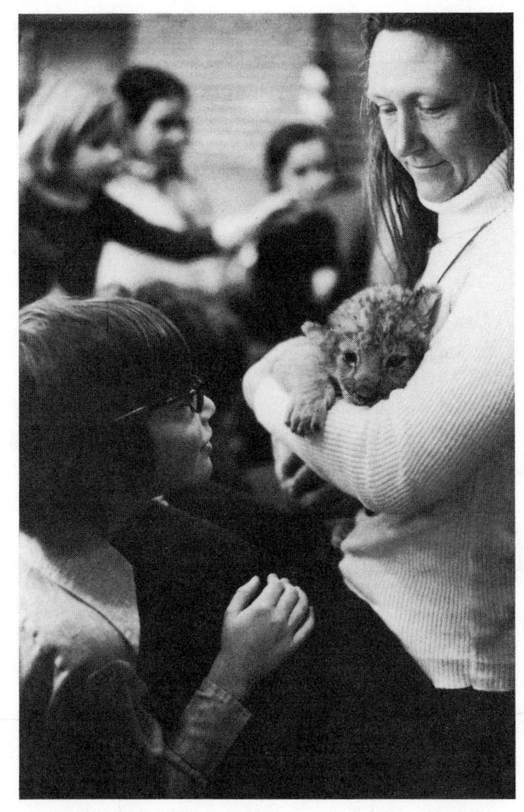

Figure 13–1 Few events generate as much enthusiasm in the classroom as the arrival of an animal.

observe animals that they might not otherwise have a chance to see, but they can be prepared to answer a variety of questions that probe the animal's habits and ways. Children can evaluate the zoo in terms of how the animals' natural habitats are replicated, how and what they are fed, how much space they are given, and how stimulating their environments are.

The youngsters in a class who develop a sense of responsibility for animals are also likely to love animals and to take these concerns and emotions with them into adulthood.

THE BROAD CONCEPTS

Children's experiences with animals can demonstrate major concepts of life as we know it. Elaborate equipment is not required for such investigations. A sidewalk, a vacant lot, a park, the schoolground, and the like are all that is really needed. The experiences children have should be related to past activities and generalized in such a way as to provide a framework for understanding and interpreting phenomena in the environment.

The Variety of Animal Types and Habitats

Life exists almost everywhere on our vast planet and perhaps in other parts of the universe. The parade of animals is diverse in terms of size, form, and the environments they inhabit. There are well over 1 million distinct kinds of animals, including more than 800,000 kinds of insects alone. Animals eat in a multitude of ways, move differently, reproduce in different ways, yet carry out the same basic life functions.

There are of course many characteristics that animals have in common and that make them distinct from plants. Animals use food from plants or other animals as their raw materials. Animals do not have chlorophyll, which is the substance plants use to make their own food. Animals do not have cell walls in their cells (the units of structure), and this too sets them apart from plants. Most animals can move about freely by means of contracting fibers in their bodies. Animals respond to environmental stimuli via a nervous system; plants do not have such a system.

How does one keep track of the vast number of animals, recognize similarities and differences, and organize them in a manageable way in order to identify and study them? *Taxonomists* are scientists who name, classify, and group living things into categories, Aristotle grouped the animals known to him on the basis of their habitats. The most widely used system of plant and animal classification was devised by Linnaeus, an eighteenth-century Swedish naturalist. This system of classification groups animals and plants according to similarity of structure and function. Each organism is assigned a two-part name, the first part being the *genus* and the second part the *species.* Human beings are assigned the name *Homo sapiens. Homo* refers to the genus and *sapiens* indicates the species. These scientific names are usually derived from Latin or Greek and are accepted by scientists throughout the world.

Living things are first grouped into two major categories—the plant kingdom and the animal kingdom.[1] Animals are further divided into those without a backbone *(invertebrates)* and those possessing a backbone *(vertebrates).* The next order of division is called the *phylum.* About ten phyla of animals are generally recognized.

[1]It should be noted that some taxonomists recognize three kingdoms—animals, plants, and *protists,* the single-celled forebears of both plants and animals. However, when working with children it is probably best to simplify things and deal with only the plant and animal kingdoms.

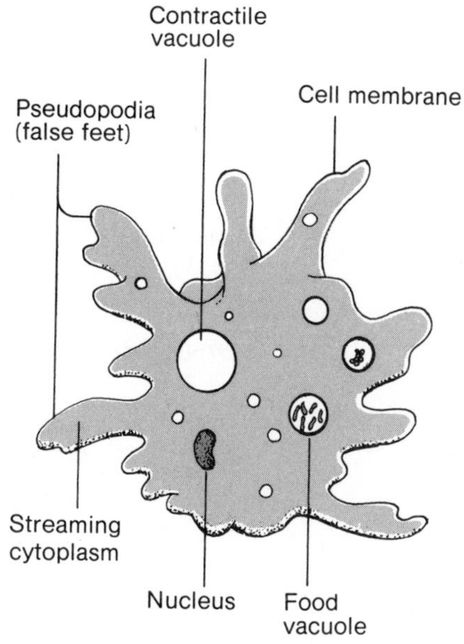

Figure 13–2 The ameba, a one-celled animal.

Protozoa. *Protozoa,* the single-celled animals, are the simplest forms of animal life. The term *protozoa* literally means "first animals." Most can barely be seen with the naked eye and live in fresh water, salt water, or soil or as parasites. The best-known examples of this group—familiar to most high school biology students—are the ameba and the paramecium. (See Figures 13–2 and 13–3.) There are approximately 35,000 known species of protozoa.

Protozoa can be found in ponds and moist garden soil. These tiny animals carry out all of the basic life functions in their single cell, with water, nutrients, and wastes passing into and out of the cell membrane. Protozoa reproduce by splitting into two, each half having all of the parts necessary to carry on its own existence!

Sponges. The *sponges* are simple animals organized as colonies of cells that inhabit warm ocean waters. There are about 5,000 species in this group. They have pores or holes all over their bodies; their phylum name, *Porifera,* means "pore bearing." This group evidences the first degree of cell specialization in the animal kingdom. Some cells attached in the colony are food gatherers; others provide structural support for the animal; and still others move water through the internal cavity of the colony to promote food use and other life functions. The natural bath sponge is the soft body skeleton that remains after the animal has decayed and dried out. (See Figure 13–4.)

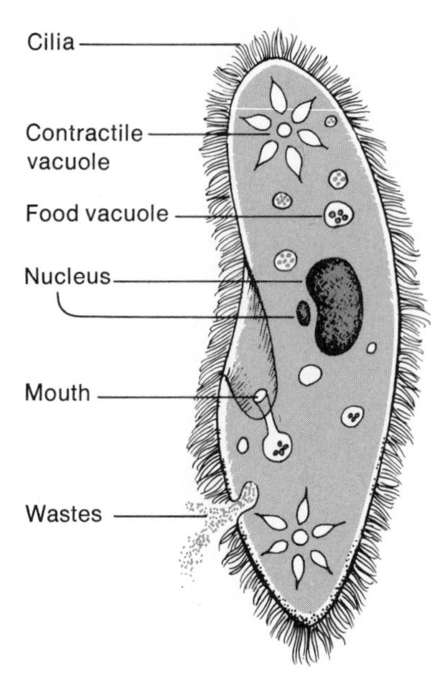

Figure 13–3 A paramecium. This one-celled animal can carry out all life functions.

Corals, jellyfish, and their relatives. This group includes the hydra, jellyfish, corals, sea anemones, and the Portuguese man-of-war. There are about 10,000 species in the group. The animals in this group are graceful, delicate-looking creatures with tentacles that aid in food getting and defense. They display an even greater degree of specialization and division of labor than the sponges. The hydra has a "mouth," digestive cells, and nerve cells.

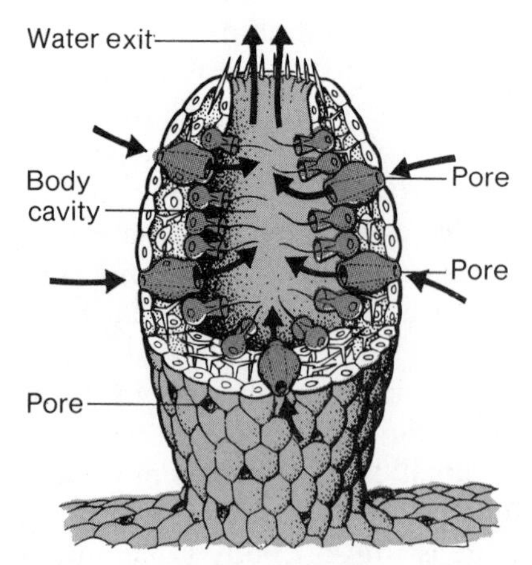

Figure 13–4 A sponge. The first example of division of labor in the animal kingdom.

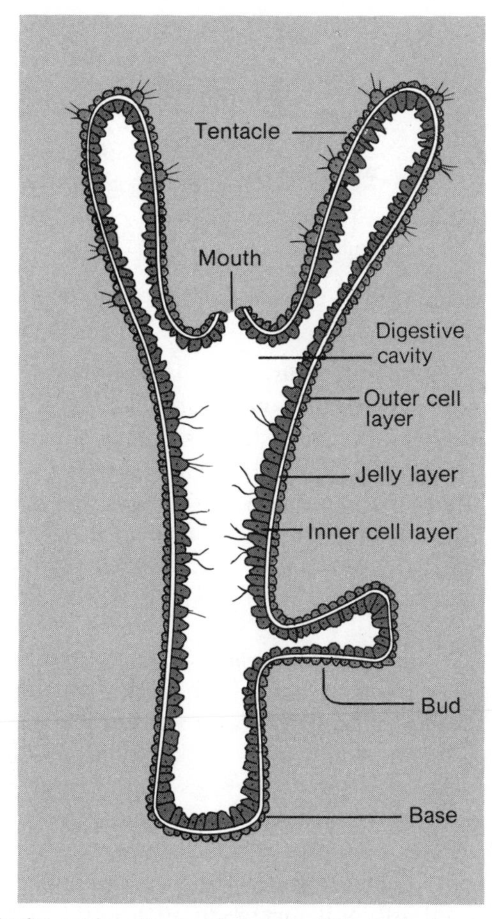

Figure 13–5 A hydra cut lengthwise. Only two of the tentacles are shown.

The familiar coral rock common in tropical waters is a collection of the skeletons of millions of these tiny animals deposited in one place. (See Figure 13–5.)

Flatworms. The *flatworm* group includes the *planaria,* sheep liver fluke, and tapeworms. Most of this phylum's 7,000 species are parasitic. The liver fluke is a dangerous parasite of sheep and cattle and also spends part of its life cycle in snails. The tapeworm is a parasite of humans, attaching itself to the inside of its host's intestine by means of hooks and suckers.

Flatworms are more advanced than the animals of the lower phyla because their bodies are composed of three distinct layers of cells, each with a specialized function. (See Figure 13–6.)

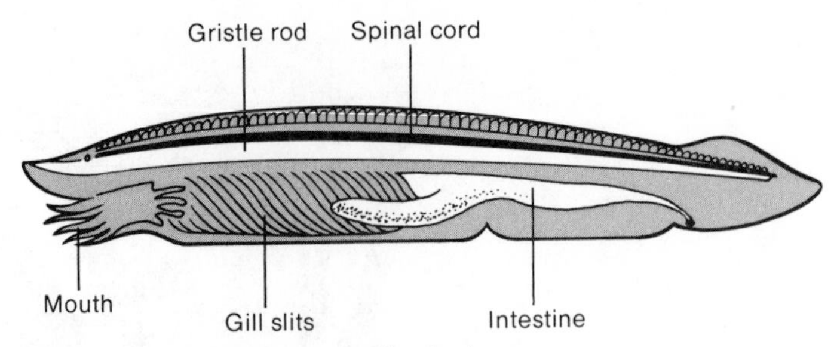

Figure 13–6 A planaria is only one centimeter long.

Roundworms. The *roundworms,* or *nematodes,* have rounded bodies and live in rich soil, fresh water, and salt water. They display a fairly sophisticated digestive system that runs from mouth to anus. Roundworms can be found in garden soil. Twenty thousand species have been identified, and at least 50 species are known parasites of humans. Roundworm parasites cause such diseases as hookworm; elephantiasis, and trichinosis; the latter can infect humans if they eat infected pork that has not been cooked sufficiently. (See Figure 13–7.)

Segmented worms. The *segmented worms,* or *annelids,* are arranged in segments, each of which is a partition of the animal's body. Earthworms, sandworms, and leeches are the most common members of this phylum, which has a total of 6,500 species. Earthworms have well-developed circulatory, excretory, and nervous systems. Some of these worms live on land, but most of the annelid worms live in tiny tubes in the sea or under rocks close to the shore.

Leeches live in freshwater ponds and streams and formerly were used as blood suckers in an attempt to rid the body of disease. (See Figure 13–8.)

Mollusks. *Mollusks* are soft-bodied creatures that display well-developed organ systems. There are three basic groups of mollusks: the double-shelled clams, oysters, mussels, and scallops; the single-shelled snails; and the shelless squids and octopuses. Mollusks are an important food group for humans. Children who visit a seashore usually enjoy finding and collecting shells—most of which come from representatives of this group, which contains some 45,000 species.

Mollusks are well adapted to their environment and have devel-

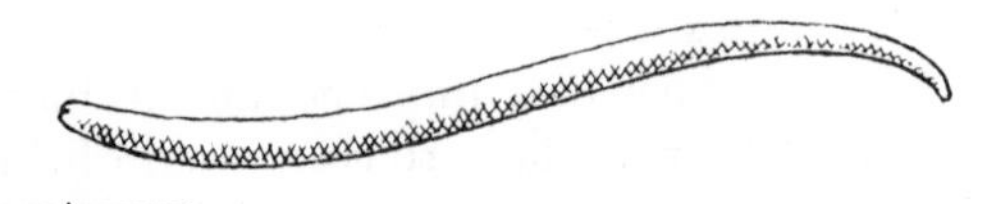

Figure 13–7 A hookworm.

Figure 13–8 An earthworm, an example of a segmented worm.

oped specialized organs for survival. The clam has a foot that protrudes outside of its shell for movement on the sea floor. It also has a filtering device that screens food out of the water carried into it by its siphons. Members of this group can be obtained at seafood stores for classroom study. (See Figure 13–9.)

Arthropods. The *arthropods* are certainly the largest phylum in terms of number of distinct species. More than 900,000 different kinds of arthropods are known. The term *arthropod* means "jointed leg." Arthropods also have a hard external skeleton made from a material not unlike that of our fingernails. The arthropods are divided into five main classes: *crustaceans,* which include shrimp, lobsters, crayfish, and barnacles; *arachnids,* which include ticks, mites, spiders, horseshoe crabs, and scorpions; and *insects,* by far the largest group, which include butterflies, dragonflies, bees, aphids, houseflies, grasshoppers, fleas, silverfish, termites, and many others. There are two other minor classes of arthropods: *centipedes* and *millipedes.*

Insects are of tremendous economic importance. They play a vital role in flower pollination. Some insects rid us of harmful parasites, and still others are used in the production of dyes, shellacs, and medicines. On the other hand, farmers seem to be constantly battling the swarms of

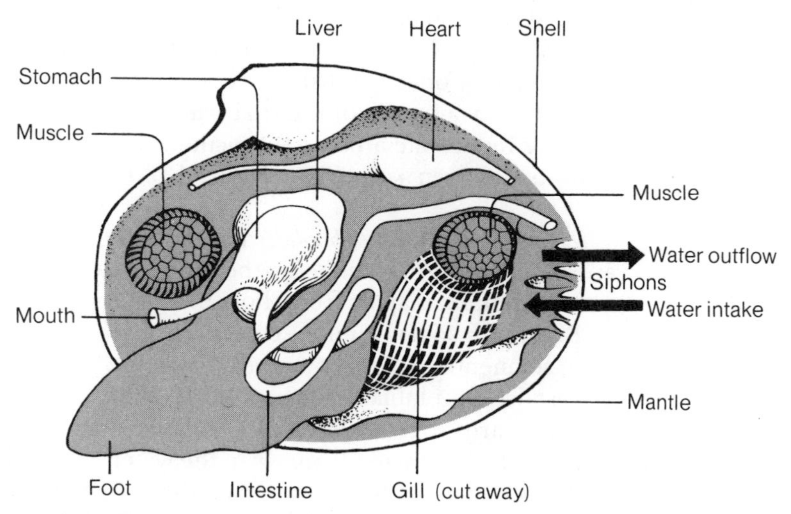

Figure 13–9 The clam has many adaptations for its life in sand or mud.

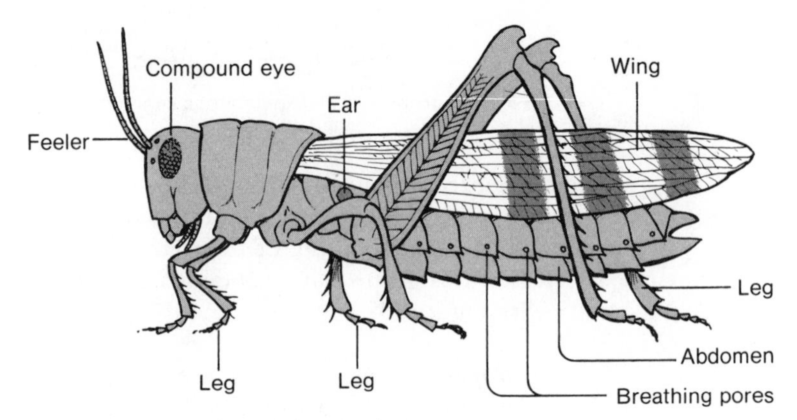

Figure 13–10 An adult insect: the grasshopper.

insects that destroy crops. Some insects carry disease or are vital links in the transmission of disease.

Working with insects in the classroom can be intriguing to youngsters. The supply of specimens is readily available, and various environments can be created for keeping them. (See pages 244–45 for activities involving insects; also see pages 202–4 for descriptions of how mosquitoes and cockroaches can be studied.) (See Figure 13–10.)

Spiny-skinned animals (echinoderms). The *echinoderms* are sea animals that have spiny skins. The most common examples are "starfish," sand dollars, sea urchins, and sea cucumbers. "Starfish" are not fish, and some scientists prefer to call them sea stars. They are supported by an internal skeleton and are probably best known for their shape and their remarkable powers of regeneration. If a starfish loses an arm or two, it can grow new ones. Starfish move about by means of tube feet that are controlled by a system of internal water canals. Through water pressure the tube feet create a powerful suction, strong enough for the starfish's arms to pull open a clam. (See Figure 13–11.)

Chordates. The *chordates* have a stiffened supporting rod, the first appearance of a backbone in the animal kingdom. This characteristic sets these animals apart from those in all of the phyla discussed previously. Except for some primitive chordates, most are *vertebrates,* animals with true, segmented spinal columns. There are five major classes of vertebrates: fish, amphibians, reptiles, birds, and mammals.

Fish are covered with scales, have fins, and use gills to obtain oxygen, which is extracted from air in the water in which they live. They are cold-blooded animals, meaning that the body temperature of the animal

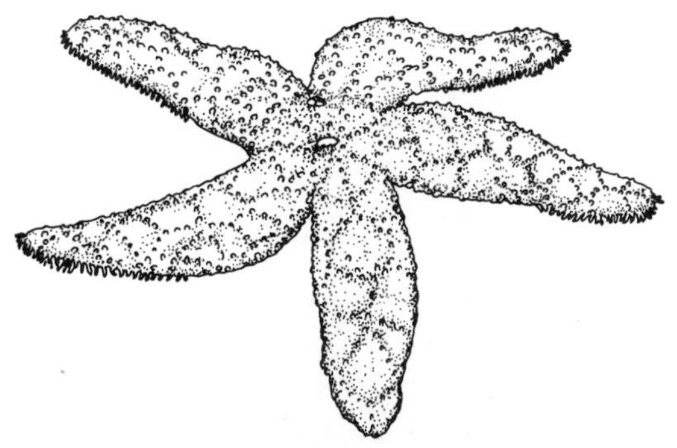

Figure 13–11 A starfish crawling over the ocean floor.

is the same as that of its immediate environment and changes as the environmental temperature changes. Fish swim by means of their paired fins, which are probably the predecessors of the arms and legs of more advanced animals. (See Figure 13 – 12.)

Amphibians generally spend the early part of their lives in the water and then inhabit land environments. Frogs, toads, and salamanders are the most common representatives of this group. Although it is difficult to raise frogs from eggs, the keeping of tadpoles is a favorite activity of elementary school youngsters and clearly demonstrates animal changes and the transition from water to land.

Reptiles dominated the earth in times past, when the giant dinosaurs reigned supreme. The present-day reptiles are snakes, turtles, lizards, crocodiles, and alligators. Reptiles evolved from amphibians and have drier skins, a better-developed heart, and a well-developed musculature for moving about. Many children fear snakes for reasons that are not entirely clear, but harmless snakes can be kept in the classroom and can live in an empty aquarium or some other large container.

Birds are the first warm-blooded animals; they have four-chambered hearts not unlike our own and are covered with feathers. They are intricately adapted for flight; for example, they have hollow bones. Birds are an important source for food not only because we eat fowl but also because of their egg production. Some birds can be kept in a classroom. Raising chicks from eggs is a popular elementary school activity.

Mammals are warm-blooded, hairy animals that nurse their young. Mammals are well distributed, inhabiting environments all over the world. Most mammals, of which the human is an example, live on land, but bats fly and whales, dolphins, and porpoises live in water. Mammals are usually familiar to schoolchildren because domesticated dogs and cats are common household pets.

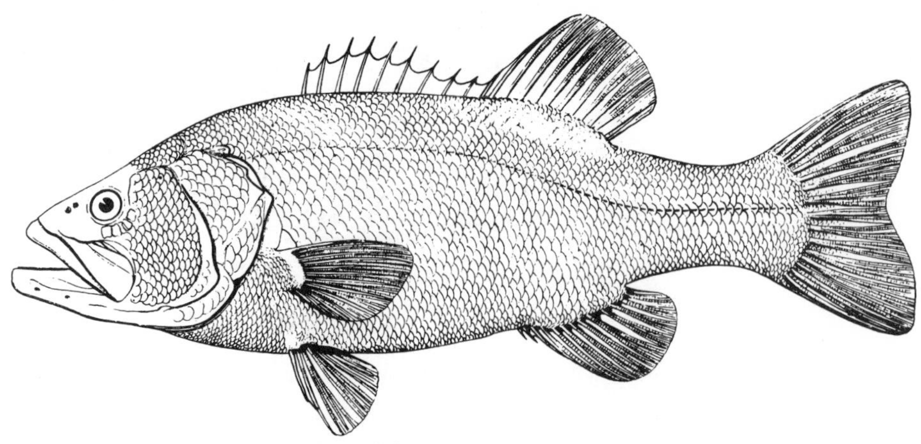

Figure 13–12 A typical bony fish.

Animal life is obviously varied, but how can children develop an appreciation for this tremendous diversity? The following are some activities for children that can promote this understanding.

Which animals live in the local community? Have your children look for and identify a variety of animals that live near their school or their homes. A wide variety of animals may be seen on the ground, in the soil, in trees, or in the air. Look for animals in such places as sidewalks, vacant lots, and the interiors of homes or schools. (See pages 192–93 for a discussion of animals that may be found in the school.) After such a survey is completed, the youngsters can list the largest and the smallest; the most colorful and the least colorful; those that live in water, on land, in the soil, or in treetops; those with the most legs and the fewest legs; those that move the fastest or the slowest; and those that eat meat, vegetables, or both. Older children can keep a list of the phyla described and find a representative of each that lives in the local community.

What can we find in a drop of pondwater? Collect a vial of water from a pond or a calm area under some rocks in a stream or small river. Examination of a drop of this water under a microscope or microprojector will often reveal a myriad of small plants and animals that children can describe, draw, and try to identify.

How many types of insects can be found locally? Insects are relatively easy to find in all seasons except winter. They may be captured by using a net made by fastening the toe of a nylon stocking to a wire loop made from a hanger. The wire loop is then attached to the end of a stick.

Grasshoppers and crickets can be kept indefinitely in a jar with moist soil and sod on the bottom. Potato, lettuce, or bits of moist bread will nourish a cricket. Children can then describe and compare the insects they have collected.

How can we quickly and painlessly kill insects for observation or study? An insect can be killed by placing it in a small bottle or jar in which a piece of solid insect repellent or "pest strip" is kept. Screw on the jar or bottle top tightly, and within a few minutes the insect should be dead. The insect can be mounted by inserting a pin through its body and into a soft piece of wood or bulletin board material. Specimens may also be placed in small boxes lined with cotton. Some older children may want to make an insect collection.

Are there variations within a species? Children can learn about variations within a single species by carefully examining two animals of the same kind. Two white mice can be compared, as can two earthworms, butterflies, or cockroaches. Are there any differences between the individuals? In most animals sexual differences account for variations in body color, shape, and size. All of the children in your class belong to a single species, *Homo sapiens;* are there any differences among them? What characteristics can be used to tell one human from another?

What similarities can be noted among groups of animals? The differences among the vast array of animal types are relatively easy to discern, but what about the similarities? Compare the skeletons of the forelimbs of a bird, a bat, a whale, and a human. Look for similarities. Can you see any commonalities in the bones of the bat's wing and those of the human arm? Do these parts have similar functions? What about the heads? Do all animals have heads? What vital functions take place in the heads of animals? (See Figure 13–13.)

Animals Change in Many Ways

It seems overly simplistic to say that animals grow and change during their lifetimes, yet these are phenomena worth studying in the elementary school. How are infants different from elderly people? While many characteristics remain the same throughout a lifetime, we can hardly chart all of the changes an individual goes through. We change physically, intellectually, emotionally; some of our needs stay the same, yet others differ at different times in our lives. Our responses to stimuli and situations change as well.

Some animals change so dramatically during their lives that at some stages of their development they cannot be recognized as the same animal. Take, for example, the case of the caterpillar and the butterfly, the tadpole and the frog, or the mealworm and the beetle.

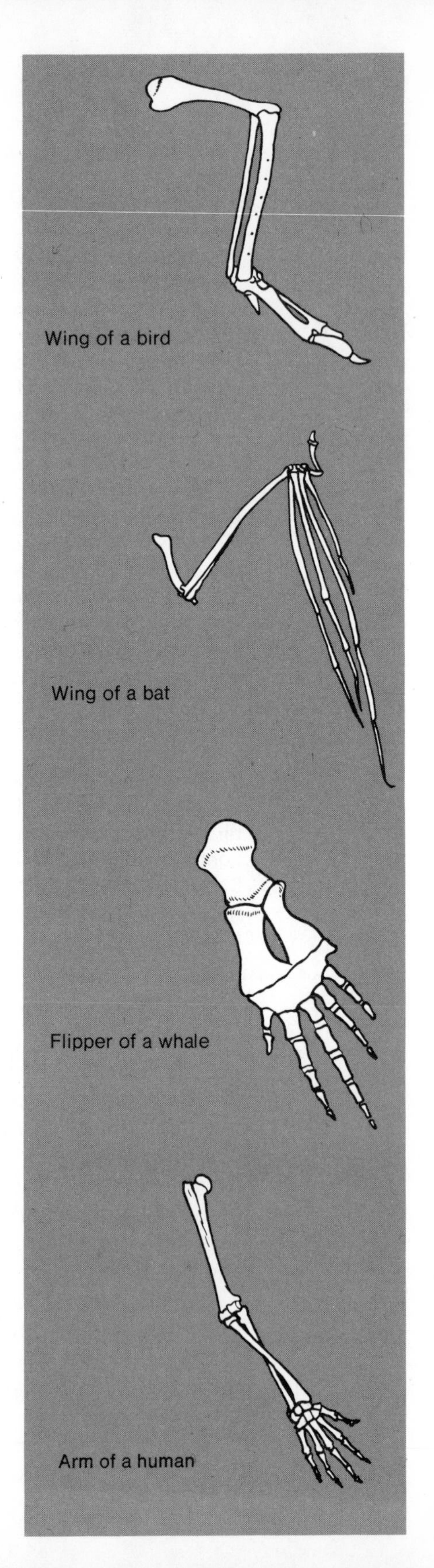

Figure 13–13 How do the forelimbs of these vertebrates resemble each other? How can this be explained?

Animals also change with the seasons by shedding a coat of fur, hibernating, or migrating to warmer climates. The study of growth and change can be undertaken with a group of children in the classroom, outdoors, and through literature.

Animal growth and development. Growth can be monitored in the classroom by keeping and observing animals from their first days of life until maturity. Tropical fish grow quickly, as do rats, gerbils, and mice. Children can be assigned to follow specific aspects of the animal's growth, such as length, body proportions, or hair development.

How can we hatch eggs in the classroom? Observing the development of an animal from the egg stage is an exciting activity. Fertilized eggs must first be obtained from a farm, hatchery, or scientific supply house. Acquiring an incubator is the next order of business. Many schools have incubators, though this may require some exploration of dusty supply rooms or school basements. If a ready-made incubator is not available, an inexpensive one can be constructed using a styrofoam cooler, a heating wire or bulb and socket, a thermometer, a thermostat, and some wire mesh.[2] (See Figure 13–14.)

Chicken eggs will take 21 days to hatch from the first day they are placed in the incubator, so the big event should be planned so as to take place during a school day. Children should be assigned various duties such as turning eggs, keeping the inside of the incubator moist, and maintaining an even temperature. You may wish to break open eggs at various times during the incubation in order to demonstrate the development of the chick embryos. While observing the progression of developmental characteristics can be interesting and instructive, the fact remains that this is a deliberate interruption of the embryo's growth. Some people feel that this indicates disrespect for life. Others feel that true life does not exist until the chick hatches

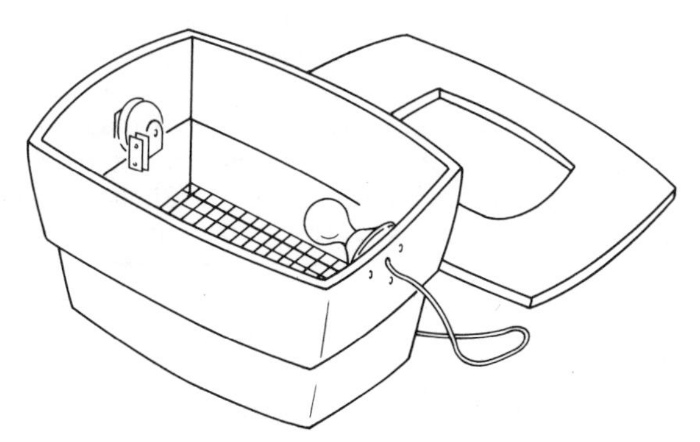

Figure 13–14 A simple incubator can be constructed using materials easily obtained at a hardware store and a styrofoam cooler.

[2]John Coulter, "An Incubator – 'Cheap, Cheep!' " *Science and Children*, September 10, 1972, pp. 21–22.

and that the experience of observing the various developmental stages is well worth the sacrifice of a few eggs. The decision must rest with each teacher and perhaps the children's opinions on the matter can be solicited. A few teachers have been able to crack a little window in a developing egg by carefully peeling a piece of shell from its underlying membrane, thereby not having to sacrifice the embryo. This procedure can be as delicate as a surgical operation and is often unsuccessful.

When the chicks hatch, they will develop rather quickly and their growth can be observed and recorded. It should be noted that not all of the eggs are likely to hatch, so it may be wise to start with at least a dozen fertilized eggs.

Animal life cycles. Profound changes occur in some animals as they go through various stages of their life cycle. *Metamorphosis* is the process by which animals change from one form to another—usually from egg to larva to pupa to adult. Some animals spend their entire life cycle on land, while others go through some stages in the water and some stages on land. The mealworm, an insect that undergoes metamorphosis, spends all of its life stages on land. The mealworm itself is the larval form of a bettle. (See pages 203–4 for a description of how the metamorphosis of mosquitos can be studied.)

Using mealworms to demonstrate metamorphosis. Mealworms are excellent animals for demonstrating metamorphosis because they are easy to keep and develop fairly quickly. Obtain mealworms from a pet shop or a scientific supply company. The animals may be kept in a plastic or glass aquarium or even a wide-mouthed jar. Fill the container about two-thirds full with oatmeal or some other breakfast cereal and provide a piece of apple or potato for moisture. Place the mealworms in the container and cover it loosely.

Children should inspect the mealworms daily and check for changes. The larva will shed its skin, and the skin cast can be seen. The larva then becomes a pupa, a dormant stage, and the children may wonder whether the animal is still alive. The number of mealworms in the container should be counted daily, for if they "disappear" they are probably changing. The adult beetle will emerge from the pupa. Similarities between the adult and the larva should be discussed. The adult beetles will lay eggs and start the cycle again.

Studying metamorphosis in frogs. Frogs spend their first stages in water and part of their adult lives on land. Frog or toad eggs may be found near the edges of ponds in the early spring, usually in gelatinous masses. The eggs may be kept in an aquarium. The water should be changed frequently or aerated, and the tank should be kept out of direct sunlight. As the tadpoles emerge, their appearance and movements should be described. What animals are they reminiscent of? Children should note periodic changes in the tadpoles as they begin to develop legs and their tails shorten. This process of

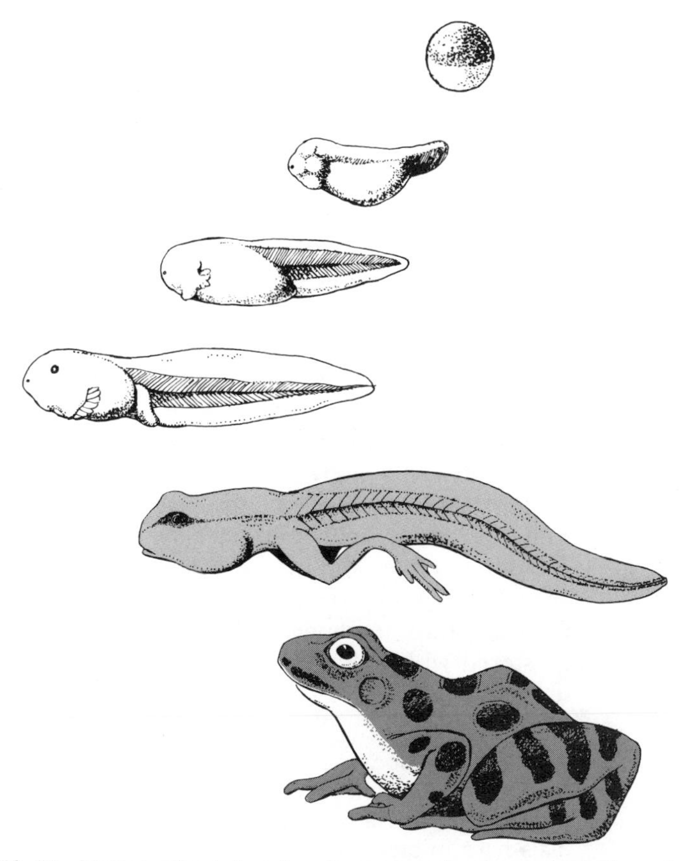

Figure 13–15 Metamorphosis in a frog is an example of an animal life cycle.

change will take several months but should be a continuing source of interest. It should be noted that many tadpoles will not survive to adulthood. (See Figure 13–15.)

There are many other animals whose metamorphosis may be studied in the classroom. Among them are moths, butterflies, caddis flies, silkworms, and houseflies. Find out the duration of the life cycle so that observations of all stages may be carried out while school is in session. Cocoons can be kept in a loosely covered jar and observed daily.

When an animal goes through the entire cycle, from egg to larva to pupa to adult, the process is called *complete metamorphosis.* Some animals, such as the grasshopper, undergo *incomplete metamorphosis;* that is, when the young emerge from the eggs they resemble the adults and can easily be recognized. The young are much smaller than the adults and will change in color and acquire some new body structures (e.g., wings in the case of the grasshopper). (See pages 202–3 for a description of how the incomplete metamorphosis of cockroaches can be studied.)

Metamorphosis may be interrupted by adverse environmental conditions. For example, if during the spring the weather suddenly becomes very cold, the development of amphibian and insect eggs may be delayed. Changes in humidity may affect some cocoons, and the oxygen–carbon dioxide balance of water can affect frog egg development.

Seasonal changes in animals. When the first balmy days of spring rush in, how many changes do we observe? Birds begin to chirp; buds begin to swell; and insects emerge and are seen wandering out of the cracks in the sidewalk. The change of seasons brings about changes in heat, light, and food supply that affect animal life.

The cold weather of winter causes many animals to seek protection. Most birds migrate to warmer climates. Amphibians and reptiles seek refuge from the elements and remain generally inactive during the winter. Many animals hibernate by spending the cold months in the ground, in caves, and in tree hollows. When an animal hibernates, all of its body functions slow down so that the energy they expend is at a minimum and their food needs will be dramatically reduced, being limited to fat stored within the animal's body. Animals in hibernation may seem scarcely alive.

How can local seasonal variations be studied? Children may conduct surveys of animals that remain in their locality during the winter. Which animals remain active during the entire year? What adaptations do these animals have in order to survive the winter? Which animals hibernate? Which ones migrate to warmer areas? Does this happen on predictable dates?

How do temperature changes affect insects? Collect some insects or flies in a jar. (See pages 244–45.) Observe the activities of the animals. Place the jar in a container of ice cubes or in a refrigerator for about fifteen minutes. Observe the insects or flies after their environment has been cooled. Are the animals' activities and behavior any different? What do you think accounts for this?

How can children keep records of animal changes? Observe a specific type of animal, such as a squirrel or a chipmunk, at various intervals during the school year. When is the animal seen more frequently? less frequently? Does the appearance of the animal change throughout the year? What factors might account for this?

Some animals, such as some eels and fish, migrate periodically to an area where they lay eggs. They swim up streams to a substantially different environment and deposit their eggs. Male members of the species are on hand to fertilize the eggs, after which both male and female swim downstream, only to die shortly thereafter.

The changes in animals are many and varied. Often we do not notice them because they are gradual. As children focus on specific animals, environments, or characteristics of animals, they will be better able to note the changes around them.

Living Things Have Changed Throughout the History of the Earth

How did we get here? Are we really related to monkeys? Why are all of the dinosaurs gone? Certainly, questions like these keep us on our toes, but without some understanding of the theory of evolution we could hardly provide honest and factual answers.

Evolution from single cells. The first organisms were probably single-celled organisms. Several kinds of bacteria, ameba, and paramecia are examples of unicellular organisms that exist today. These single-celled organisms were probably formed by the joining together of large numbers of nucleic-acid and protein molecules. Perhaps lightning provided the energy necessary for these molecules to combine and form these first types of life.

All living material must have energy to maintain itself. The earliest cells probably obtained their energy directly from the sun. Over a long period the substance called *chlorophyll* evolved in some cells. In the presence of light, chlorophyll can cause water and carbon dioxide to unite to form carbohydrates, which can be used to sustain the cell. Some bacteria contain chlorophyll, and the first cells to manufacture their own food may have been somewhat like these bacteria. Organisms that contain chlorophyll are usually greenish in color.

Oxygen is an important by-product of *photosynthesis,* the process by which plants manufacture food. This oxygen is usually released into the atmosphere, making possible the evolution of plants and animals that use oxygen to turn food into energy. Most present-day plants and animals use oxygen in the process of producing energy from food. The supply of available oxygen is replenished through photosynthesis.

Animal cells must have some external source of food. In general, this food comes from green plants. Also, there must be some system of processing food that enters the organism. The small hydra has the beginnings of a digestive system. The hydra is really a hollow tube. Unlike the digestive systems of higher organisms, which are open at both ends, the hydra's tube is open at only one end. It ingests food and egests wastes through the same opening. With this tube it is able to process food for all of its cells. In a similar but much more highly developed digestive system, food is processed for the billions of cells that make up the human body. Food materials still diffuse through membranes into the body. In unicellular organisms this diffusion takes place through the outer membrane covering the cell. But there is a limit to how large organisms can evolve and still depend on diffusion through the outer membranes for all the materials needed for life. By evolving a hollow tube, which in many organisms, including humans, is twisted and coiled so that it is many feet long, the surface area of the tube is greatly increased.

With the evolution of larger organisms came specialization among cells and the structures of cells. Some cells and structures perform digestive functions. Other structures process the oxygen from the atmosphere, distribute food and oxygen to all the cells, remove waste materials, and carry on the many functions necessary to sustain life. But this specialization made possible greater variety among living organisms, from the giant whale and enormous kelp to the tiny diatoms and microscopic hydra.

From water to land. Life originated in an aquatic medium. Some forms of life are almost entirely water; in fact, about 70 percent of the weight of the human body is water. The gases and minerals needed by the primeval organisms were dissolved in the water and could diffuse through the membranes into the organism to be used for food and growth. And the water medium provided external support for the jellylike microorganisms.

But another environment was available—the terrestrial environment, the dry land of the continents. The first organisms to invade the terrestrial environment may have been the primitive mosses, which remained very small and could live only in places like the shores of seas and rivers that remained moist all the time. Since there were advantages to life on land, such as increased sunlight and absence of competition, plants may have evolved that could live in environments where there was less moisture, although all organisms need access to water in some way or other. The organisms that successfully invaded the terrestrial environment needed such developments as the following:

1. *Outer coverings to reduce the loss of water to the atmosphere.* The waxy coatings of plants and the outer shells of turtles are examples of such coverings.
2. *A source of water and minerals.* Many organisms get the water and minerals they need from the soil.
3. *A transportation system for water, food, and minerals.* The circulatory systems in animals carry these necessities throughout the organism.
4. *A system of support for the organism.* Fibers in plants and skele-

tons and exoskeletons in animals provide the support that prevents terrestrial organisms from collapsing.

The invasion of the terrestrial environment led to the evolution of a great diversity of forms among land plants and animals. The evolution of sexual reproduction among both plants and animals was a mechanism for increasing variety. During successive periods in evolutionary history, different kinds of plants and animals became dominant. Although certain kinds of organisms ceased to be dominant, many of them remained on the scene. Some kinds of organisms, such as the giant ferns and the dinosaurs, have become extinct; yet other kinds of ferns and reptiles are still with us. It is noteworthy that probably no major basic *type* of animal became extinct, and most types of plants also remain with us. New kinds of organisms have evolved while the older types continue to exist. All of this leads to great variety among living organisms.

Adaptations among living things. Living organisms adapt to their environments. However, environments and conditions are extremely diverse. Contrast, for example, the humid and sweltering tropical rain forest with the arid and windblown desert. There are living organisms in both of these environments, but these organisms have different adaptations. Most organisms that thrive in the rain forest would die in the desert, and desert organisms would not survive in the rain forest.

The chameleon—the animal that is noted for its ability to change color—can live in the rain forest. In addition to being able to blend with the color of its environment, the chameleon is adapted for life on the swaying branches and twigs of trees. It has a tail with which it can grasp a twig or branch; in fact, it can hang suspended only by the tail. Also, it can place its four legs in line on a branch and grasp it firmly with its toes. Apparently, it can maintain its grasp on a branch even when it is asleep. The chameleon can swivel its eyes so that one is looking back while the other peers ahead. Since the chameleon is rather slow moving, these independent eyes make it possible for it to observe a delectable fly in front while watching a potential enemy to the rear. Since most animals are most sensitive to the sight of motion, the chameleon can postpone capturing the fly until the enemy is no longer to be seen by its rearward-looking eye. Then it shoots out its remarkable tongue—a tongue that can be as long as its body and tail combined. The fly adheres to the tip of the tongue and in a split second is pulled back into the mouth to be devoured. The chameleon, like other organisms, has adaptations that help it survive and live in its particular environment.

Organisms have a variety of adaptations for protection. With *protective coloration* animals such as the plaice and the grasshopper blend in with background colors. In *mimicry,* harmless organisms look like other more dangerous or inedible organisms. Some snakes mimic the deadly cobra; to birds, the viceroy butterfly looks just like the bad-tasting monarch butterfly. The opossum and certain snakes play dead until their enemies leave. The skunk can eject an odorous liquid at its attackers. Hard outer coverings protect animals such as the turtle and crab. The zebra and the deer, on the other hand, protect themselves with their speed, which enables them to escape most pursuers.

There are many adaptations for getting food. The hawk has the equivalent of an eight-power magnifier with which to see mice and other small animals from high in the sky. The hummingbird has a long tube through which it extracts nectar from flowers. Since animals do not manufacture their own food, almost all of them have some kind of adaptation for getting food.

Organisms with these various kinds of adaptations have not always existed. Instead, organisms with certain kinds of characteristics were more likely to survive in a particular environment. Their offspring tended to have those characteristics. Consideration of the process of evolution helps explain how these adaptations gave rise to new kinds of animals over long periods.

The process of evolution. Charles Darwin, whose *Voyage of the Beagle* enabled him to study and draw conclusions about natural phenomena on the Galapagos Islands, is considered the father of modern evolutionary theory. In 1858 Darwin and a fellow naturalist, Alfred Russel Wallace, presented a joint paper to the Linnaean Society of London outlining their views on evolution. (Darwin built a firmer base of observation to support the statement, so these ideas came to be known as Darwin's theory of natural selection.) Darwin's theory has five major aspects: (1) mass production, (2) struggle for existence, (3) variation, (4) natural selection, and (5) origin of new species through inherited variations.

Mass production is a characteristic of populations of living things. All populations of living things tend to overproduce at such a rate that the population outstrips the food supply. A single female codfish may lay nearly 10 million eggs each year. If all of these eggs were fertilized and the offspring survived, the oceans would be virtually solid with codfish in a few years. Naturally, a large fraction of the codfish eggs never hatch and become adults; as a matter of fact, a great many never become fertilized. But the number of eggs produced is so high that the small fraction that do hatch usually produce a total population of such a size that it has an inadequate food supply.

Children may see evidence of this mass production in the environment. How many tadpoles emerge from the masses of frog eggs in a pond? Trees produce thousands of seeds that are dispersed in a variety of ways. What percentage of the seeds produced become mature trees? What would our neighborhood look like if each seed became a tree?

The struggle for existence arises as a result of overproduction. In short, the living things that are produced must compete for food, shelter, and other necessities of survival. The struggle for existence is not necessarily a dramatic event like the warring of two ferocious dinosaurs; it may simply be the competition between two seedlings for water, minerals, and protection from strong winds. Organisms that are adapted to survive in their environment until they reproduce will be most likely to continue to exist.

Variation within a species is a biological fact. No two tadpoles that emerge from the numerous masses of frog eggs in a pond will be exactly alike. There may be differences in size, vigor, agility, and so on. As pointed out earlier, some members of a species have developed adaptations that make them better suited for survival than others.

Natural selection is the basic principle of the evolutionary theory. Stated simply, it means that some organisms in a species have adaptations that give them an advantage over other organisms. The organisms with the advantage are better able to compete in the environment and are more likely to survive than those without such an advantage. In nature the most fit tend to survive; others that do not have these adaptations tend to die and never have a chance to produce offspring. Since the surviving organisms are selected by these natural mechanisms, this is called *natural selection.*

The origin of new species can be traced to the continued selection, over many generations, of variations that have proved successful. These variations are passed on from one generation to the next. If a type of organism is produced that differs greatly from the original type, it is considered a new species.

The mechanism of evolution. Just how are adaptations and variations passed on from one generation to the next? This is an extremely important question and one that Darwin was not able to answer. Some of the laws of heredity were expounded by Gregor Mendel not too long after Darwin presented the theory of evolution. Basically, Mendel described the ways in which physical traits are passed along from parent to offspring. We now know that traits and characteristics of living things are transmitted by *genes* within their cells. In the case of sexual reproduction, the genes are carried by the male sperm and the female egg. When a sperm and an egg unite, the genes that code information for specific traits and characteristics pair and combine. As a result of this pairing, the offspring will have some traits like the male parent's, some like the female parent's, and some that are blends of the traits of both parents. Mendel isolated specific characteristics in pea plants and figured out the likelihood of observing particular traits in the offspring that would result from the combination of defined parent types.

Hereditary traits are transmitted from one generation to the next;

acquired characteristics are not passed along. For example, if a man loses a finger in an accident he will not pass this trait along to his own children because his genetic material has not been altered. There are, however, occasional changes in the structure of genes that are passed along to offspring. Such changes are called *mutations*. Mutations can be promoted by X-rays and cosmic rays bombarding cellular material from the environment. These rays emanate naturally from radioactive materials. Most mutations are harmful to the organism and usually result in death before the individual can reproduce and pass on the bad trait. However, a small percentage of mutations are helpful in that they cause the individual to be better adapted to its environment. This positive adaptation gives the individual an advantage in the struggle for existence and the new trait, since it results from a change in the genetic material, will be passed along to succeeding generations. For example, sometime in the distant past an ancestor of *Homo sapiens* may have had a mutation that produced a thumb that was slightly opposed to the other fingers of the hand. Over long periods other mutations led to our present opposable thumb. The opposable thumb was very helpful in grasping and manipulating objects. The natural selection, over millions of years, of helpful variations and adaptations produced by mutations is apparently the basic mechanism by which evolution operates.

Animals Live in a Wide Variety of Environments

All organisms live in an environment that may include a variety of materials and, of course, other living things. Consider the woodchuck that browses beside the asphalt highway. It has its burrow in the soil, where it hibernates in the cold winter. It has interrelationships with other woodchucks for reproduction and to care for its young. But it also has interrelationships with other organisms, not the least of which is the fellow mammal who has cleverly invented a steel-and-chrome vehicle that can travel at great speeds and kill thousands of unwary woodchucks each year.

Systems of living organisms. Living organisms can be studied in terms of their relationships with other organisms and their environment. In fact, a strong case can be made that an organism cannot be understood without some comprehension of these interrelationships. These interrelationships can be studied in terms of several different kinds of systems.

Populations include the individuals of a particular kind or species. For example, the size of the human population and its relationship to the available food supply is a critical matter and a subject of study. In a somewhat similar manner, game biologists will study and be concerned about the deer population in a region, and fish and wildlife specialists will monitor the population of a certain kind of fish in a stream or lake. A population consists only of the individuals of a particular kind.

A *community*, on the other hand, consists of the various populations that live together in a region. On the grasslands of the plains, for example, the grass may be the dominant population. But there are the cattle and other animals that graze on the grass, the insects that pollinate the flowers, the worms and microorganisms in the soil, and the man who raises the beef cattle for the big city markets. The community is made up of all the living organisms in a region that interact with each other.

The *ecosystem* includes such physical elements of the environment as air, water, and soil as well as the living organisms. A city may be considered an ecosystem in which the human population is dominant. There are also grass, trees, birds, mice, rats, insects, and many other organisms in the city. However, the soil is covered with concrete; the water is polluted with wastes, and the air is contaminated with the outpourings of smokestacks and exhaust pipes. These physical conditions affect living organisms, and only certain kinds of organisms can survive the harsh environment of the city ecosystem.

Interrelationships among living things. Living things depend on one another in a variety of ways. The basic factor underlying most of these relationships is the acquisition of food. Animals may be divided into three groups depending on how they are involved in food transfer. There are predators, scavengers, and symbionts.

Predators kill another animal and then eat it. Predators tend to be adapted for their life style by having strong senses of smell, sight, and hearing. They may also have strong jaws, sharp teeth, or long claws. Their prey, in turn, develop adaptations to help them escape from their predators, and thus a "balance" is maintained. (See pages 253–54 for a discussion of mimicry.)

Scavengers prey on the bodies of animals or plants that are already dead. Vultures are common scavengers that can see carcasses from far above the ground.

Symbionts are organisms that live together in some way. Symbiotic relationships fall into three main categories: commensalism, mutualism, and parasitism. In *commensalism* one organism benefits by using the unused food of the other organism, which is unharmed by the relationship. Barnacles attach themselves to whales and are provided with "free transportation" to various feeding grounds. The barnacle benefits, but the whale is unaffected. *Mutualism* is a relationship between two organisms in which both members derive benefits. Termites, which eat wood but cannot digest cellulose, are aided by protozoa that live in their digestive tracts. The protozoa digest the cellulose for the termite and in return are afforded food, water, and a place to live. The members in the relationship could not exist without each other. *Parasitism* is a relationship in which one organism lives at the expense of another but does not necessarily kill or devour its host. Some humans are parasitized by tapeworms that live and eat in their intestines. The hosts, though, are adversely affected.

Categorizing the relationships among animals can be an interesting project for children. Start with a list of ten unrelated animals. The eating habits of each animal may be researched in books or guides, and then each animal may be categorized as a predator, scavenger, or symbiont. The youngsters may also start with these three categories and try to identify five animals that exhibit each type of relationship. How would humans be categorized?

Limiting factors. Organisms have certain requirements for life. They need water, food, temperatures within a certain range, shelter, and so forth. On the other hand, the presence of too much of certain substances can limit life and growth. Often, for example, there are too many toxic substances in the atmosphere in a large city for some plants to grow. All of the factors that control or regulate the growth of organisms in an environment are called *limiting factors.* The nature and extent of limiting factors help determine what kinds of organisms will thrive in a region. Water is often a limiting factor. In a desert only plants and animals with adaptations that allow them to live under arid conditions can survive. The camel, which has the ability to store water, is adapted to desert life.

Temperatures can be critical. While some forms of life can exist within a temperature range of nearly 100°C, (180°F) most organisms can survive only within a much narrower range of temperatures. Some animals burrow into the ground or find refuge in the shade of rocks to escape the high temperatures of the midday sun in a desert.

Nutrients are needed for survival. All animals are ultimately dependent on green plants for food, and if there is a short supply of certain kinds of food owing to lack of water, nutrients, or light, the animal population may be limited. A severe drought that limits rice production, for example, may be a limiting factor for the human population in regions where rice is the staple food.

Cover can also be a limiting factor. Rabbits, field mice, and many kinds of birds depend on bushes, brambles, and grass for protection from hawks, owls, and other predators. If fire or the farmer's plow removes too much of this cover, the populations of these animals will be threatened.

All around us we see factors limiting the growth and multiplication of plants and animals. A path on the schoolgrounds becomes barren as it is trodden by young feet. The grass under a board left out on the lawn becomes yellow and ceases to grow. The bluebird population diminishes as starlings move into the bluebirds' former nesting places.

How can youngsters study limiting factors? Children can analyze ecosystems in their neighborhoods to try to determine the factors that are limiting the growth of various organisms. What would be the effect of the introduction of a natural predator of a bird population that is thriving on the schoolgrounds? How have humans altered the supply of limiting factors in the environment? How can we avoid depleting certain resources that are limiting factors for the existence of certain populations?

Studying animal environments in the classroom. Various kinds of environments can be simulated in the classroom. As was suggested for plants, we can try to replicate certain habitats (e.g., desert, woodland, etc.) and observe animals as they live in their environments. Naturally, the children ought to discuss how their simulated environment differs from that in which the animal usually exists.

How can we keep ant colonies? Studying the ways of ants is a fascinating activity for children. Ants are found nearly everywhere—in backyards, schoolgrounds, gardens, or sidewalks. An ant "cage" can easily be constructed by obtaining a large glass jar and placing a tin can (sealed end up) inside of it so that there is not more than 3 or 4 cm between the outside of the can and the sides of the jar. Fill the jar around the can with soil. This apparatus requires the ants to stay close to the glass as they build their tunnels. Find an anthill. Dig down with a trowel or shovel to remove the entire colony of ants. Try to get the queen (a much larger ant), as she is necessary for the maintenance of the colony. (Ants for classroom use are also available through scientific supply houses or, as complete kits with an appropriate cage, in toy stores.) Place the ants in the jar. For food, provide bread crumbs, jelly, and sugar. Try small quantities of other foods to see what the ants like. Place a wet sponge on top of the soil and cover the jar with a cloth secured with a rubber band or string. Cover the outside of the jar with aluminum foil or black paper when no one is observing, so that the ants can avoid light as they build their tunnels.

The children may obtain books about ants from the school library and study the fascinating relationships among soldiers, workers, and their queen. Ant behavior can be particularly interesting as children try to find out what ants do with their antennae, how they carry particles, how they eat, and how they rest.

How can earthworms be kept in the classroom? Earthworms can be found by digging in moist soil. Any large jar or plastic container will provide an adequate home. Fill the container three-fourths full of soil that has been mixed with decaying leaf material. For food, place some cornmeal in a corner of the container. Keep the soil moist by covering the container with a wet cloth.

Children can make specific observations of earthworms and study their ways. Do they prefer light or darkness? Do they prefer wet or dry environments? Do they prefer warm or cold spots? Children may conduct investigations to find the answers to each of these questions.

How can the aquarium be used as a teaching and learning tool? A simple aquarium with a goldfish or two may be appropriate for use with very young children. As children get older, though, their observations and studies of an aquarium can become relatively sophisticated.

To set up an aquarium, obtain a glass container, a large jar, or a fish tank. At the bottom of the aquarium, place 3 to 5 cm (1 to 2 in) of clean sand or gravel. (This bottom material should be washed by pouring it into a pail of water and then pouring off the water. Continue until the water is clear.) Pour in clean water to a height of 10 to 12 cm. (3 to 4 in.). Now set a few aquarium plants such as Vallisneria, Sagittaria, Cabomba, or Elodea in the sand. Fill the aquarium to within 3 to 5 cm of the top with clear water. (Do not use water drawn from copper pipes, as such water may be harmful to the fish. If possible use spring water, which is available at most supermarkets.) Allow the water to stand for two days to allow any chlorine dissolved in it to escape. Place fish, snails, and other water animals in the aquarium. Because of their hardiness and ability to withstand changes in temperature, goldfish are an excellent choice. (See Figure 13–16.)

Most aquariums require relatively little care. The aquarium should receive some light, but not too much. Usually it is best to place the aquarium against the wall opposite the classroom windows. If an aquarium turns a greenish color (the green is due to the growth of algae), it is probably receiving too much light. Fish food can be obtained from pet shops, supermarkets, or department stores and should be given to the fish three or four times a week. Children have a tendency to overfeed fish. A good rule of thumb is to never feed fish more than they can consume in ten minutes. The temperature of the aquarium should be kept fairly constant. If the classroom temperature goes down too low at night, a special aquarium heater may be needed. Usually the aquarium should be covered with a glass pane to reduce loss of water and prevent dirt and dust from contaminating the water.

After the children have some experience in maintaining a simple aquarium, it may be desirable to have them set up and maintain a more sophisticated aquarium. A pump will help ensure that sufficient air will be dissolved in the water. A thermostatically controlled heater will make it possible to introduce tropical fish and other organisms that are sensitive to changes in temperature.

As a tool for learning, the aquarium can provide the opportunity for many observations:

1. How do the fish move? What parts of the body are involved? Observe the movements of the gill covers. How many times per minute do the gill covers open and close? Take a fish and put it in water

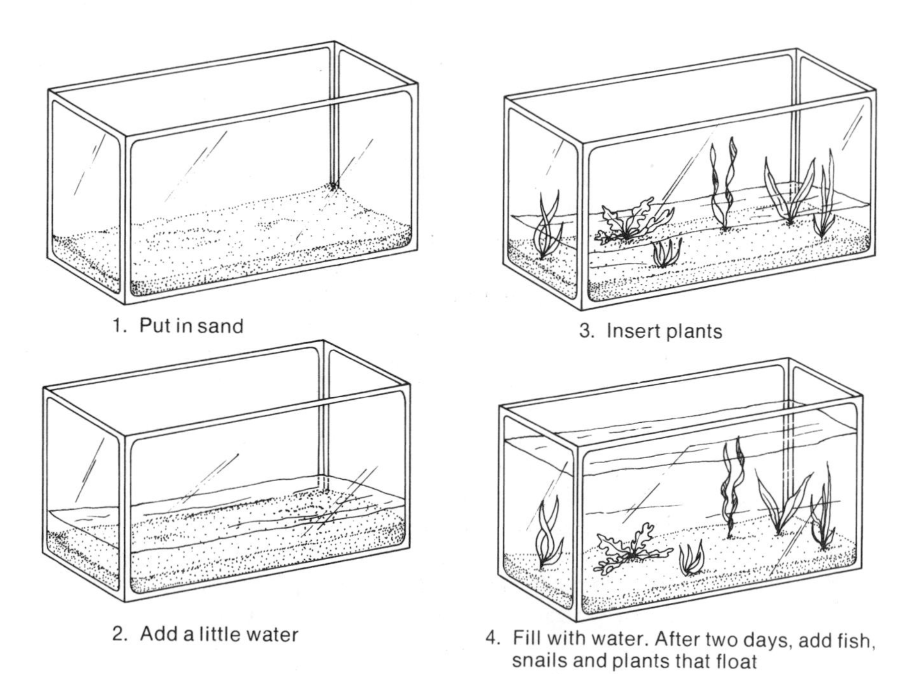

Figure 13–16 Setting up an aquarium helps show some of the ways that organisms live in water.

cooled with ice cubes. Now how many times do the gill covers open and close per minute? What do you think accounts for the change in rate?

2. Where do air bubbles appear in the aquarium? By what process might these bubbles have been formed? Do the fish come to the surface to gulp air? What could be the reason for this? What would happen if the number of animals in the aquarium were changed?
3. Does the number of animals in the aquarium change? Periodically the children should take a census of the various kinds of animals in the aquarium. They may be able to see eggs, very small fish, and snails.
4. Look at the movement of the snails across the aquarium glass. What are the steps involved in the snail's movement? Place a piece of glass or clear plastic over graph paper and have a snail move along its length. What is the snail's rate of movement in kilometers (miles) per hour? What is meant by the expression "moving at a snail's pace"?
5. Can fish learn? Children can try tapping the sides of the aquarium before feeding. Do the fish eventually learn to come to the top in search of food whenever the aquarium is tapped?
6. Do any kinds of materials form on the bottom of the aquarium? If so, what might be the source?

Maintaining a bird feeder. A simple bird feeder may be constructed by fastening a board to the outside ledge of the classroom window, on a post, or to branches on a tree within sight of the classroom. A frame or lip of wood molding should be nailed around the edges of the board to prevent food from blowing away. A water dish may be placed on the feeder as well as grains, dry cereals, all types of seeds, cracker crumbs, and the like. All of these foods may also be rolled into a ball using peanut butter as the "glue."

Bird feeding stations can have disappointing results, so the children should be prepared for this possibility. When birds do visit the station, the children should be cautioned not to make sudden movements that might frighten them. The birds that visit the station should be carefully described and identified by using pictures and manuals.

How can mammals be kept in the classroom? Mammals are perhaps the most popular classroom animals. Gerbils, hamsters, mice, rats, and guinea pigs are relatively easy to care for and observe. Large glass or plastic aquariums (even old ones that have leaks) make fine cages for these animals. Wire cages are probably less desirable, since active mammals are likely to throw the groundcover out of the cage. The groundcover should be cedar chips mixed with some chlorophyll chips to absorb odors. Gerbils, hamsters, mice, rats, and guinea pigs may all be fed a mixture of dry dog food, sunflower seeds (particularly for gerbils), with occasional strips of lettuce or carrot.

Mammals can usually learn to run a maze, and the construction of a maze can be an interesting activity. Children may enjoy focusing on specific habits of the animals, for example, watching the animal eat, play, respond to various (but harmless) stimuli, mate, and so forth.

Keeping animals in the classroom can and should be one of the most exciting activities children engage in. The youngsters should be helped to appreciate that animals need gentle and responsible care. The rewards and benefits are well worth the preparation of food and equipment for keeping classroom animals. Firsthand observation of the basic phenomena of animal life can hardly be accomplished in a better way.

RELATED ACTIVITIES

Caring for and studying animals, p. 106.
Animal training, p. 112.
How do parts of organisms vary? p. 220.
How can we study cells in the classroom? p. 270.
Rocks and shells, p. 110.
Where did fossilized organisms live? p. 351.
Chicken bones, p. 108.
How can a housefly carry wastes? p. 202.
How do cockroaches live and change? p. 202.

How do mosquitoes live and change? p. 203.
How are humans similar and different? p. 267
What organisms live within a square meter of schoolground? p. 193.
What birds are to be seen in our neighborhood? p. 204.
How can blood cells be seen as they pass through capillaries? p. 276.
The Law of Mass Production, p. 73.
How are flowering plants pollinated? p. 225.
How are seeds spread? p. 218.
The eye color of children in a class, p. 128.
How has the human population grown? p. 158.
What is the population profile of the members of the families of children in a classroom? p. 123.
How can terrariums be set up and used as tools for learning? p. 225.
How does an aquarium heater work? p. 31.
How does a sensing device work? p. 191.
What is the "black stuff" on the bottom of an aquarium? p. 8.

REFERENCES

Books for Teaching

HENNINGS, GEORGE AND DOROTHY GRANT HENNINGS. *Keep Earth Clean, Blue and Green: Environmental Activities for Young People.* New York: Citation Press, 1976. An activities book to help elementary school teachers guide their pupils in projects dealing with environmental education.

HILL, KATHERINE E. *Exploring the Natural World with Young Children.* New York: Harcourt Brace Jovanovich, 1976. An activities book for use with children in the early elementary grades. Appropriate topics and experiences with age-level indications.

ORLANS, F. BARBARA. *Animal Care from Protozoa to Small Animals.* Menlo Park, Calif.: Addison-Wesley, 1977. An excellent resource for classroom teachers and older students on keeping a wide variety of animals. A comprehensive volume including humane and ethical considerations.

For Young Children

BEHNKE, FRANCES L. *What We Find When We Look Under Rocks.* New York: McGraw-Hill, 1971. There is a great deal of life in the dark caves under rocks. Young children can read about some of the animals that live under rocks.

BROWNER, RICHARD. *Everyone Has a Name.* New York: H. Z. Walck, 1961. A book for young children dealing with the diversity and variety of animal life.

PODENDORF, ILLA. *Animals and More Animals.* Ill. by Elizabeth Rice. Chicago: Children's Press, 1970. A book about animals and their relative sizes and other characteristics.

MCCLUNG, ROBERT M. *Ladybug.* New York: William Morrow, 1966. A fascinating tale about a ladybug and its environs.

SIMON, SEYMOUR. *Animals in Your Neighborhood.* New York: Walker, 1976. This book advises children to find the many small creatures that undoubtedly live in their neighborhoods.

For Older Children

BENDICK, JEANNE. *How Animals Behave.* New York: Parents Magazine Press, 1976. A well-written and attractively illustrated book on basic mechanisms of animal behavior, including instinct, imprinting, reasoning, and higher learning.

BUSCH, PHYLLIS S. *At Home in Its Habitat.* Ill. by Arline Strong. New York: World, 1970. A book about animals and their habitats, with photographs.

COHEN, DANIEL. *Animals of the City.* Ill. by Kiyo Komoda. New York: McGraw-Hill, 1969. An account of animals commonly seen in city streets and parks.

GERGELY, TIBOR. *Animals: A Picture Book of Facts.* New York: McGraw-Hill, 1974. A lively book about the size, speed, and geographic distribution of animals.

HESS, LILO. *Animals That Hide, Imitate and Bluff.* New York: Charles Scribner's Sons, 1970. An interesting account of mimicry and protective coloration, with photographs.

LESKOWITZ, IRVING AND A. HARRIS STONE. *Animals Are Like This.* Ill. by Peter Plasencia. Englewood Cliffs, N. J.: Prentice-Hall, 1968. An attractive book about animal behavior.

MUSSELMAN, VIRGINIA. *Learning About Nature Through Pets.* Harrisburg, Pa.: Stackpole Books, 1971. A guide for the use of pets to study the basic concepts of life.

PRINGLE, LAURENCE. *Cockroaches: Here, There, and Everywhere.* Ill. by James and Ruth McCrea. New York: Harper & Row, 1971. A fascinating and accurate account of these familiar animals.

———, ed. *Discovering Nature Indoors.* New York: Natural History Press, 1970. A collection of articles including science activities with small animals.

SELSAM, MILLICENT E. *Animals as Parents.* New York: William Morrow, 1965. A book about how animals care for their young.

WATERS, J. AND B. *Salt Water Aquariums.* New York: Holiday House, 1968. All you need to know about establishing and maintaining saltwater aquariums.

We Study Ourselves

CHAPTER 14

Take a look in the mirror! What do you see? What are the most prominent features on the human body? When you look at another individual, what features do you notice first—hair, eyes, height, clothing? As you look in the mirror, think of yourself as a mammal. How is the human different from other mammals? How are our bodies adapted for our erect, walking-on-two-feet life style?

A great convenience in teaching and learning the big ideas of the human body is the fact that each child has his or her own laboratory. Some activities or investigations require physical demonstrations with chemicals, glassware, or equipment, but most of the activities in this chapter require only a human body.

EXTERNAL FEATURES

External features should be studied first. There are literally thousands of vital life functions taking place within the human body at any given moment. Most of these cannot be seen or observed by looking at a person. As children learn about the human being, they can begin by studying the easily observable external features.

Have your children take that look in the mirror and examine their gross characteristics; hundreds of observations can be made at this time. How many fingers do we have? How extensive is our hair coverage? How many eyes, ears, teeth do we have? Are there any external evidences of the many vital processes taking place that help keep us alive? Is breathing noticeable? Do any reactions to stimuli occur as we look at ourselves?

One of the most important prominent features of the human body is the pairing of body parts. We have two arms, two legs, two ears, two nostrils, and so on. An imaginary line may be drawn down the middle of the human body dividing it into two fairly even parts. This line is appropriately called a *line of symmetry*. Many of the body parts found on one side of the line have a corresponding part on the other side (see Figure 14–1.) This type of symmetry is called *bilateral symmetry, bi* meaning "two" and *lateral* meaning "sides." A butterfly exhibits a striking degree

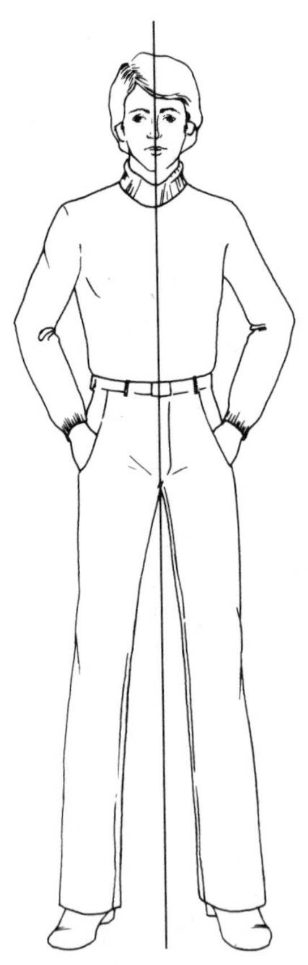

Figure 14–1 An imaginary line may be drawn down the middle of the body, dividing it into two fairly equal parts.

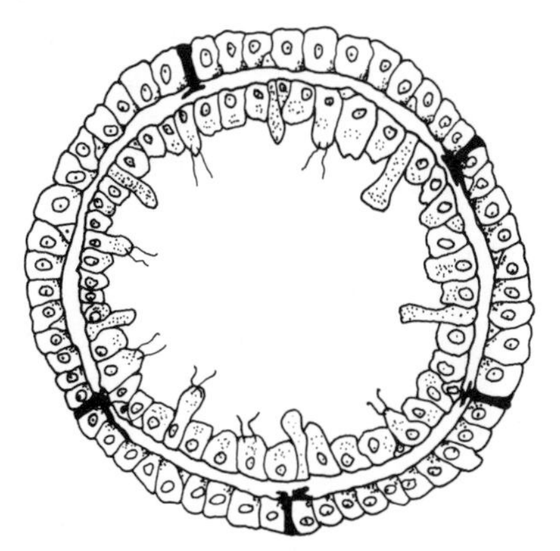

Figure 14–2 A cross-section of a hydra—an example of radial symmetry in nature.

of bilateral symmetry, as each wing is practically a mirror image of the other. Children can see evidence of their own symmetry as they place their two hands on their desks and note the similarities. What parts of the body are not symmetrical? This, too, should be studied. The heart lies near the midline of the body, slightly to its owner's left side. There is no counterpart on the right side. What other parts of the body are exceptions to the general symmetry?

In nature other types of symmetry are displayed. Among some of the invertebrates are animals that exhibit *radial symmetry:* All their body parts are arranged in a radiating fashion around some central point, not unlike the spokes of a wheel. A cross section of a hydra demonstrates how the parts seem to be arranged around some central axis (see Figure 14–2.) Some sponges, all coelenterates, and some echinoderms exhibit radial symmetry.

Children may collect objects from nature and the classroom and determine whether each demonstrates bilateral symmetry, radial symmetry, or asymmetry.

How are humans similar and different? Humans, all being members of the same species, are alike in thousands of ways. We have two arms, two legs, two eyes, two ears, two kidneys, two lungs, a four-chambered heart, and so on. Yet despite all of these similarities, we are all different. We exhibit a variety of hair and skin types. Hair may be straight and black, blond and curly, black and curly, brown and wavy. Have the children list the various hair types exhibited in their class and make a frequency count to determine which type is the most common. Which hair type is the most common in the world?

There are many other features that distinguish one individual from another. A feature that is widely used in identifying individuals is fingerprints. Every human has his or her own trademark; no fingerprint is exactly like another. Children may take their own fingerprints by placing a thumb (or other finger) on an inked stamp pad and then pressing it on a clean slip of paper. The children may compare their fingerprints and, by making careful observations, discuss the features that distinguish one fingerprint from another. What other characteristics may be used to identify individual humans?

Children should study external characteristics of the human body before any attempts are made to investigate internal processes and functions. They may begin by listing the gross features of their own bodies (e.g., hair type, skin type, height, weight, lengths of arm and legs, size of waist, head circumference, etc.).

How can young children begin to explore their body characteristics? Some of the observations and measurements mentioned earlier may be too sophisticated for children in the early elementary grades. These younger children may begin to discover body parts and relationships by having their bodies traced. Obtain some butcher paper or a large roll of brown wrapping paper. Have a child lie on top of the paper and trace around him or her with a crayon or marker. After the tracing is completed, each child should cut around his or her own outline and then draw, as accurately as possible, his or her prominent external features: eye color, hair color, skin color, clothing fabric. When each child has completed his or her own likeness, all of the "bodies" may be discussed and compared. Height differences will be immediately noticeable. Discussions regarding the variety of individual types will naturally ensue, and positive values regarding the nature of human differences may be reinforced. Why should people consider differences in skin color any more important than differences in height, hair type, or clothing worn?

ORGANIZATION OF THE BODY

After children have investigated their external features, attention may be turned to the investigation of bodily functions. However, the general organization of the human body can also be considered.

Cells

Cells are the building blocks of living material. The external features of the human body are, for the most part, easy to see and examine. If we take a part of the body and keep breaking it down into smaller and smaller parts, we will eventually encounter the basic unit of structure of all living material, the *cell.* Some organisms, such as amebas or paramecia, are composed of a single cell. The human body, however, is composed of millions of cells. Although there are some differences, there are remarkable similarities between cells in various kinds of living matter.

Most cells cannot be seen with the naked eye; microscopes are used to see and study them.

The cell is enveloped by a *cell membrane,* a thin covering which permits the passing of materials into the cell (e.g., food, oxygen, water) and out of the cell (e.g., waste matter). In addition, plant cells have a cell wall surrounding the cell membrane that helps protect and provide support for the cell.

The *nucleus* is a spherical body usually located near the center of the cell. The nucleus controls all of the cell's vital metabolic and reproductive functions. Inside the nucleus are *chromosomes,* long, thin, coiled fibers that contain all of the cell's hereditary information. It is by means of the chromosomes that cell characteristics are passed along from one generation to the next. Body cells reproduce by simply splitting in two, a process called *mitosis.* As a cell splits, the chromosomes themselves appear to split, so that each new half of the original cell contains a complete set of chromosomes and, thus, all of the information necessary for the continued life of the new cell and its subsequent "reproduction." (See Figure 14–3.).

The bulk of the rest of the cell is composed of *cytoplasm.* The cytoplasm's streaming, flowing movements help materials circulate throughout the cell. Nourishment is brought to all parts of the cell by the movements of the cytoplasm. There are other cell parts that carry out specific

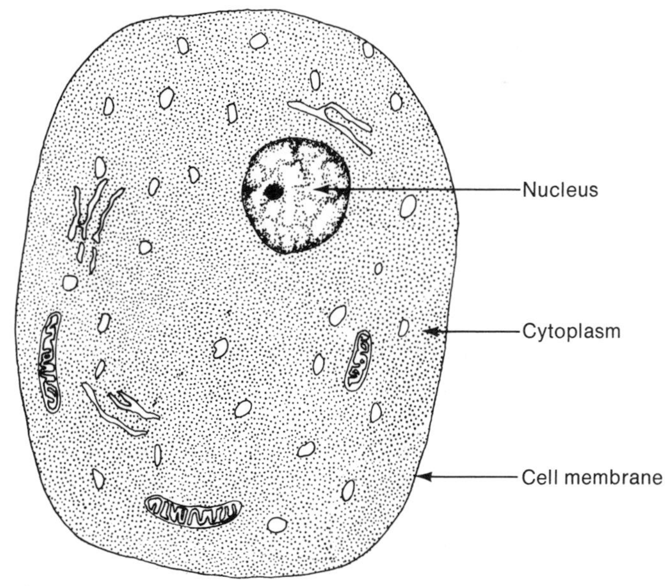

Figure 14–3 A typical body cell. It should be noted that this is a two-dimensional drawing of a three-dimensional structure. Children often think cells are flat, but they are not.

functions for the maintenance of the cell, but these are usually beyond the scope of an elementary school science program.

How can we study cells in the classroom? Samples of a particular kind of cell *(epithelial cells)* can be obtained by scraping the inside of the mouth with a spoon or a tongue depressor. Place the material obtained on a glass microscope slide and apply a drop of tincture of iodine on top of the scrapings. Place a cover glass over the drop.

Examine the slide under a microscope and try to find the nucleus and the cell membrane. (The iodine stains the nucleus dark brown, and this should be the darkest part of the cell.) A slide of the thin membrane of the inside of an onion peel may be stained with iodine to permit the children to compare the human epithelial cells with the plant cells. The cell wall of the plant cell should be conspicuous.

Although cells are best examined under a microscope, an image of cells can be projected onto a screen with a microprojector.

Tissues

Tissues are large groups of similar cells that are specialized for a particular function. Epithelial tissue, for example, serves as a covering for the body surfaces on both the inside and the outside. Other types of tissues include muscle, connective tissue, nerves, bone, cartilage, and blood.

Epithelial tissue covers the outer parts of the body. Our skin is epithelial tissue. It also forms the inner and outer coverings of the respiratory and digestive tracts. Epithelial cells are very close together, so that there is little room for germs to pass between them. Some epithelial cells form *glands,* which produce secretions such as digestive juices. Sweat glands, essential for the regulation of body temperature, are also made up of epithelial cells.

Hair and nails are parts of the skin that children can examine. However, they are not living tissue. The nails are horny plates formed from the outer skin layers. Hairs are also developed from the outer skin layers. As children examine a hair, they will notice a bulb at the end of the root of the hair. This hair bulb is set in the inner skin layers.

How does human hair appear under the microscope? Place a human hair on a slide, cover it with a cover glass, and examine it under a microscope. Is the color concentrated in any particular part of the hair? Are there scales in the hair? Comparison of a human hair with a wool fiber, which is a hair from a sheep, will provide additional information.

Get a cross-section of a human hair by having a man shave closely for a second time and then rinse the razor in clear water. Put a little of the water containing the shaving on a slide and examine it under a microscope. What is the shape of the hair? Are there distinct features to be seen in the cross-sectional view of the hair?

Muscle tissue is composed of cells that are capable of contracting. One kind of muscle tissue helps move arms, legs, and other segments of the skeleton. In general, the movement of these muscles, called *voluntary* muscles, can be controlled. Other muscles, which help operate such organs as the kidneys, lungs, and intestines, are *involuntary* muscles because they cannot be controlled at will. The heart is an extremely important muscle that pumps blood throughout the body.

Connective tissue is composed of ligaments, tendons, and fatty tissue. *Ligaments* connect bones to each other; *tendons* connect muscles to bones. The cells of fatty tissue store fats and oils. The cells of connective tissue tend to be long, narrow, and widely separated. Connective tissue is strong, as it must withstand the great stress put on it by bodily movements.

Nerve tissue is made up of nerve cells or *neurons.* The extensions of nerve cells that carry impulses from the body of the nerve cell to another place may be very long, often as long as a meter. Nerve cells are covered by a protective sheath and have branched end regions that help carry the impulses from one cell to another. The brain is thought to consist of trillions of nerve cells.

Bone tissue is composed of bone cells, which secrete substances (basically calcium carbonate and phosphate) that fill in the spaces between the cells. These hard substances form the bones, which are necessary to support the human body. *Cartilage tissue* is composed of cells that are embedded in spongy masses. Cartilage is strong and flexible, resists shock, and is found between bones in the spine, in the tip of the nose, in the outer ear, and in the walls of the windpipe.

Blood tissue is made up of three basic kinds of cells: *red blood cells, white blood cells,* and *platelets.* These cells are carried in a fluid called *plasma* that can transport the cells to all parts of the body. Red blood cells carry oxygen to the many places in the body that need it. White blood cells help fight infection and disease. Platelets are essential for the clotting of blood.

Organs and Organ Systems

Organs consist of one or more types of tissue working to perform a specific bodily function. The small intestine, for example, is an organ that performs certain digestive functions. The outside of this organ is covered with epithelial tissue; below this covering are layers of muscle tissue and then a layer of connective tissue; lining the inside of the small intestine is a layer of epithelium. Nerve and blood cells are in close association with these component tissues. Each of the tissues performs a specific job, but together they act in concert to accomplish the vital task of digestion and absorption of some foods. The entire process of digestion and absorption cannot be performed by the small intestine alone. It also involves the actions of the mouth, teeth, esophagus, stomach, large intestine, liver, pancreas, colon, and anus. The entire job of digestion, then, is performed by a group, or *system,* of organs. Other organ systems that carry out body functions are the respiratory system, the circulatory system, the nervous system, the reproductive system, the endocrine system, and the excretory system.

The digestive system. *The digestive system prepares food so that it can be used by the body.* The food we eat must undergo a series of changes so that it can be broken down to a form that can pass through the walls of the digestive system and be absorbed into the bloodstream. In this simpler form the food can be carried by the bloodstream and either stored or used to nourish body parts.

Food enters the mouth, where it is chewed and torn into small pieces. It is mixed with *saliva,* which softens it for its journey down the digestive tract. The saliva carries *ptyalin,* an enzyme produced by the salivary glands, which changes some starch into sugar. The food is then swallowed by a series of muscular movements called *peristalsis.*

The food next enters the stomach, where hydrochloric acid and the enzymes *pepsin, rennin,* and *lipase* are secreted. Proteins are broken down into simpler compounds called *amino acids.* Milk proteins are also partially digested in the stomach. This partially digested food passes into the *small intestine,* where most digestion occurs. The small intestine of humans is a narrow, wavy tube nearly seven meters (22 feet) long. Protein digestion is completed here, resulting in the formation of more amino acids. *Bile,* secreted by the liver, enters the small intestine to *emulsify* fats (i.e., break them up into small globules so that they may be acted upon by lipase). Lipase, along with other enzymes, is secreted by the pancreas and completes the digestion of fats, amino acids, and starch. Most of these end products of the digestive process pass into the bloodstream in the small intestine through the process of absorption. The small intestine is lined with a great many small, finger-like projections called *villi.* Among other things, the villi increase the surface area through which absorption can take place. In each of the villi are tiny blood vessels. The digested food, in the form of simple sugars, amino acids, and fatty acids, passes through the membranes of the villi and into the bloodstream.

The food that has not been digested or absorbed passes on to the *large intestine,* a much wider but shorter tube than the small intestine. Here water is absorbed from the remaining food material, returning to the body much of the liquid added along the way. Bacteria that live in the large intestine are instrumental in manufacturing certain vitamins. The remainder of the undigested food material, mainly the cellulose of plants, is mixed with bacteria and forms *feces,* which are excreted from the body through the *anus.*

Conducting food tests. The concept of a chemical test for the presence of certain foods may be developed in an elementary school classroom. Such tests use indicators to detect the presence of the substances under study.

A simple test may be used to check for the presence of fats and oils. Rub some of the material to be tested on a piece of brown paper bag. Hold the test spots up to the light. What are your observations? If the spot becomes translucent, it means that fats or oils are present. Try testing a peanut, butter, margarine, meat, and so on.

A test for starch may be demonstrated by cutting a thin slice of potato. Place a drop of iodine solution on the potato slice. What changes do you see? The dark stains confirm the presence of starch. Test bread, celery, crackers, halved lima beans, and other substances for starch. Have the children list the results.

What are the sounds of food passing down the esophagus? One child may listen to food as it passes down another child's esophagus by pressing a cardboard tube or stethoscope on the child's throat. The swallower should eat a cracker and sip some water. What does the food sound like as it passes down the esophagus? What might account for the sounds?

The respiratory system. *The respiratory system makes oxygen available to cells all over the body.* Oxygen is needed by cells for the release of energy from foods. Energy is, of course, needed for muscular movements and many other necessities of life. The respiratory system brings oxygen to the blood cells, where it is carried to other parts of the body. The respiratory system also carries carbon dioxide and water (the products of food use and energy release) out of the body.

Air enters our bodies through the nose or mouth. The nasal passages are moist with mucous and contain tiny hairlike projections called *cilia.* The cilia and mucous filter the inhaled air and capture some dust and bacteria. The air then passes through the *trachea* or windpipe, which branches into *bronchi,* and then enters the *lungs.* The lungs contain air sacs, which are in close association with finely branched blood vessels. These tiny vessels, just one cell thick, allow oxygen to pass into them and enter the bloodstream. Similarly, carbon dioxide is able to pass out of the bloodstream and into the lungs, where it is expelled during exhalation.

To inhale air the body must reduce air pressure in the lungs. This is done largely by moving the *diaphragm*, a sheet of muscle just beneath the lungs, downward so that the size of the chest cavity is increased. When the pressure within the lungs is decreased, the greater air pressure outside the body forces air into the nose or mouth, through the trachea and bronchi, and into the lungs. To exhale, the diaphragm moves upward, decreasing the size of the chest cavity, increasing the pressure within the lungs, and in this way forcing the air containing carbon dioxide out of the lungs.

How does exercise affect the breathing rate? Have the children count and record the number of times they breathe in a minute. Then have them jump up and down in place twenty times. Have the children count and record their breathing rates again. How do the two rates compare? Why do you think there is a difference in the breathing rates? How long does it take the breathing rate to return to the initial rate?

How can the presence of water vapor in the breath be demonstrated? Have a child breathe against a cold glass window or chalkboard. Can you see anything form on the window or chalkboard? Does this provide any evidence that water vapor is exhaled?

Demonstrating the presence of carbon dioxide in exhaled air. Limewater can be used as an indicator of carbon dioxide as it turns milky or cloudy in its presence. Obtain some limewater from a druggist. Add about 2 cm (1 in) of limewater to each of two test tubes. Using a drinking straw, blow into the limewater in one test tube. Draw air in through the other tube, taking care to keep the bottom of the straw just above the level of the limewater. What conclusions may be drawn about the relative carbon dioxide content of inhaled and exhaled air?

The circulatory system. *The circulatory system transports food and oxygen to all parts of the body.* It also carries waste products from the body cells to the lungs, where they are released from blood cells and exhaled. The circulatory system is a closed system consisting of the heart, which acts as a pump, and a network of arteries, veins, and capillaries. *Arteries* carry blood away from the heart. *Veins* return blood from all over the body to the heart. *Capillaries* are finely branched blood vessels with walls one cell thick that connect small arteries and veins.

The *heart* is essentially a muscle that pumps the blood through the circulatory system. It consists of four chambers. (See Figure 14–4.) A partition divides the heart into two parts, the right side and the left side

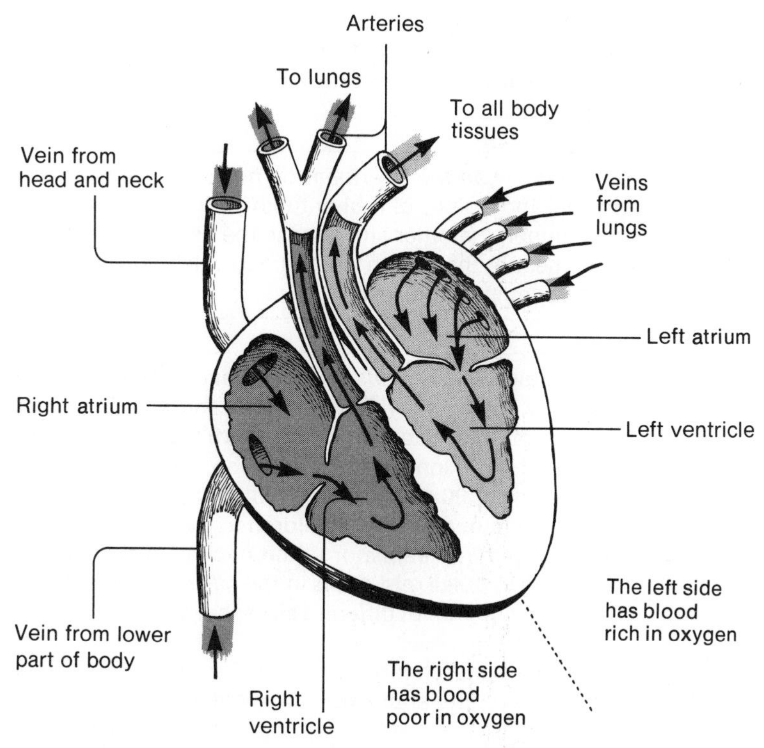

Figure 14–4 Diagram of the human heart.

(the owner's right and left side, not the observer's). On each side veins empty blood into the thin-walled *atria,* which then contract, forcing the blood down into the *ventricles.* The ventricles are thick-walled, muscular structures that, upon contraction, force blood away from the heart through the arteries. Between the atria and the ventricles there are valves that control the direction of blood flow.

Blood returning from the different parts of the body is brought to the right atrium by large veins. From here it flows down into the right ventricle, which contracts. The valve between the right atrium and ventricle closes, and the blood is forced through the pulmonary artery into the lungs. In the lungs the blood gives up carbon dioxide and absorbs oxygen. The oxygenated blood leaves the lungs through the pulmonary veins and enters the left atrium. It flows from the left atrium into the left ventricle. As the left ventricle contracts, the valve between the ventricle and the atrium closes and the blood is forced out through a large blood vessel, the *aorta,* into the various parts of the body. The aorta branches into smaller arteries, and these smaller arteries branch into still smaller ones and finally connect to capillaries. Here the oxygen and digested food materials pass through the thin capillary walls to the body cells and waste products (carbon dioxide, urea, and others) are picked up. These waste products, now in the blood, are returned to the heart through the veins.

How can the heartbeat be heard? A stethoscope may be made from a funnel and a piece of rubber tubing. Fit one end of the tubing over the neck of the funnel; the other end goes in the listener's ear. Press the funnel firmly on the chest in the region of the heart. Listen to the heartbeat. Can more than one sound be identified?

How can the pulse rate be taken? Have the children take their own pulse by lightly applying their index and middle fingers to an area below the thumb (on the underside of the hand) 2 or 3 cm (1 to 1½ in) from the base of the hand. After locating the pulsation, have the children count the number of pulsations in 30 seconds and multiply by 2 in order to derive the number of beats per minute. Ask the children to try to locate other areas of pulsation, for example, in the temple or in the neck near the hinge of the jaw. How much variation in pulse rate exists in the class? Is there any difference in pulse rate between children of different sizes, between boys and girls, between children and adults?

Have the children exercise for one minute and compare their pulse rates before and after exercise. What accounts for the differences observed?

How can blood cells be seen as they pass through capillaries? Remove a goldfish from a fish tank and wrap it in a wet gauze pad, leaving the tail exposed. Place the fish in a shallow dish (such as a petri dish) containing water. Place a microscope slide above and below the tail to prevent movement and examine it under a microscope or microprojector.

Can the arteries and veins be distinguished? Look for capillaries; how can they be identified? Can blood cells be seen? If so, watch them carefully as they pass through the capillaries. What do you observe? Trace a drop of blood from a small artery, through a capillary, and into a small vein. Are there any differences in the rate of flow in each type of vessel? Why do you think this is so?

The excretory system. *The excretory system removes from the body the waste products of metabolism.* While the skin and lungs remove some of the wastes from the body, wastes such as urea and uric acid are removed by the *kidneys.* The human body has two kidneys; they receive blood directly from the aorta and discharge it into a major vein. Urine is extracted from the blood by tiny filters in the kidneys that are in close association with capillaries. Blood flows through these filters, and the waste products of food breakdown are removed. The process of filtration can be illustrated by having the children filter some muddy water through filter paper or paper toweling and pointing out that in the kidney some filtered material is reabsorbed into the bloodstream. This may seem like a slow process, but in the course of a day, more than one liter (1.06 qt.) of urine is removed from the blood by the kidneys and passed out of the body. (See Figure 14–5.) As urine is removed from the blood, the filters pass it into a tube, the *ureter*, which empties it into the bladder, where it is stored until it is voided.

The nervous system. *The nervous system is the central control system of the body.* The nervous system coordinates actions and responses so that the organism can function as a whole. It also integrates the work of the various organs and systems of the body. For example, the heart pumps blood to all parts of the body, and all parts of the body are dependent on

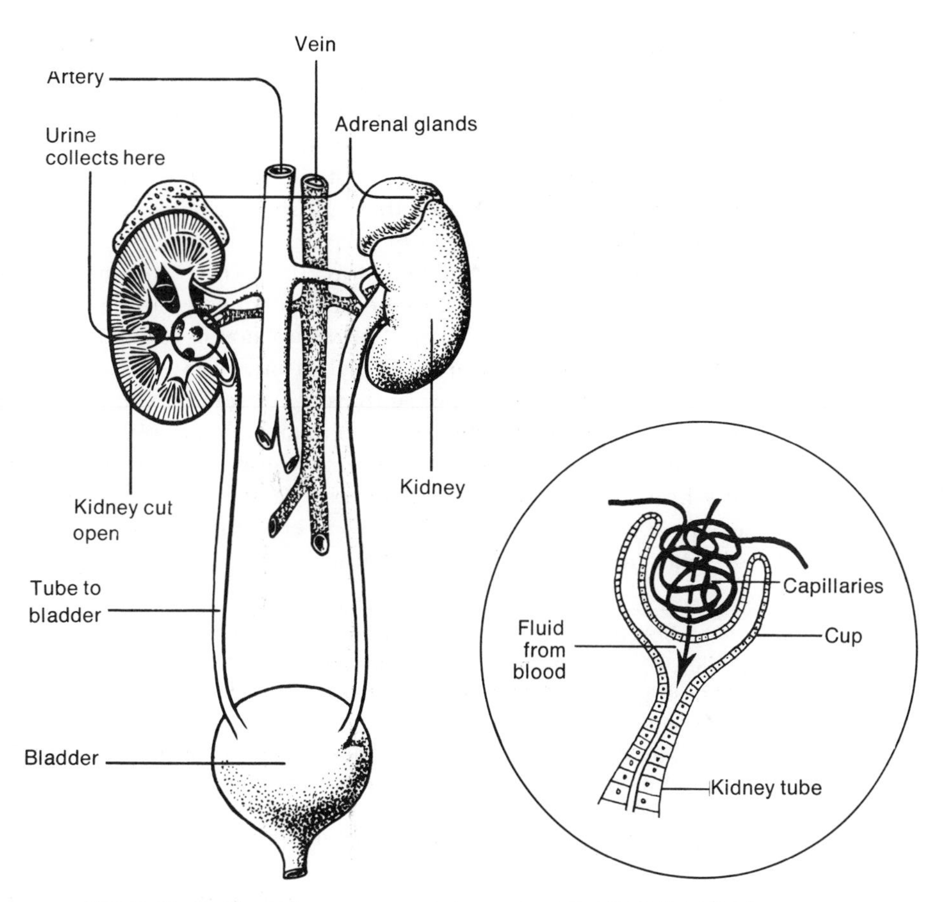

Figure 14–5 The kidneys remove poisonous wastes from the blood.

it. However, the heart is dependent on food from the digestive system and oxygen from the respiratory system for the energy needed to keep it beating. The nervous system helps keep all parts of the body working smoothly together.

The nervous system also helps the body respond to changes in the environment. For example, if the weather suddenly becomes colder, this information is gathered in by receptors in the nervous system; it is then interpreted, and the nervous system activates the muscles that may do the work of putting on a coat or sweater.

The nervous system is composed of the *brain,* the *spinal cord,* and 43 pairs of *spinal nerves.* The brain is located at one end of the spinal cord. It may be considered an outgrowth of the brain stem found in lower animals. A highly developed brain is probably the most important physiological characteristic of humans. The brain is connected to the spinal cord, which extends down the spinal column. The 43 pairs of peripheral nerves extend from the brain or spinal cord to different parts of the body.

To illustrate one of the ways in which the nervous system functions, it may be desirable to describe what occurs in a reflex action. In a *reflex action* a nerve impulse is set up in a sensory nerve. For example, if a person accidentally touches a hot stove, a nerve impulse is set up. This nerve impulse is transmitted through several neurons to a neuron connected to a muscle. The muscle is activated and the part of the body touching the hot stove is pulled away. In this case the nerve impulse travels from the sensory nerve to the nerve connected to the muscle without having to travel to the brain. This reflex action is one of the important ways in which the body protects itself from injury.

In other cases sensory impulses are sent to the brain, where the impulses are interpreted and voluntary (consciously controlled) responses are made. The details of how these interpretations are made in the brain are not known. However, the fact that many actions can be voluntary and may be performed after considerable thought is one of the most important "human" characteristics.

How can nervous reflexes be investigated in the classroom? There are a variety of reflex actions which may be easily demonstrated. Children should work in pairs with one child serving as the observer. Roles should then be reversed.

The *pupillary reflex* may be observed by looking into the pupils of the eyes of a child who has covered his or her eyes for one minute. What happens? What controls the response? This observation can be even more impressive if one child turns on bright lights and then observes the action of the pupils.

The *blinking reflex* can be demonstrated as one child holds a sheet of clear plastic in front of his or her face and another throws a piece of crumpled paper against the plastic, aiming at the eyes. What is the reaction?

The *knee jerk reflex* may be observed by having one child cross his or her legs and another child lightly chop the first child's free leg just below the kneecap with an open hand. What happens to the leg? This method is used by many physicians to test the integrity of the nerves.

The endocrine system. *The endocrine system controls many bodily functions by chemical means.* The system consists of such glands as the *pituitary, thyroid, parathyroid, thymus, pancreas,* and *adrenal* glands, as well as *ovaries* in the female and the *testes* in the male. These glands secrete hormones directly into the bloodstream which carries them throughout the body.

The hormones secreted by the endocrine system have critical control functions, so that either a lack or an oversupply of a hormone can have serious consequences. The secretions of the thyroid gland, for example, exercise general control over the metabolic rate. If there is a lack of thyroid hormone, the heart rate is reduced and the body tends to become fat and sluggish. An oversupply of thyroid hormone leads to an increased metabolic rate, nervousness, muscular tremors, and loss of weight. The adrenal gland is sometimes called the emergency gland. In times of stress and danger, the secretion from the adrenals stimulates body activity. In similar ways the other endocrine glands carry out control functions.

The reproductive system. *The reproductive system of humans provides the means by which we can perpetuate our species.* In the male, the testes produce *sperm.* Sperm are transported in a fluid called *semen.* The testes also produce male sex hormones, which are responsible for the development of such characteristics as the pitch of the voice and the growth of facial and pubic hair. During sexual intercourse the sperm are passed from male to female through a tube in the *penis.*

In the female, the ovaries produce *eggs.* An egg is produced approximately every 28 days. After it breaks out of the ovary wall, a process called *ovulation,* the egg moves slowly down a tube called the *oviduct* until it comes to the *uterus.* This journey from ovary to uterus usually takes 3 to 6 days. The *vagina,* the external opening of the female reproductive system, receives the male sperm during sexual intercourse. The sperm swim through the uterus and often up into the *oviducts.* The union of sperm and egg is called *fertilization.* Millions of sperm are produced; however, only one fertilizes an egg. If the egg is fertilized, it will implant on the uterus wall and begin to divide; it is now termed an *embryo.* If fertilization does not occur by the time the egg reaches the uterus, the inner lining of the uterus (which was prepared for the implantation of a fertilized egg) will slough off and the blood-rich lining will be discharged through the vagina. This is called *menstruation.* The entire menstrual cycle in females is controlled by hormones secreted by both the pituitary gland and the ovaries.

The muscular and skeletal systems. *Our bodies move as components of the muscular and skeletal systems interact.* The human skeleton has a main axis (the skull, vertebral column, and rib cage) and appendages (arms and legs). These bones provide support for the body, serve as levers for bodily movement, and protect the internal organs. *Ligaments,* which connect bones to bones, and *tendons,* which connect muscles to bones, join the parts together for movement.

The main movement performed by a muscle is *contraction.* Referring to Figure 14–6, it can be seen that as the biceps muscle contracts, it pulls the lower arm bones up. This is so because the biceps muscle has two attachments, one at the shoulder and one at the lower arm bones. The triceps muscle relaxes at the same time, loosening the tension on the lower arm so that it may respond to the biceps' contraction. When the arm straightens out, the triceps contracts and the biceps relaxes. Thus, muscles act as opposable pairs. The biceps is a *flexor* muscle; it bends a joint toward the body. The triceps is an *extensor* muscle; it straightens out a joint. Using many pairs of opposable muscles, humans can move their limbs in many directions.

How can we feel the action of muscle pairs? Ask the children to sit up in their chairs and place one hand, palm up, inside their desks, pushing up on the desk top. With their other hand have them feel the front and back of their upper arm. Is the flexor (biceps) or the extensor (triceps) contracted (hard)? The muscle that feels hard is contracted; the one that is soft is relaxed. Next ask the pupils to press down on their desk tops. As they touch the two upper arm muscles, ask them to discern which one is hard now. Which muscle's

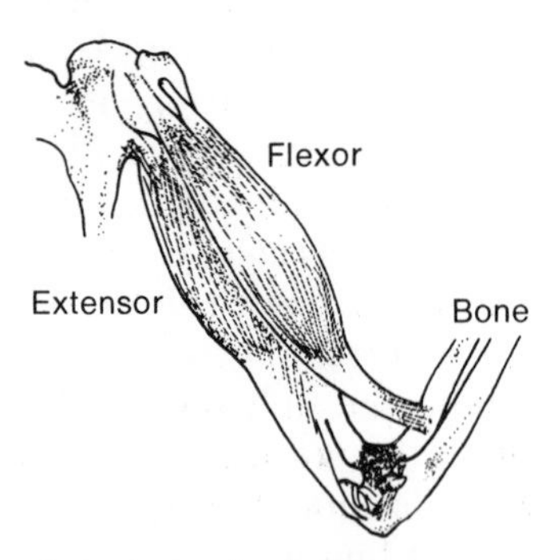

Figure 14–6 Flexor and extensor muscles in the arm.

contraction is responsible for the straightening of the arm? How does this demonstrate the operation of opposable pairs of muscles?

While the pupils are sitting in their chairs, have them alternately slide their feet back and forth on the floor. With their hands on the undersides of their thighs as they slowly slide their feet, have them see whether they can feel any muscle contraction. Then ask the children to place their hands on top of their thighs and feel the muscle movements as they slide their feet. Which muscle (upper or lower thigh) contracts as the foot is moved forward? Which muscle contracts as the foot moves back?

Ask the pupils to locate other muscle pairs and analyze their movements. Actions of the arms, fingers, tongue, eyelids, and wrists may be explored.

The senses. *The senses are externally oriented receptors that are sensitive to such stimuli as light, sound, taste, smell, and touch.* These specialized receptors are ultimately the basic means by which we receive information from the environment. The senses are essential in our daily lives; when one or more is lost, the individual must make dramatic accommodations to successfully meet the demands of living.

We are perhaps more dependent on our eyes than on any of our other senses. It has been estimated that 85 percent of human learning is derived from the sense of sight. The eye receives light and transmits the information perceived to the brain for interpretation. Light passes through the *pupil* of the eye (the dark, round dot in the center of the eyeball), and the *lens* focuses an upside-down image on the *retina.* (See Figure 14–7.) The image is perceived as being right side up because the brain mediates and interprets it for us. The retina contains a network of tiny nerve cells called *rods* and *cones.* The rods are sensitive to objects in subdued light; they are particularly important for night vision. The

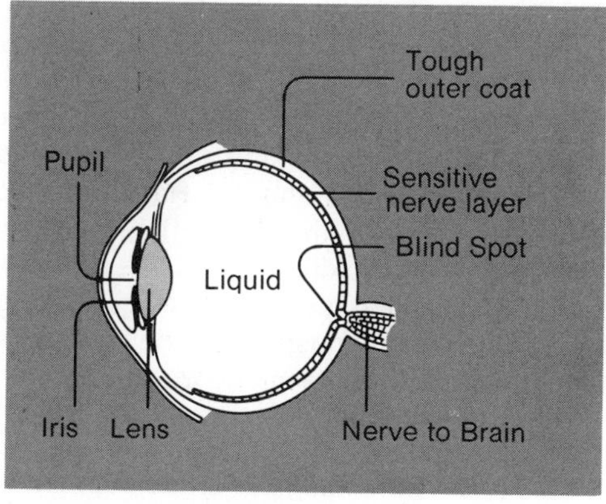

Figure 14–7 Cross-section of the human eye.

cones are primarily sensitive to bright light and colors. The rods and cones are connected by the *optic nerve* to the brain.

The eye works very much like a camera, and studying the similarities between the two may be useful in helping children explore interactions through the use of models and analogies. The size of the pupil is automatically controlled by the *iris.* The iris opens the pupil in the presence of dim light and closes it somewhat in the presence of bright light. Similar adjustments are made by the user of a camera in opening or closing the diaphragm. The lens in the eye focuses the image on the retina. In a camera, the image is focused by moving the lens back and forth. In the eye, however, the focusing is controlled by muscles at the side of the lens that vary its thickness. In the camera, the light-sensitive material is on the film. In the eye, it is the retina. When light falls on the retina, the image, picked up by the rods and cones, is sent to the brain.

Our perception of depth and three-dimensionality is due to the fact that we have two eyes; each eye receives a slightly different scene, and the brain's interpretation of these different scenes allows us to judge distances. Cameras with two lenses (called stereo cameras) essentially do the same thing as our two eyes; they take two pictures, and when the developed slides are shown on a screen or viewed in a stereopticon there is an illusion of three-dimensionality.

Demonstrating the effects on perception of the loss of sight in one eye. Have the children play a game of "catch" in groups of two using a small rubber ball. One child in each pair should blindfold one eye. Have each pair stand about three meters (one yard) apart and toss the ball back and forth, keeping a record of catches and misses. After about 20 or 25 tosses, have the blindfolded child remove the blindfold and continue for another 20 or so trials. How does the record of tosses and catches when one eye is used compare with the record when both eyes are used? What might be the cause of the difference in results?

How can the blind spot be located? Where the optic nerve is attached to the retina, there is a little blind spot in the eye. Each child can find his or her own blind spot by marking a heavy dot on a blank index card and marking an X 5 cm to the left of the dot. Have the children rest their cards on their desk tops and focus on the X with their right eye. Their left eye should be closed. While they are looking at the X, they should bring the card slowly toward their eyes while still focusing on the X. The dot should disappear when its image is exactly on the blindspot. As the card is brought still closer, the dot should reappear because its image was moved.

How is the image of an object inverted? The image of an object on the retina is inverted. A pinhole camera can be used to demonstrate how this comes about.

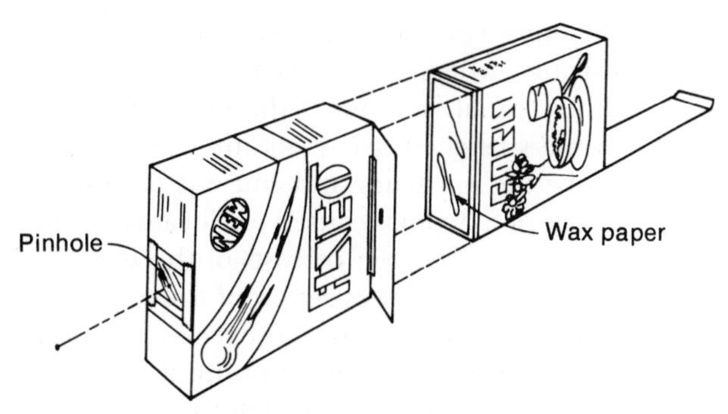

Figure 14–8 A pinhole camera may be made from two cereal boxes.

Obtain two cereal boxes. One should be slightly smaller than the other so that it can slide back and forth within the larger box.

Cut a square in the bottom of the larger box and tape a piece of aluminum foil over it. Make a small pinhole in the aluminum by sticking a small pin through it.

Cut the bottom off the smaller cereal box and tape a piece of wax paper over it. Insert the end covered with wax paper into the larger box and slide it back and forth. (See Figure 14–8.)

Darken the room and light a candle. Have the children point the pinhole toward the candle and look through the open end of the smaller box. When their eyes become "dark adapted," they should be able to see an image of the candle on the wax paper. They can slide the smaller box back and forth to change the size of the image. Is the image of the candle right side up or upside down?

What happens when the retina tires? Shine a bright light on a piece of red paper taped onto a white screen. Ask the children to stare intently at the paper for two or three minutes and then quickly remove the paper. What do they see?

After their eyes have rested, have the children stare intently at a piece of green paper for two or three minutes and then quickly remove it. What do they see now? It is believed that the cones that are sensitive to a particular color tire after a few minutes of exposure. When exposed to white light, the retina is not as sensitive to that part of white light as it is to the complementary color. (*Complementary colors* are two colors that give white light when mixed.) Therefore, the complementary color is usually seen until those cones are rested again.

If 85 percent of our learning is visual, then the bulk of the remaining 15 percent is auditory. The ear receives sound waves and transmits them to the brain by means of nerve impulses. The parts of the ear are generally grouped into three sections: the outer ear, the middle ear, and the inner ear. The *outer ear* gathers sound waves, operating like a funnel through which the waves are concentrated. (See Figure 14–9.) The auditory canal represents the neck of the funnel and ends with a membrane stretched across its inner end, called the *eardrum.* Sound waves strike the eardrum and cause it to vibrate. In a similar way the membrane of a drum is caused to vibrate by a nearby loud sound.

The *middle ear* consists of three tiny bones called the *hammer, anvil,* and *stirrup.* The movement of the eardrum is transmitted through these bones, where the vibration received is magnified. The pressure of this vibration is then exerted on fluid in the *cochlea* of the inner ear, where the fluid's movement is detected by auditory nerve endings. Electrical impulses are set up in the auditory nerve and transmitted to the brain, where they are interpreted. Also in the inner ear are the semicircular canals, which are involved in balance and the detection of our position along the three planes of space.

Behind the eardrum there is a tube—the *Eustachian tube*—that leads down into the mouth. It is through this tube that the air pressure on the inside surface of the eardrum is equalized with that on the outside. Swallowing or chewing gum allows air to enter or leave the middle ear through the Eustachian tube.

It is important that children learn to protect their ears from damage. The eardrum can be pierced by pins, pencils, and other objects. It has been said that "no object smaller than your elbow should be placed in your ear."

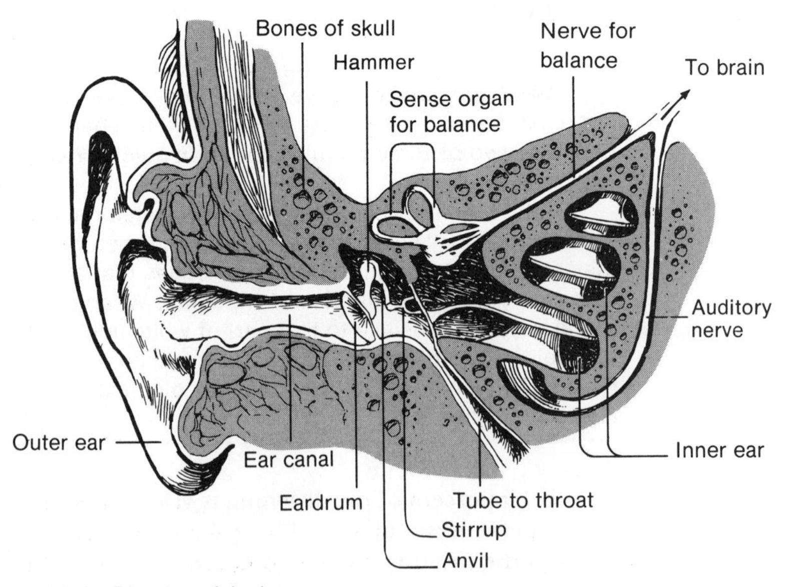

Figure 14–9 Diagram of the human ear.

How well do we hear? Hold a watch near a child's ear. Then move the watch out from the ear until you reach the point at which the sound can no longer be heard. Measure and record this distance. Do the same for the other ear. Have the children use this method to check each other's hearing. If there are children who have difficulty hearing the watch, refer them to the school nurse or a physician.

How can auditory fatigue be demonstrated? Strike a tuning fork and place it beside a child's left ear. When the sound is no longer heard, remove the tuning fork and a few seconds later bring it back beside the child's ear. Repeat this, but as soon as the tuning fork is no longer heard by the child's left ear, place it beside his or her right ear. What are the results? Ask the children to try to account for the results.

How do we locate sounds? Place one student in the center of the room and cover his or her eyes with a blindfold. Arrange a number of other children in a circle around the outside of the room. Provide them with some way of making a short, sharp sound. (One way is to give each child two small sticks or stones to strike together.)

Ask the child in the center of the room to close one ear by placing a hand over it. Have one youngster hit his or her sticks or stones together. Have the blindfolded child point in the direction of the sound. After having done this several times, have the blindfolded child uncover his or her ear and again point in the direction of the sound.

Was the child better able to locate the sounds in some directions than in others? Was he or she better able to locate sounds with two ears than with one?

Have several children try it. Are the results always the same?

The senses of smell and taste are interrelated. When you hold your nose, many substances seem to have no taste. A blindfolded person who holds his or her nose cannot taste the difference between an apple and an onion.

The tongue, besides helping with eating and speech, is the site of our *taste buds.* These are hidden among the tiny bumps on the surface of the tongue. Taste buds contain the endings of nerve cells that transmit taste information to the brain. The four basic sensations of taste are sweet, sour, bitter, and salty. Other tastes are due to various combinations of these four. The taste buds for sweet and salty are located near the tip of the tongue. The taste buds for sour tastes are found near the edges of the tongue, while those for bitter tastes are at the base of the tongue.

Both smell and taste are said to be chemical senses. To be tasted, the substance must be in solution or become dissolved in saliva. For example, if the surface of the tongue is wiped dry, some substances, such as salt or sugar, cannot be tasted. They have to be dissolved before they can be tasted. Similarly, materials in the air that are said to have odors have to become dissolved in the fluid that covers the mucous membrane.

The nose is the major receptor of odors. In the upper, moist portion of the nostrils are tiny hairlike nerve endings that transmit odor information to the brain. The sense of smell is a very sensitive one. In the case of certain chemicals, as little as one part in 30 billion parts of air can be detected. Like hearing, sensitivity to odors can become fatigued. This is why we often smell an odor when we first enter a room but soon become accustomed to the odor and can no longer smell it. The sense of smell often serves vital protective functions, as the early detection of certain poisonous fumes (e.g., natural gas and refrigerator gas) can save lives.

Demonstrating the interrelationship between taste and smell. Prepare small cubes of raw potato, pear, apple, and onion. Have the children work in pairs. One member of each pair should be blindfolded; the other member presents the blindfolded child with the food samples. The blindfolded child (i.e., the taster) should be given each sample, one at a time, while holding his or her nose. Can the samples be identified by taste alone? Also have one child hold a piece of paper under the taster's nose and at the same time give him or her a piece of apple to eat. What does the taster think he or she is eating? Why do foods seem tasteless when one has a heavy cold?

Mapping the tongue. Dip separate toothpicks in salt water, a sugar solution, vinegar (or lemon juice), and black coffee (or tea). Touch each toothpick to spots all over the tongue, rinsing the mouth after each test. Which areas are sensitive to sour, sweet, salty, and bitter substances? Have the children sketch their tongues and mark the areas where they received the best response for each taste.

How do odors permeate a room? With the pupils evenly distributed throughout the classroom, open a bottle of ammonia in one corner of the room. Ask the children to raise their hands as soon as they smell the ammonia. Chart the path of the odor. In which area of the room was the odor sensed last?

For more activities involving these senses, see Chapter 4.

The skin is the major receptor of tactile sensation. The skin is laden with sensitive nerve endings that detect touch, pressure, pain, heat, and cold. Each of these sensations can be stimulated separately. For example, the touch sensation may be felt by rubbing a soft material against the skin;

the pressure sensation may be felt as a pencil is pressed against the skin; the sensation of pain may emanate from internal or external sources and can serve as an indicator of danger to the body; hot and cold sensations are stimulated by hot or cold objects or environmental temperature conditions. Different feelings may also occur simultaneously, as when you brush up against a hot kettle. The brain registers touch, pressure, pain, and heat all at once. In different parts of the body, the nerve endings for sensation are situated with varying degrees of density. For example, nerve endings that are pressure sensitive are very close together in the fingertips and much farther apart in the arms and legs.

How far apart must two points be to be distinguished from each other? If two points such as the points of two toothpicks are applied close together on the skin, they will be felt as one point. As the distance between the points is increased, the sensations are eventually recognized as being made by two points. The distance required to distinguish between the two sensations varies in different regions of the body.

Have the children work in pairs. With one member of each pair blindfolded, have the other child move the points of two toothpicks farther and farther apart in the palm of the blindfolded child's hand until he or she can detect the two points. Have the child doing the investigation measure the distance at which the two points can be detected and record it. Repeat for a fingertip, the back of the hand, the forearm, the back of the neck, and the leg.

Have the children in each pair change places and repeat the investigation. How much variation is there between the measurements taken on different parts of the body? How much do the measurements for the various body regions differ among the children?

NUTRITION EDUCATION IN THE CLASSROOM

It is one thing to know how different body systems operate but quite another to know how to keep the body working properly. A healthy body is not merely one that is free from disease; to be healthy we must take care of ourselves so as to maintain proper bodily functioning. One of the keys to healthful living is good nutrition. A good diet in childhood is essential for optimum growth and development and for good health and well-being throughout life. This is reason enough to devote some of our energies to developing successful programs of nutrition education.

Ask the youngsters in your class what they had for breakfast. How many did not have any breakfast? Eating habits are established early in life and, like most other habits, are difficult to change. An objective for working with children in nutrition is to change habits or patterns of behavior. This is almost always difficult, but if we begin early enough in a child's school career we can have some impact on the child's dietary habits. Children's general health and well-being also affect their work in school. If a nutrition education program is to be effective, the cooperation and support of school lunchroom personnel, parents, and parent groups must be enlisted.

Food is both the fuel and the building material of the body. Foods provide the energy to operate our essential organs and all bodily processes. The energy used by muscles to lift a book or pedal a bicycle is also produced by food. Food is also the building material used in growth and repair.

In the study of foods and the energy requirements of the body, energy is usually stated in terms of *calories.* In nutrition the *large Calorie* (or *kilocalorie*) is the unit that is used. A large Calorie is the heat (or energy) required to raise the temperature of one kilogram of water one degree Celsius (A *small calorie* is the energy required to raise the temperature of one gram of water one degree Celsius.)

A certain amount of energy is needed just to maintain essential bodily processes (e.g., the beating of the heart, the expansion and contraction of the lungs, etc.). The energy required for these vital processes is called *basal metabolism.* If the energy expenditure of an individual is measured just after awakening and about 15 hours after eating, the energy that is expended is essentially just that required to operate the body. Among young adults the basal metabolism is about one Calorie per kilogram per hour. To maintain the metabolic processes for 24 hours, a young adult weighing 70 kilograms (154 lbs.) would require about 1,680 calories ($1 \times 70 \times 24$). The basal metabolism varies, depending on age, body weight, and among individuals.

Various physical activities, from pushing a pencil to running up stairs, require energy, and the amount of food needed depends on the amount of activity engaged in. After foods have been eaten and digested by the body, they are taken to body cells, where they are *oxidized* (i.e., essentially "burned" for body use). Scientists have been able to study the amount of energy given off by foods as they are oxidized, since the energy given off by food oxidation in the body is equal to the energy given off when it is oxidized outside the body. Therefore, it is possible to determine the caloric value of all foods. When people try to lose weight by "counting calories," they are essentially keeping track of all that they eat and trying to consume just a little less food than they need to maintain their bodily functions and other energy requirements. If the body uses more calories than it consumes, the difference is provided by stored food or fat. Thus, weight loss occurs. (All dieting to lose weight should be carefully planned with a physician.)

Almost every living thing undergoes growth during some part of its life span. The raw material needed for this growth is derived from the food eaten by the organism. There is continual replacement of old protoplasm by new. It has been estimated that over any seven-year period almost all of the material in our bodies is replaced by new material. When the body is injured, new tissue is formed to repair the injury. The material for the body repair also comes from the food we eat. With growth and replacement both dependent on the foods we eat, it may not be an exaggeration to say, "We are what we eat."

Proteins are the basic food materials utilized in bodily growth and repair. Milk, meat, poultry, fish, and eggs are major sources of protein. Foods from plants such as corn, beans, and wheat also contain protein.

Carbohydrates are foods that contain some form of sugar or starch. They are important for quick energy supply and are a "cheap" source of calories. Strenuous activity, for example, does not necessarily require more protein than is needed for growth and repair, but the body does require a relatively high caloric intake. Carbohydrates are the major fuel for our bodies in that they supply most of the calories in our diets. Among the foods rich in carbohydrates are bread, cake, macaroni products, potatoes, and fruits. *Fats* are another good source of calories. Butter, margarine, lard, and vegetable oils are sources of fat.

Certain mineral elements are also essential for growth. A little more than one kilogram (2.2 lbs.) of the total weight of the body consists of minerals. Calcium and phosphorus are needed for the proper development of bones and teeth as well as for the contraction of muscles. Milk and milk products are the most important sources of calcium. Phosphorus is found in so many foods that sufficient amounts of this mineral are almost always consumed. Iodine is essential for the proper functioning of the thyroid gland, which, in turn, is an important regulator of basal metabolism. Goiter, a condition in which the individual's thyroid gland becomes so enlarged that a large lump hangs from the neck, is due to a deficiency of iodine.

Mineral elements help control the movement of substances within the body. Iron is essential for the transport of oxygen and carbon dioxide in the blood. The movement of digestive foods into the bloodstream is partially controlled by the concentration of certain mineral elements.

Vitamins are also essential to good health. The stories of the discovery of vitamins read like interesting novels. English sailors suffered from scurvy; Japanese sailors suffered from beriberi. As humans limited their intake of fresh foods and began to rely on stored or dried foods, deficiency diseases became apparent. Often chance played as much a part in the discovery and treatment of vitamin deficiency illnesses as legitimate scientific research. It is well known how sailors in the British navy would develop a sallow complexion, tire easily, lose teeth, and feel general pain. When a ship docked, though, and fresh fruits were eaten, the sailors "miraculously" recovered. Now it is known that the sailors suffered from scurvy, a disease caused by a lack of vitamin C. Table 14–1 is a sum-

Table 14–1 The Essential Vitamins

Vitamins	*Important Sources*	*Functions*	*Symptoms of Deficiency*
A	Liver, fish liver oils, green and yellow vegetables, tomatoes, butter, milk, egg yolk	Important for healthy eyes, skin, and mucous membranes and for normal growth	Night blindness, dry skin and eyes, susceptibility to infection
B_1 (thiamine)	Meat (especially liver), fish, fowl, milk, whole-grain cereals, peas, nuts	Promotes normal growth, good appetite, health of nerves, muscles, heart; necessary for proper metabolism of sugar	Beriberi: retarded growth, loss of weight, poor digestion, nervous disorders and fatigue
B_2	Milk, lean meat, fresh vegetables, yeast, egg yolk	Important for growth, health of skin and mouth, sugar metabolism, functioning of eyes	Retarded growth, inflammation of tongue, cracks in lips, premature aging, dimness of vision, intolerance to light
Niacin	Meat, fish, fowl, nuts, tomatoes, potatoes, whole-grain cereals and breads, green leafy vegetables	Important for growth, metabolism of sugar, health of stomach and intestines, health of nervous system	Pellagra: skin eruptions, digestive disturbances, mental disorders
B_{12} (cobalamin)	Green vegetables, liver	Helps control pernicious anemia	Reduction in number of red blood cells
C (ascorbic acid)	Fresh fruits (especially citrus) and green vegetables, tomatoes	Important for growth, strength of blood vessels, development of teeth, good skin, healing	Scurvy: Sore gums, hemorrhages around bones, tendency to bruise easily
D (calciferol)	Liver, fish liver oils, irradiated foods, fortified milk, eggs; also produced in body when sunlight strikes skin directly	Needed for building and maintaining bones and teeth, growth, regulation of body's use of calcium and phosphorus	Rickets: soft bones and poor teeth
E (tocopherol)	Wheat germ, green leafy vegetables, milk, butter	Prevents sterility in rats; effects in human uncertain	Undetermined
K (phylloquinone)	Green vegetables, tomatoes	Important for clotting of blood and normal liver function	Hemorrhages

mary of the essential vitamins, together with their sources and importance.

Water, while not a nutrient, is essential for life. We can live without food for several weeks, but we cannot go without water for more than a few days. Lack of water leads to fatigue. Complete deprivation of water leads to death. More than half of the body's weight is water. Water is also used to remove nitrogenous wastes from the kidneys. As water evaporates on skin surfaces after perspiration, it cools the body. This is an essential aspect of body temperature control.

The Basic Food Groups

A well-planned diet can help maintain good health. The human body must get about fifty different substances from food. Among these essentials are fats, proteins, and carbohydrates. These foods supply the energy and substances needed for growth and repair of the body. In addition, the body needs a variety of vitamins and minerals.

Variety is also a key to good nutrition. However, the body does have basic requirements, and different foods vary in their nutritional contribution to a diet. A plan based on the following basic food groups has been suggested by the U.S. Department of Agriculture and can serve as the foundation of a good diet:

1. *Milk, cheese,* or *ice cream* (three to four cups of milk a day for children) should be eaten each day. Without milk, it is difficult for the body to receive the calcium that it needs. Therefore, it is especially important that children drink ample quantities of milk. Milk also provides other minerals, vitamins, and high-quality proteins that supply energy for active young bodies.
2. Two or more servings of *beef, veal, pork, lamb, poultry, fish,* or *eggs* should be eaten each day. These foods are the main source of high-quality protein. They also supply some minerals and vitamins.
3. Four or more servings of *dark green* and *deep yellow vegetables, citrus fruit,* and *other fruits and vegetables* should be eaten each day. These foods are the major sources of vitamins. Almost all vitamin C comes from fruits and vegetables. Since the body does not store vitamin C, it is important that some good source of vitamin C, such as citrus fruit, be eaten every day. American diets are most often weak in this group of foods, and a special attempt should be made to encourage children to eat vegetables and fruit.
4. Four or more servings of *bread, breakfast cereals, wheat, rice,* or *oats* should be eaten each day. The cereal foods, such as wheat, wheat germ, rice, and soybean products, are the most economical sources of energy. Wheat, corn, or rice serves as the base for diets in many parts of the world. The cereal foods are also important for protein, iron, and several of the B vitamins. In addition, most of the cereal foods sold in the United States have been enriched with vitamin supplements and other nutrients.

These four food groups provide a framework for planning diets. Additional food is needed, but if the suggested amounts from each of the food groups are eaten, sufficient quantities of the essential nutrients will be taken into the body.

Studying Foods and Nutrition

The study of nutrition may begin as children are helped to find the answers to their own questions. These questions may emerge into central themes which naturally lead to activities and investigations. Children may begin with questions like these:

Where does our food come from?
How and where are our foods grown?
What animal meats do we eat?
How do we get our foods?
Why are different foods important?

Each group of children will undoubtedly ask questions like these as well as some unique ones. In helping children to answer their questions, a variety of resources may be tapped. Books, charts, films and filmstrips should be used. Trips and excursions to the grocery store, bakery, supermarket, or food organizations may be appropriate. Food-processing companies may conduct tours or send resource people to the schools. school menus may be analyzed and discussed with lunchroom personnel.

Children (and teacher) should summarize and evaluate their investigations to make sure the original questions have been answered adequately. The results of the children's findings may be presented in reports, tape/slide shows, assembly programs, or colloquia.

What foods do we eat? Since one of the most effective ways of improving diets is to help children become aware of what they are and are not eating, a survey approach may be used to find out what foods the children eat. As this study is carried out by the children, they should become aware of some of the pitfalls in this kind of research. "Are the records that are kept accurate?" "Are the records that are kept representative of the foods that are really eaten?" "What do the records mean?"

For the teacher, the results of a diet survey may be suggestive of matters that should be stressed in teaching. For example, if the survey shows that many youngsters in a class do not eat enough green and yellow vegetables, additional attention may be given to the contributions of these foods to the diet. The diet survey procedures described here do not give a complete picture of any individual child's pattern of eating, but they do give an indication of the general patterns of eating to be found in a group.

In the diet survey a record is kept of the foods eaten over a period of three or more days. If possible, the period should include at least one day that is a holiday.

In the upper elementary grades, the children should keep their own records. In the morning they can be asked to record the food that they ate after school the day before, at dinner, and for breakfast. In the afternoon they can record foods eaten at lunch and for snacks.

If the study is carried out with children in the lower grades, the information will probably have to be obtained through interviews. One approach is to interview every third child listed on the class rolls.

To get information about the amount of food consumed, some attention will have to be given to common measures. The children can be shown what is meant by one liter, one tablespoon or teaspoon, one slice, and so on.

An effective way for the children to analyze their diet records is in terms of the four basic food groups. How many children ate sufficient amounts of each of the four groups each day? In which food groups did the children tend to be deficient?

How can children learn to enjoy a variety of foods? Variety is one of the keys to a balanced diet, and children who regularly eat a wide variety of foods are likely to be including all the essential nutrients. Some children, though, are "picky eaters" and tend to limit their intake of foods to those that they like or are most familiar with.

Children ought to be encouraged to taste new and unusual foods. Their diets can be analyzed to find out whether they have consistently selected foods from only one or two of the four basic food groups. By setting up classroom snack bars with foods planned and prepared by the youngsters, children can expand the list of foods they eat.

When children study foreign lands, it is often effective to have them sample some of the foods from those places. Also, it is often helpful to investigate the ingredients of various prepared foods. For example, some children are more likely to eat a chef's salad if they know what goes into it.

Lectures, demonstrations, and discussion groups involving teachers, parents, school administrators, lunchroom personnel, and children can contribute greatly to a nutrition education program.

RELATED ACTIVITIES

Learning body parts, p. 95.
Self-exploration, p. 108.
How do parts of organisms vary? p. 220.
How many paper clips tall are you? p. 154.
What's inside? p. 82.
The eye color of children in a class. p. 128.

Are there variations within a species? p. 245.
Identifying and locating major organs in a manikin, p. 54.
Rhythmic counting: The human pulse clock, p. 133.
A camera and the eye are alike in many ways, p. 408.
Introducing basic taste words, p. 95.
Identifying the sweeter of two drinks, p. 86.
Sorting liquids by taste, p. 88.
Sorting by odor, p. 87.
Smelling objects, p. 81.
How reliable is our sense of temperature? p. 396.
Working with tactile sensations, p. 96.
An orange is an orange, p. 82.
A tasting party, p. 81.
Where does our food come from? p. 198.
How are foods preserved and stored? p. 198
Selecting foods that will make up a nutritionally sound diet, p. 54.

REFERENCES

For Teaching

ASIMOV, ISAAC. *The Human Body.* New York: Houghton-Mifflin, 1963. An interesting account of *Homo sapiens,* its genealogy, and the operation of the human body.

JACOBSON, WILLARD ET AL. *The Human Organism.* New York: American Book – Van Nostrand, 1969. A factual and concise study of the human organism's growth and well-being.

KAUFMAN, JOE. *How We Are Born, How We Grow, How Our Bodies Work and How We Learn.* New York: Golden Press, 1975. Essentially a book for children, this account can be used as a teacher's reference. This comprehensive book is about the human body and its functions, with easy activities for the classroom.

For Young Children:

GANS, ROMA AND FRANKLYN M. BRANLEY, EDS. *Let's Read-and-Find-Out Science Books: The Human Body.* New York: Harper & Row. This series includes useful and interesting information for young children. Titles in the series include the following:

A Baby Starts to Grow
Before You Were a Baby
A Drop of Blood
Find Out by Touching

Follow Your Nose
Hear Your Heart
How Many Teeth
How You Talk
In the Night
Look at Your Eyes
My Five Senses
My Hands
The Skeleton Inside You
Sleep Is for Everyone
Straight Hair, Curly Hair
Use Your Brain
What Happens to a Hamburger
Your Skin and Mine

SIMON, SEYMOUR. *Finding Out with Your Senses.* Ill. by Emily McCully. New York: McGraw-Hill, 1971. A book of activities for children about the five senses.

STEIN, SARA BONNETT. *Making Babies.* New York: Walker, 1974. An accurate account with many fine photographs dealing with major concepts of human reproduction. A separate text for adults provides suggestions for discussing this material with children.

For Older Children

ALLISON, LINDA. *Blood and Guts: A Working Guide to Your Own Insides.* Boston, Mass.: Little, Brown, 1976. An intriguing book about the human body with related experiments and projects.

ASIMOV, ISAAC. *How Did We Find Out About Vitamins?* Ill. by David Wool. New York: Walker, 1974. An interesting collection of stories about the discovery of vitamins and deficiency diseases. Appropriate for children in upper elementary and middle-school grades.

GALLANT, ROY A. *Me and My Bones.* Garden City, N.Y.: Doubleday, 1971. A book in words and photographs dealing with the skeletal system and its operation.

JONES, HETTIE. *How to Eat Your ABC's: A Book About Vitamins.* New York: Four Winds Press, 1976. A collection of interesting stories about vitamins and vitamin deficiencies. Nutritional planning and good eating habits are explored.

MCGOMEN, ANN. *The Question and Answer Book About the Human Body.* Ill. by Lorelle M. Raboni. New York: Random House, 1965. Detailed information regarding the human body is provided through a series of questions and answers. Well illustrated.

ROGERS, FRED B. *Your Body Is Wonderfully Made.* New York: G. P. Putnam's Sons, 1974. A short book with easy text that concisely summarizes the major systems of the human body. Contains several helpful and accurate illustrations.

Children and the Sun, Moon, and Stars

CHAPTER 15

A group of fourth-graders were viewing the night sky. Several parents were also peering at the sky as we pointed out some of the features that could be seen. We were hampered by the bright lights of the city, but still there was much to be seen.

In the west there was a very bright light. The children had checked the star map that afternoon. "That's Venus. It's a planet and one of the brightest objects in the sky. Sometimes, it is seen in the evening soon after sunset. At other times it is seen in the morning before sunrise." One of our children told her father, "It is sometimes called the morning star or the evening star. But it is not a star. It is a planet!"

To the north we saw the Big Dipper. We asked the children to find the two stars opposite the handle. "These are the 'pointers.' Draw an imaginary line five times the distance between the pointers. See that star? It's Polaris—the north star!" This group of children had become keenly interested in astronomy. One of the boys remarked, "Polaris is supposed to be at the end of the handle of the Little Dipper. But I can't see it. I guess the lights of the city are too bright."

It was winter, and we saw the belt of the winter constellation Orion. We noted the reddish star with the wonderful name Betelgeuse. Near Orion we saw the brightest star, Sirius. Then we looked to the east and watched as the moon rose. "It's so red!" "What makes it so red?" "It's moving so fast! I didn't know it moved so fast."

Then, the unexpected. One of the girls saw a bright object—it looked like a star—moving across the sky. We watched it. It moved quite rapidly against the background stars. It seemed to be quite low. It was a manmade satellite reflecting sunlight as it moved on its orbit around the earth.

Children should have a chance to see the stars, moon, sun, constellations, and artificial satellites. We have always found children to be keenly interested. They wonder. They raise questions. As teachers, we can help them explore these questions. But we should also do everything we can to see that they have the firsthand experience of seeing something that they will continue to find out about throughout their lives.

THE SOLAR SYSTEM

Our solar system consists of the sun, the planets that revolve around it, their associated satellites, asteroids, meteors, comets, and the vast space between these objects. The sun at the center of our solar system is a star much like the ones seen on a clear night. The sun dominates our sky because of its closeness, but it is quite an ordinary star in terms of size and brightness.

The Sun

Our sun is a medium-sized star. Its diameter is approximately 1.4 million km, (870 thousand mi) nearly 110 times that of the earth. The volume of the sun is so great that more than 1 million earths could fit inside of it. If a circle representing the sun were 30 cm (11.8 in) in diameter, then a proportionately sized earth would be only 2.7 mm (.1 in) in diameter, barely the width of a chalk dot. Some other stars (e.g., Betelgeuse and Antares) have diameters that are 200 or 300 times as large as our sun's. There are also millions of stars that are considerably smaller than our sun.

The sun is composed of hot, glowing materials. About 92 percent of the sun's mass is hydrogen and about 7 percent is helium. This has been determined by spectroscopic observations. A *spectroscope* is an instrument in which light passes through a narrow slit onto a prism, which breaks up the light into separate parts and colors. By comparing the spectrum of the sun with the spectra of various chemical elements when they are heated, the component elements of the sun (and other stars) may be determined.

When the sun is viewed with properly equipped instruments (never with the naked eye, because the eyes can be seriously harmed by its very bright light), it is apparent that the bright disk of the sun is not uniform. There exist dark spots of varying size called *sunspots.* These dark spots are not devoid of heat and light but appear dark when compared with surrounding parts of the sun. Sunspots appear periodically and seem to occur in 11-year cycles. It has been theorized that sunspots are due to large-scale magnetic disturbances within the sun.

The gases in the sun give off tremendous amounts of heat and light. But this is not due to the burning of the sun's gases. Actually, the sun's energy output is a result of massive atomic explosions within the sun. In these reactions atoms of hydrogen combine to form helium, resulting in the release of enormous quantities of energy. These conversions are in compliance with Einstein's famous equation:

$$E = mc^2$$
$$\text{Energy} = \text{mass} \times (\text{speed of light})^2$$

The speed of light is a very large number. Therefore, in such nuclear reactions small masses of matter are transformed into relatively enormous amounts of energy. Actually, the sun's mass is consumed at the rate of nearly 5 billion kilograms (11 billion pounds) per second! But since the sun's mass is so large, it will continue to emanate heat and light in much the same way for hundreds of millions of years.

The type of reaction that takes place in the sun is called *nuclear fusion;* this is also the type of reaction that takes place in a hydrogen bomb. The controlling of fusion reactions and harnessing of the resulting energy hold great promise as possible sources of energy for our everyday needs on earth.

The earth actually receives only a small fraction of the sun's light and heat. The energy is released in all directions and mostly passes into the vast empty spaces of our solar system and beyond. Yet if it were not for the small fraction of the sun's light that we *do* receive, life as we know it could not exist.

The sun is the ultimate source of most of the energy we use. Nearly every form of energy can be traced to the sun either directly or indirectly. As we work we expend energy, which must be replaced by the foods we eat. Green plants synthesize food from the sun's energy. We eat plant products and meats from animals. These animals were dependent on plants and grains. The sun's energy evaporates water from oceans; clouds are formed; and when water from these clouds falls as precipitation it runs off in streams and rivers. Turbines placed in the path of running water can convert some of this energy into electricity. The fossil fuels we use were originally the remains of organisms that once depended on the sun for their existence. All of these forms of energy are derived from solar energy.

Do all objects absorb the sun's heat equally? Materials on the earth become differentially heated when exposed to the sun. Children may recall walking barefoot on different surfaces in the summer. How does it feel to walk on soil, grass, concrete, or sand on a hot, sunny day?

Children can determine the effects of the heating of several materials on a school playground on a sunny day. Are metal objects warmer than wooden ones? What are the apparent differences as stone, brick, concrete, metal, grass, and so forth are heated?

Older children can obtain more precise data if they place samples of several materials on a sunny windowsill for a few hours and then touch a thermometer to each of the materials. Can any generalizations be made? Objects of the same material but different colors may also be investigated to see whether they absorb heat equally.

What is the warmest time of the day? Children may make a record of hourly temperature readings, using a thermometer located within view of a classroom window. The thermometer should not be in direct sunlight. Repeat the procedure for several days. Have the children examine their records. What is the warmest time of day? What might account for this?

Examining light from the sun. Sunlight can be examined by using a prism in much the same way that sunlight and starlight are examined with a spectroscope.

Have a beam of sunlight pass through a thin slit in an index card. Hold a glass prism so that the light coming through the slit passes through the prism. (See Figure 15–1.) By turning the prism you can produce a spectrum that may be projected onto a wall or a paper screen. What colors are in the spectrum of sunlight? By adjusting the distance from the prism to the wall or paper, the colors may be sharpened. What is the width of the total spectrum? What is the width of each band of color?

Do green plants need sun? Obtain two small green plants of the same kind. Place one on a sunny windowsill and the other in a dark closet. Water both plants two or three times per week. Have the children *predict* what changes they think will occur. After two or three weeks of periodic examination, have the children note as many differences between the two plants as they can. What may have accounted for the differences?

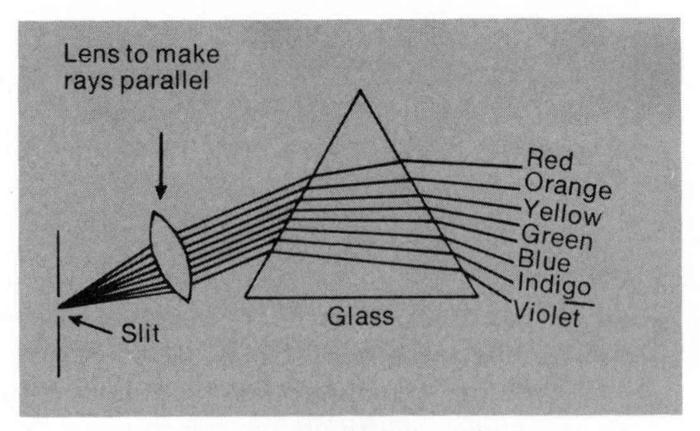

Figure 15–1 White light is made up of light of many colors.

How does the length of shadows change in the course of the day? Obtain a tall, slim object such as a thick pencil, a statue, or a stick. Find a place where sunshine can be seen for most of the school day. Place the object in the sun early in the morning. With the object on a piece of paper, trace its shadow on the paper. Measure the length of the shadow as projected on the paper. Return to the site at one- or two-hour intervals, tracing and measuring the shadow each time. Also note the sun's apparent position relative to where the children are standing. How does the direction of the shadow change? When was the shadow longest? shortest? What relationship seems to exist between the length of the shadow and the position of the sun in the sky?

The Planets

There are nine known planets in our solar system. These planets revolve around the sun in elliptical orbits. All of the planets can be seen with telescopes, and several of them can be seen with the unaided eye. The term *planet* comes from Greek and means "wanderer." This name arose because planets' "wandering" movements are distinctly different from the movements of stars. (See Figure 15–2.)

Mercury is the closest planet to the sun. Its revolution around the sun follows the shortest path of any planet and takes only 88 days. (See Table 15–1.) It rotates on its axis once every 59 Earth days. Since it is so close to the sun, Mercury is extremely hot, certainly too hot to support life as we know it.

Venus, sometimes considered a "sister planet" of Earth, is very similar in size to Earth. It is the brightest object in the sky aside from the sun and our moon. Since Venus is fairly close to the sun, it is seen in the west after sunset and in the east just before sunrise. Space probes have provided us with much recent information about Venus. The surface temperature of Venus is extremely hot, so hot that any form of life that requires liquid water could not exist. The atmosphere around Venus is

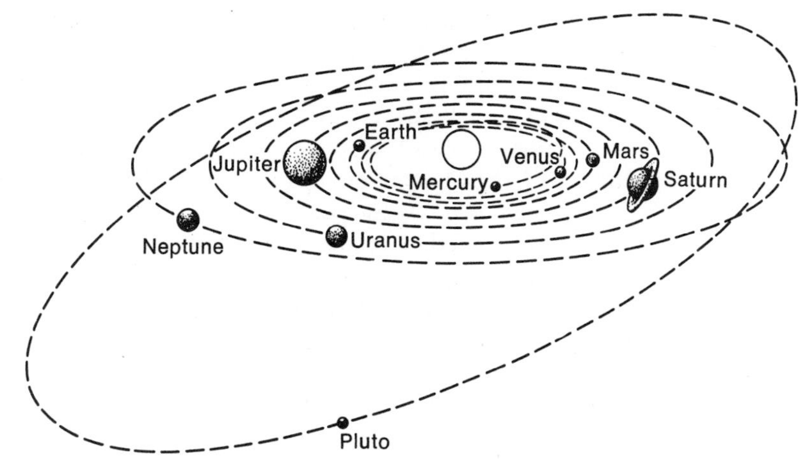

Figure 15–2 A modern model of the solar system showing the planets and their orbits.

mostly carbon dioxide, with minuscule amounts of water vapor. It takes Venus 225 Earth days to revolve around the sun; its rotation period is relatively long, about 244 days.

Earth is our home. It is the third planet from the sun and receives just enough solar energy to make life possible. The atmosphere enveloping Earth is composed largely of nitrogen, oxygen, and small amounts of water vapor. The Earth revolves around the sun once every 365 days (i.e., an Earth year) and rotates on its axis once every 24 hours. Earth has one natural satellite.

Mars has been the subject of much speculation in science and science fiction alike. It is not as close to us as Venus, but it has little atmosphere to obscure our view of its surface. Mars usually appears as a red

Table 15–1 Facts about the Planets (as of 1979)

Planet	*Diameter (in km)*	*Distance from Sun (in millions of km)*	*Time it takes to revolve around Sun.*	*Time it takes to rotate on axis*	*Number of Moons*
Mercury	4,827	58	88 days	59 days	0
Venus	12,228	108	225 days	244 days	0
Earth	12,872	150	365 days	1 day	1
Mars	6,919	227	687 days	1 day	2
Jupiter	139,983	777	12 years	10 hours	13
Saturn	115,848	1,426	29 years	10 hours	10
Uranus	56,315	2,867	84 years	11 hours	5
Neptune	51,488	4,494	165 years	16 hours	2
Pluto	3,000(?)	5,900	248 years	6 days	?

planet, but has white polar caps that extend and recede with the seasons. The surface of Mars has craters and valleys whose origin is still unclear. A Martian day is almost identical to our own, being 24 hours and 37 minutes long. We have gathered a great deal of information about Mars from space probes that have been landed on its surface. Compared to the Earth, the Martian environment is a harsh one for life.

Jupiter, the largest planet in the solar system, has a cloudy atmosphere composed of methane and ammonia. When seen through a telescope, Jupiter appears to have a deep red spot that has puzzled astronomers for a long time. Jupiter is an extremely cold planet because of its great distance from the sun. It rotates on its axis with great speed, making one turn in about 10 hours. This causes the planet to bulge at the equator and to be flattened at the poles. (The same is true of Earth, but this effect is much more pronounced in the case of Jupiter.)

Saturn has fascinated youngsters and adults alike because of its famous rings. The rings extend out from the planet nearly 140,000 km (86,800 mi.) but are only about 15 km (9 mi) thick. They are thought to be composed of frozen particles of ammonia and carbon dioxide. Saturn has 10 moons, one of which seems to have an atmosphere.

Uranus is a dark planet compared to those described previously. Discovered in 1781 by William Herschel, Uranus was long thought to be a star. One of the unique features of Uranus is that it rotates almost at right angles to the plane in which the planets revolve around the sun.

Neptune was discovered in 1846 as a result of observed deviations in the expected path of Uranus. Apparently, these deviations are due to the gravitational attraction between the two planets.

Pluto was discovered in 1930 by Clyde Tombaugh. It is a relatively small planet, being considerably smaller than Earth, and is so far from the sun that the sun would appear to be nothing more than a bright star if we were standing on Pluto.

Asteroids are relatively small planetary bodies revolving around the sun between Mars and Jupiter. They number in the thousands and range in diameter from about 1 to 800 km (½ to 500 mi). It is believed that the asteroids were originally part of a small planet (or planets) that was broken up into bits, perhaps by a collision.

Comets are composed of small particles of matter that travel in extremely elliptical paths around the sun. The head, the most prominent part of a comet, is followed by a long tail composed of such widely spaced particles that stars can be seen through it. The most famous comet, Halley's comet, will be seen again in 1986.

Meteors are particles of iron, stone, or combinations of the two that travel through space and occasionally enter the earth's atmosphere. Commonly referred to as "shooting stars," almost all of the meteors that enter the atmosphere burn up as a result of friction. Those that do strike the earth's surface are called *meteorites.* The largest meteorite in North America fell in Quebec Province and its impact crater is over 3 km in diameter.

Making a model of the solar system. Constructing models of the solar system is a favorite activity of elementary school children. Using the data provided in Table 15.1, children can build scale models of the planets. This will help them conceptualize the relative sizes of the planets. Models may be made from clay or styrofoam balls or carved from soap and similar materials. An appropriate scale might be 1 cm = 10,000 km (1 in = 4000 mi). Thus, the earth's diameter would be approximately 1.3 cm, (½ in) or about the size of a large marble. If the sun is to be included, it will be extremely large compared to the planets, but this may help the children realize its enormity. If a schoolyard or field is available, the relative distances between the sun and planets may be measured. This will help the children realize that our solar system consists primarily of empty space.

The following is a scale model that one group of youngsters developed to show both planet sizes and distances:

Sun	Slightly larger than a basketball
Mercury	Grass seed 83 feet (25 meters) from sun
Venus	Bean 155 feet (47 meters) from sun
Earth	Marble 215 feet (65 meters) from sun
Mars	Small pea 330 feet (99 meters) from sun
Jupiter	Small grapefruit at a distance of 1/5 mile (317 meters) from sun
Saturn	Apple 2/5 mile (634 meters) from sun
Uranus	Large marble 4/5 mile (1.3 kilometers) from sun
Pluto	Small pea 1-3/5 miles (2.5 kilometers) from sun

Day and Night

The rotation of the earth on its axis causes day and night. In ancient times it was believed that the earth was the center of the universe and that all celestial bodies, including the sun, moved around the earth. This explanation, though a faulty one, was convenient in that it accounted for many of the observations made by the ancients.

Today of course, we know that we are not stationary. Aside from our revolution around the sun and the movement of our entire solar system through space, the earth spins on its axis, and this accounts for day and night. One side is in the sunlight while the other is away from it. The earth rotates from west to east. If an observer could look down on the earth from the north star (Polaris), it would appear to be spinning in a counterclockwise direction. The stars are stationary relative to each other, but because the earth is rotating they appear to be moving relative to the earth. The sun also seems to move across the sky as it rises in the east and sets in the west, but we are actually moving relative to the sun.

The spinning of the earth probably accounts for the fact that the earth is flattened at the poles and bulges at the equator. It is theorized that when the earth was in a more fluid form the forces exerted on it as it rotated caused this slight distortion in shape.

The Seasons

The revolution of the earth around the sun and the tilt of its axis account for the seasons. The warmth that we experience in summer is not due to the earth's being any closer to the sun than it is in the winter. Rather, it is due primarily to the directness with which the sun's rays strike us. The earth's orbit around the sun is elliptical; in fact, in the Northern Hemisphere we are closer to the sun in the winter than we are in the summer. Obviously, this cannot account for seasonal differences.

The earth's *axis* (i.e., the imaginary line from pole to pole around which it spins) is inclined about 23.5 degrees from a vertical plane. (A vertical plane would be perpendicular to the plane of the earth's orbit about the sun.)

The earth revolves around the sun once every 365¼ days. On December 21, the *winter solstice,* the North Pole is pointing away from the sun. It is winter in the Northern Hemisphere and summer in the Southern Hemisphere. (See Figure 15–3.) By March 21, the *vernal equinox* (spring), the earth has moved one-fourth of its orbit. On June 21, the *summer solstice,* the North Pole is pointing in the direction of the sun and the Northern Hemisphere has its summer and the Southern Hemisphere its winter. By September 23, the *autumnal equinox,* the earth has

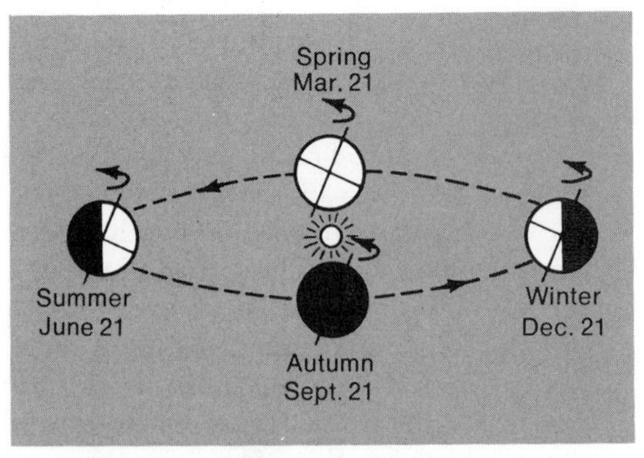

Figure 15–3 The earth's axis always tilts the same way as it revolves around the sun.

Figure 15–4 The direct rays of the sun (a) give more heat than slanted rays, (b) because they are not spread over such a large area.

completed three-fourths of its annual orbit. Then the North Pole again begins to point away from the sun.

The basic cause of the seasons is the change in the amount of solar energy received by any given area on the earth. When the sun appears high in the sky and not low (toward the horizon), the sun's rays are most direct. A comparatively large amount of energy strikes a given area. When the sun's rays strike the surface at an oblique angle, the solar energy is spread out over a larger area than when the incoming sunlight is perpendicular to the surface. (See Figure 15–4.) A similar effect can be observed when a flashlight is beamed directly onto a surface. The light covers a relatively small area. This is similar to the sun's effect in the summer (in the Northern Hemisphere). When the flashlight is shined at an oblique angle to the surface, the same amount of light is spread out over a larger area. This is the case during our winter. Each unit of surface area receives less solar energy than in the summer. This effect (i.e., the directness of the sun's rays), combined with the fact that there are more hours of sunlight during summer days and fewer hours of sunlight in the winter, is the basic reason for the seasons.

Demonstrating day and night in the classroom. Using a globe and a flashlight or the beam of a projector, shine the light onto the globe. With the classroom lights dimmed, it should be apparent that only half of the earth receives the sun's light (i.e., the flashlight or projector beam) at any given moment. Children can follow a specific city as the globe is slowly turned on its axis and determine when the city would be having day or night. They should be able to understand why the sun cannot be seen at night.

Explaining the seasons. The seasons are caused by the earth's annual revolution around the sun and the tilt of the earth's axis. The effect of these two factors can be demonstrated by carrying a globe (with an inclined axis) around a light representing the sun in a darkened room.

A bulb and socket should be set up in the middle of a large table or on the classroom floor. Prop up the light with books so that its center is at the same level as the center of the globe. The bulb should not be shaded, so that it will shine in all directions. As the globe is carried around the light, stops should be made at points representing each of the four seasons. For example, when the North Pole of the globe is pointing away from the sun it is winter; when the North Pole is pointing toward the sun it is summer. (See Figure 15–3.) Between these two positions are spring and fall. At each of the stops, the children should consider such questions as the following:

Where would the sun appear to be during this season if one were standing at the North Pole?

Where would the sun appear to be if one were standing at the South Pole?

Where would the sun appear to be if one were at "home"?

Would the sun's rays at "home" be slanted or direct?

Where on the earth's surface would the sun be seen directly overhead at noon?

Where would you look to see the sun at noon if you were at the Tropic of Cancer or the Tropic of Capricorn?

Demonstrating the effects of the angle of the sun's rays. Remove the transparency from a cardboard projection slide frame and replace it with wire screening such as window or door screening. Place the slide of wire screening into a projector and focus it on a chalkboard. Have a child outline the projected squares with chalk. Are all of the squares about equal in area? Now project the slide of wire screening onto a globe. What happens to the shape and area of the projected squares? How can this be used to explain the difference in solar energy received at different places at different times of the year?

The Moon

The moon, which revolves around the earth, is our natural satellite. The distinction "natural satellite" has become necessary in the past 25 years, since there are now many manmade satellites circling the earth in varied orbits. The diameter of our moon is roughly one-fourth that of the earth. The distance from the earth to the moon varies (because of the moon's elliptical path around the earth), but the average distance is about 380,000 km (236,000 mi).

The moon lacks an atmosphere. There are no clouds around it and thus no precipitation. The moon therefore is extremely hot during the day (hotter than the boiling point of water) and extremely cold at night, since there is no atmosphere to absorb the daytime heat. Such extremes would make lunar life impossible for humans without protection. Moon missions and probes have not revealed any evidences of life on the moon.

When the moon is observed with the unaided eye, dark patches, called *maria*, can be seen. With binoculars or a small telescope, *craters* and their light-colored *rays* can be seen. Some of the moon's surface is fine and powdery. Some moon rocks have been found to be over 4.5 billion years old. In the highlands, *anorthosite* is most common. In the lowlands and plains, *basalt* is the most common rock. The moon has mountain ranges, and the highest mountains in those ranges are higher than the highest mountains on earth. Moon craters are believed to be formed by the impact of meteoroids or by exploding volcanoes.

As the moon revolves around the earth, it rotates on its axis. This occurs in such a way that as we view the moon from the earth we always see the same side facing us. It is somewhat like walking around another

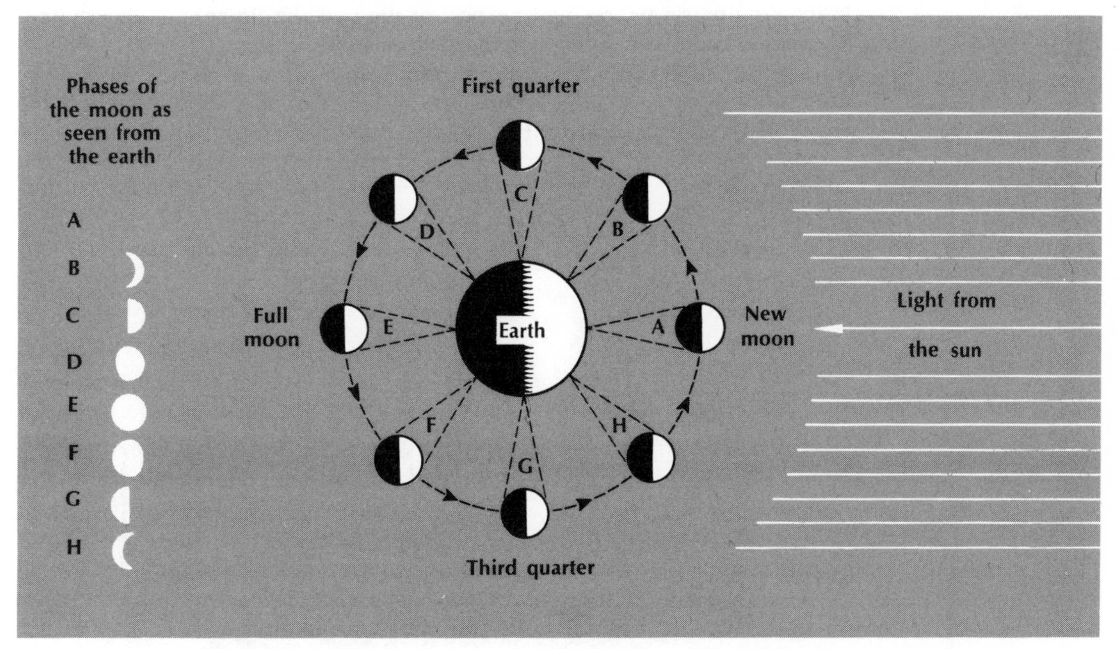

Figure 15–5 Phases of the moon. The diagram at the right shows the actual illumination of the moon at all phases. The lines drawn from the earth to the moon show the part of the moon that can be seen from the earth. The sketch at the left shows how the moon looks from the earth.

person while always keeping your face in the direction of that person. In the process of walking around the person once, you will have rotated once on your axis. In earth terms, the moon's "day" and "year" are of nearly equal duration, each lasting about twenty-eight days.

The moon appears in different shapes as it revolves around the earth. (See Figure 15–5.) When the moon is between the earth and the sun, the side we cannot see is illuminated and the side facing us cannot be seen because it receives no light. This is the new-moon phase. Soon a tiny sliver of the moon can be seen, and in about seven days, one-quarter of the moon will be illuminated. At this phase, called the first quarter, half of the illuminated side of the moon can be seen. In another seven days the full moon (all of the illuminated side of the moon) is visible. In another seven days the moon will be at its last quarter, when half of the illuminated side is again visible. Finally only a thin sliver of a moon can be seen, and then the moon becomes invisible again—the new moon.

Explaining the phases of the moon. Place a projector at one side of the classroom; its beam is to represent the sun. Obtain a ball (about the size of a volleyball) and paint it with white or silver paint. (See Figure 15–6.) Group some children in the center of the room. They will represent the earth in this demonstration. The ball should be held at the same height as the projector beam. Start with the ball between the children and the projector. This is the new-moon phase, and no part of the ball should appear illuminated. The ball should be moved around the children counterclockwise (as viewed from above). When do the children representing the earth first see a little light reflected from the ball? Have the "moon" stop at each quarter and ask the children to report how much of the illuminated ball they see.

Figure 15–6 Children can demonstrate how the revolution of the moon around the earth leads to our seeing phases of the moon.

Gravitation

All bodies in the universe are attracted to one another by a force called gravitation. It is this force that makes an apple fall to the earth and holds the moon and the planets in their orbits.

In the seventeenth century Sir Isaac Newton described how the force of gravitation between two objects could be determined. The magnitude of the gravitational force depends on the masses of the objects. The larger the masses, the greater the force of attraction. The force of gravitation varies inversely as the square of the distance between the objects; that is, if the distance between the two objects is doubled, the gravitational force between them will be reduced to one-fourth.

Newton's law of gravitation is expressed in equation form as follows:

$$F = \frac{G \times m_1 \times m_2}{d^2}$$

where F is the force of gravitation; m_1 and m_2 are the masses of the objects; d^2 is the distance between the objects, squared; and G is the gravitational constant used to keep the units of measure in the equation the same. Newton's law is of tremendous value in that it can be used to explain observations and known facts.

It might seem from the preceding discussion that the planets and the sun, which are mutually attracted to one another, ought to move directly toward one another and collide. This is not the case because of the effects of another law explaining the motion of objects. This law, called the law of *inertia,* states that *bodies in motion tend to stay in motion in a straight line, and bodies at rest tend to stay at rest, unless they are acted upon by other forces.* Of course, a ball rolling down a street may eventually stop because of the friction between the street's surface and the ball, but this holds true in ideal or frictionless situations. When an automobile is accelerated rapidly, you feel the back of the seat pushing against your back, since your body tends to stay at rest. However, if a fast-moving automobile is braked suddenly, your body tends to continue moving forward.

The laws of gravitation and inertia explain why planets do not fly off into space but maintain nearly circular orbits around larger heavenly bodies. The earth, which is moving rapidly through space, ought to move at a constant speed in a straight line, as the law of inertia predicts. But because of the mutual gravitational force between the earth and the sun, the earth is also pulled toward the sun. It is the resolution, or net result, of these two forces that keeps the planets in their nearly circular paths. These laws of motion also explain why the moon (as well as man-made satellites) revolves around the earth and why some stars (of comparable mass) revolve around a common point.

How can the force of gravitation be explored? Have the children each drop an object to the floor. Ask them why their objects stopped falling. If all obstacles in the object's path were removed, how far would they fall? Use a globe to explain the direction of the gravitational force. (Objects fall toward the center of the earth.)

Obtain a bathroom or spring scale. How does such a device measure the amount of gravitational pull on objects? What is the gravitational force acting on each child? How is the operation of a scale different from that of a balance that measures mass?

As objects are brought higher and higher above the surface of the earth, their weight is reduced. Why is this so? Why would we weigh less on the moon than on the earth? If an astronaut took a balance to the moon with him to check his mass, would it be more, less, or the same as his earth mass? Why?

How can the forces acting on the earth as it revolves around the sun be demonstrated? Tie a rubber stopper to a string. Have a child swing the rubber stopper around in a continuous circle in a vertical plane. (The child represents the sun and the stopper represents the earth.) What forces are acting on the rubber stopper to keep it in orbit? Ask the child to let go of the string when the stopper reaches the top of one of its orbits. (Make sure the observers are not in a dangerous position.) What path does the stopper follow? Which force acting on the earth was removed when the string was let go? Which force took over? How does this demonstration help explain the laws of gravitation and inertia?

Eclipses

Eclipses occur when the sun, moon, and earth are positioned in a straight line. When the moon passes between the sun and the earth so that its shadow falls on part of the earth, that part of the earth's surface has a *solar eclipse.* On the regions of the earth having a solar eclipse, part or all of the sun cannot be seen because the moon is between those regions and the sun. (See Figure 15–7.) A *lunar eclipse* occurs when the earth passes between the moon and the sun. Here we see the earth's shadow as it is cast upon the moon. During a lunar eclipse anyone on the surface of the moon would be unable to see the sun because the earth would be between the moon and the sun.

Basically, shadows, and therefore eclipses too, occur because light travels in a straight line. The shadow of an object has some resemblance to the shape of the object casting the shadow. An interesting observation for children is to note the edge of the earth's shadow as it moves across the moon. The edge of the shadow is curved, and this is part of the evidence that the earth is round.

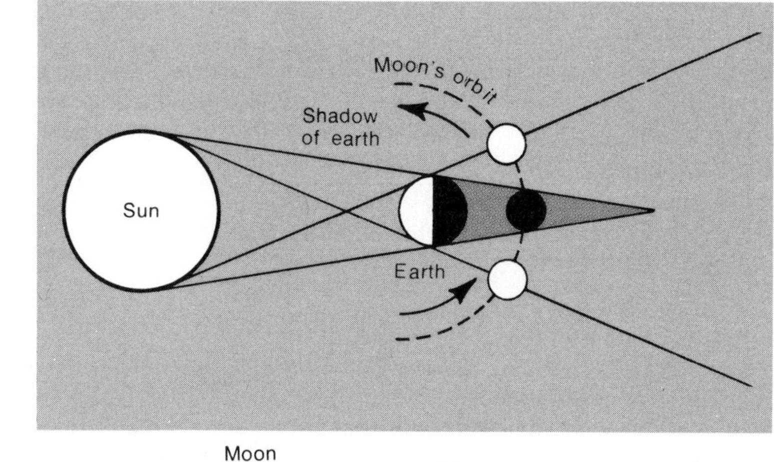

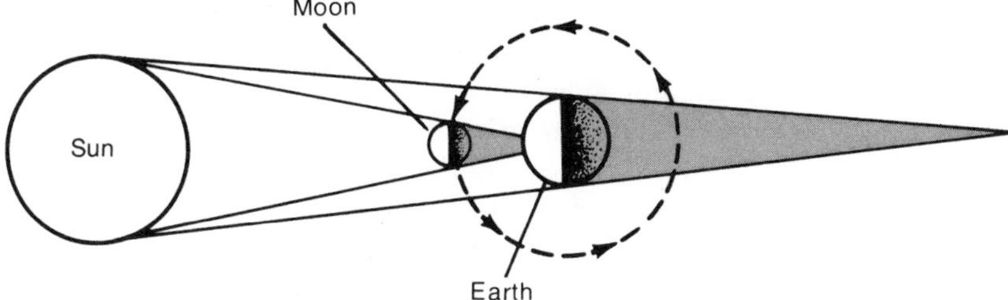

Figure 15–7 A lunar eclipse (top) occurs when the moon moves into the shadow of the earth. A solar eclipse (bottom) occurs when the moon moves between the sun and the earth, casting its shadow on the earth.

A total eclipse of the sun is an important astronomical event, and often scientific expeditions as well as pleasure cruises go to the places on land or sea where it can be viewed.

Care should be taken in viewing solar eclipses in order to protect the eyes. One way is to view the sun through enough layers of dense photographic negatives so that the sun can just barely be seen. Another way is to view the sun at the back of a pinhole camera. (See pages 282–83 for the construction of a pinhole camera.)

How can an eclipse be simulated in the classroom? An eclipse can be simulated by having a child move a softball (the moon) around a globe (the earth) that is illuminated by a distant projector beam (the sun). The softball should be much closer to the globe than to the projector beam. Why? Ask the children to predict what will happen as the softball passes between the globe and the projector bea.n. Can the shadow of the softball be seen on the globe? This is the area experiencing an eclipse of the sun. Why are only some regions of the earth affected?

Next have the child holding the softball move it around the globe, just as the moon orbits the earth. Ask the children to *predict* how the moon will appear as it passes on the other side of the globe (i.e., the side opposite the projector beam). What would an observer from the earth at nighttime (i.e., facing away from the sun) see? What would be preventing a person on the earth from seeing the moon? How would the sun appear to anyone on the moon during an eclipse of the moon?

The Formation of the Solar System

Several hypotheses have been advanced to explain the formation of the solar system. As children wonder about the origin of the solar system, they are engaging in thoughts that have provoked many individuals for centuries.

Any hypothesis that attempts to explain the formation of the solar system must be consistent with known characteristics and features of the solar system. It is important to emphasize how hypotheses are developed concerning an event that took place so long ago and could not have been witnessed by human observers.

Most of the hypotheses concerning the origin of the solar system fall into one of two major categories: those that describe a gradual, formative process and those that depict a more explosive, cataclysmic origin.

The major hypothesis describing gradual formation of the solar system is called the *dust cloud hypothesis.* This hypothesis suggests that throughout space there are vast amounts of dust and gas. Particles of dust and gas are pulled together by their mutual gravitational attraction into a huge cloud. Starlight (which does exert some pressure on matter) could also have helped bring the particles together.

Within this cloud of dust and gas there probably was some swirling movement, and whirls of dust and gas could have formed. Such swirls can be observed as light is shined on smoke in a jar. Usually, the entire cloud of dust and gas begins rotating, and within the cloud there are smaller cloudlets that also revolve.

As the cloud and smaller cloudlets continue to revolve, they contract and the rate of their spinning increases. This is like a figure skater who starts spinning slowly but then places her arms and hands close to her chest and begins to rotate much faster. It is hypothesized that the smaller cloudlets contracted to form the planets and some of the satellites. The remainder of the original cloud is thought to have formed the sun.

The other major hypothesis about the formation of the solar system is the *planetesimal hypothesis.* Proposed at the turn of the century, this hypothesis suggests that another star came close to our sun and that great bulges formed because of the strong attraction between the two bodies. These bulges, it is supposed, formed long streamers of gas that became planets.

The latter theory explains fewer of the realities we observe today and is considered to be far less likely to have occurred than the circumstances described in the dust cloud hypothesis. Perhaps some of the children in the classroom could suggest hypotheses that are consistent with what they know about the solar system.

BEYOND OUR SOLAR SYSTEM

We know less about the far reaches of outer space than about the solar system. It has been said that considering the billions of stars that exist in the universe, and the fact that our sun is an ordinary star, it is *statistically impossible* that another solar system could *not* have formed in a manner similar to our own. Perhaps many such solar systems exist. Some of these solar systems must have planets with conditions similar to ours. Could life have evolved on such planets? Or do other forms of life completely unfamiliar to us exist? These questions are particularly intriguing and can capture the interest of children.

Stars and Space

There are billions of stars in the universe. Many characteristics of stars are studied by scientists. Among these are color, brightness, size, and distance from our solar system.

When stars are viewed in the sky, they *appear* to be part of a huge umbrella surrounding the earth at a uniform distance. This, of course, is not the case, but we lose our sense of depth perspective when we look at the stars. A bright star may be assumed to be large compared to others. Actually, there are three major factors that influence how bright a star appears: its size, its distance from the earth, and its color.

Distances in space are measured in light years. A *light year* is the distance that a ray of light, traveling at a rate of 300,000 km (186,000 mi) per second, would travel in one year. This is an enormous distance, but it can give a clue to the tremendous distances we are concerned with as we study the universe.

Children sometimes enjoy calculating the length of a light year. The problem would look like this:

$$300{,}000 \times 60 \times 60 \times 24 \times 365\text{-}1/4 = 1 \text{ light year} = 9{,}467{,}280{,}000{,}000 \text{ km}$$

Some concept of distance can be gained when it is realized that it takes 8 minutes for light to reach us from our sun, and that the next-nearest star, Proxima Centauri, is about 4.3 light years away. Polaris, the north star, is approximately 650 light years away.

It is intriguing (although sometimes confusing) when dealing with youngsters to ponder the relationships among light, distance, and time. When we view a star 650 light years away, we are receiving the light emitted from that star 650 years ago. That star may have changed and be somewhat different today, but we will not see light from the star as it is today for another 650 years.

Astronomers classify the *apparent* brightness of stars according to their *magnitude.* The lower the number, the brighter the star. First-magnitude stars are the brightest. Examples of first-magnitude stars are Sirius, Arcturus, Rigel, and Betelgeuse. Second-magnitude stars are the next brightest, and in general, first-magnitude stars are 2-1/2 times as bright as second-magnitude stars. Each order of magnitude is about 2-1/2 times brighter than the next-higher number. It should be remembered, though, that this scale measures apparent brightness and *not* actual color, size, or distance.

The actual size or diameter of stars is also categorized. Some stars are as small as 16,000 km (10,000 mi) or less in diameter, about the size of our planet. These very small stars are called dwarfs or subdwarfs. The largest stars (e.g., Betelgeuse or Antares) are called giants or super giants and may be at least 400 times larger than our sun.

Stars also vary in their color, and color, in turn, is related to the temperature of a star. For example, if a nail is held in a flame it will eventually become red. As the nail is kept in the flame, it will change from red to orange to yellow to white as its temperature increases. This is also the case for stars; the coolest ones appear red and the hottest ones white or blue–white. Rigel, a blue–white star, is one of the hottest, whereas Betelgeuse, a red star, is relatively cool.

How can we help children develop concepts of astronomical space? Astronomical space is so vast that it is difficult for children (or adults) to comprehend. One of the best ways to help children gain some conception of the great distances in space is through the use of analogies. One analogy would be to consider how long it might take to travel to various parts of the known universe.

If a newborn infant were placed in a conventional jet traveling at a speed of 1,250 km (750 mi) per hour, how long would it take him to get to the moon? How long would it take him to get to the sun? What if he were traveling on an SST at twice the speed of sound; how old would he be when he got to the sun?

Suppose another baby set out for the nearest star flying in a jet plane at 1,250 km (750 mi) per hour. She would have to live to the ripe old age of at least 3 million before she could approach it!

Demonstrating the relationship between color and temperature. With a pair of pliers, hold a wire above a lighted Bunsen burner. Have the children record the changes in the color of the wire. Our sun is a yellow star. Is a red star hotter or cooler than our sun? Which color appears to be the hottest?

Constellations

Constellations are groups of stars that, when viewed from the earth, form patterns or figures in the sky. The stars in a particular constellation may represent a variety of sizes, colors, and distances. However, the study of constellations has provided a convenient framework for identifying and locating individual stars.

Observers from different points on the earth see different parts of the sky. Since we cannot see through our planet, it necessarily limits the view. We in the Northern Hemisphere see different constellations than can be viewed in the Southern Hemisphere. Also, because of the earth's yearly trip around the sun we see different parts of the sky at different times of the year. It is for this reason that different star charts are used for different seasons of the year.

Since stars are most readily seen at night, field trips or evening meetings have to be arranged. Parents can be asked to bring the children back to the schoolgrounds for a guided view of the heavens.

It has been said that one needs an awful lot of imagination to recognize the objects, people, or animals depicted by the stars in a constellation. For this reason it is advisable to review charts of the constellations before attempting to view them in the field. This can also orient the children to the locations of what they will be looking for.

Locating constellations. The constellations can best be seen on a clear, moonless night. If possible, it is desirable to find a spot where there is a minimum of distracting light from cities and highways. Obtain a star map from a local planetarium or observatory, or use one found in a textbook or journal.[1]

[1]*Science and Children,* an official publication of the National Science Teachers Association, contains, as a regular feature, a current sky calendar and map in each of its issues. SCIENCE AND CHILDREN (Washington, D.C., National Science Teachers Association).

Current sky maps may also be found in the journal *Natural History* (New York, American Museum of Natural History).

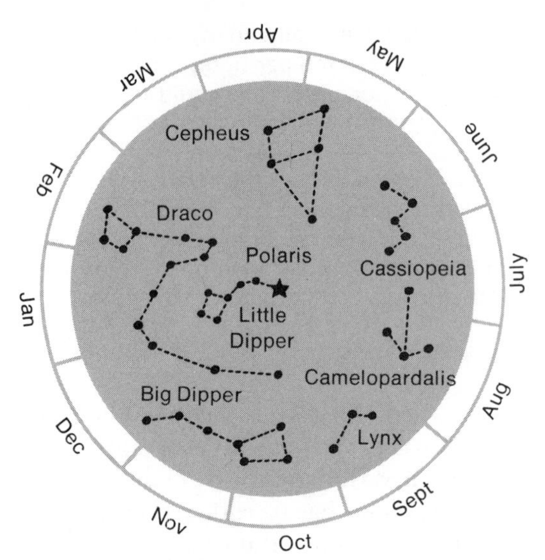

Figure 15–8 The polar constellations as they would appear in the sky during each month of the year at 8:00 P.M. To use this map, turn it until the month in which you are interested is below the pole star.

Orient your map so that it is in the correct position in terms of date and time of night. (See Figure 15–8.)

Perhaps the best way to start is to find an easily recognizable constellation such as the Big Dipper, which appears in the northern sky. Many other constellations can be found in terms of their location with respect to the Big Dipper.

Be sure to point out to the children how Polaris can be located. The two stars in the Big Dipper opposite the handle are called the "pointers." Imagine a line between the two pointers and extend it to a distance about five times the distance between the pointers. The star at the end of this imaginary line is Polaris (the north star). Having found Polaris, you have the direction of true north.

Polaris is at the end of the handle of the Little Dipper. On the side of Polaris opposite the Big Dipper is a W-shaped constellation called Cassiopeia. With these constellations as reference points, many other constellations can be found.

If possible, view the constellations at different times of the night. How do they appear to move during the night?

Galaxies

Our sun is but one of more than one hundred billion stars in our galaxy—the Milky Way. A *galaxy* is a huge system of stars, gas, and dust. Our galaxy is shaped like a flattened spiral and is believed to be about 80,000 light years from edge to edge. The galaxy has a core region where a great many stars are concentrated. Radiating from the core are three or more complicated spiral arms.

Our sun is located in one of the spiral arms at a distance of about 27,000 light years from the center. This means that the sun, one of the medium-sized stars in the galaxy, is about two-thirds of the way out from the center.

The entire galaxy rotates around its center. The central regions are believed to rotate faster than the trailing spiral arms. In the spiral arms there are dark, cloudy patches. These are believed to be composed of interstellar gas, which is probably the raw material from which new stars are formed. No such clouds are visible near the core of the galaxy. Here the stars are older, and apparently no new stars are being formed. In the spiral arms, however, many of the stars are younger. In fact, some of the stars are probably still in the process of being formed.

Observing the Milky Way. On a clear, moonless night, and far away from city lights, the Milky Way can be seen. Our sun is located about two-thirds of the way out from the center of the galaxy. Therefore, we are looking out from the inside of the galaxy; we can never get a full view of it. Early in the evening during late summer, it can be seen as a hazy, white band stretching across the sky from northeast to southwest. In late winter the Milky Way stretches across the sky from northwest to southeast in the early evening. The stars seem to be concentrated along an imaginary line that can be drawn across the sky through the center of the Milky Way. Spreading out from this line, the number of stars diminishes, suggesting that the galaxy is flattened in shape. It is believed that an observer who looks at the bright section of the Milky Way is looking toward the center of the galaxy. When the observer looks toward the part of the Milky Way that has fewer stars, he or she is looking away from the center and toward the outside. This observation also suggests that our sun is located at some distance from the center of the galaxy.

Our galaxy is but one of billions of galaxies in the universe. Galaxies that can be observed are distributed fairly evenly in space about 2 million light years apart. At present the 200-inch (508 cm) telescope at the Mount Palomar observatory can detect galaxies to a distance of about 2 billion light years. The vast majority of the observable galaxies are shaped as either open or closed spirals. Others are shaped like clouds, spheres, or other forms. The average diameter of a galaxy is between 10,000 and 15,000 light years.

Our galaxy is part of a cluster of galaxies called the *local group.* The Andromeda and the Triangulum are other major galaxies in the local group. Other galaxies beyond our local group can also be seen. A galaxy outside of our local group can be observed in the bowl of the Big Dipper.

This galaxy, designated M-81, is about 8 million light years from the earth.

Since the known universe is so large, is there any limit to it? There is some evidence that suggests that the universe is expanding. The galaxies are moving away from each other. The planets remain in their respective orbits and the stars in their relative positions in the galaxies, but the galaxies are rushing away from each other at great speeds. It should not be assumed that all the galaxies are rushing away from the Milky Way galaxy alone. This expansion has been compared to that of an expanding balloon. Make a series of dark marks on the surface of a balloon. Imagine yourself as an observer at any one of these points. As the balloon is inflated, regardless of which point you have chosen, it would appear that all other points are moving away from you. Similarly, regardless of the galaxy from which the universe is viewed, it would appear that all the other galaxies are moving away.

Does this expansion have a limit? Is our universe finite or infinite? Undoubtedly, one of the most fascinating questions with which scientists deal is that of whether or not there is a limit to the universe.

Studying the Universe

Astronomers use indirect methods to measure distances in space. A meter stick cannot be used to measure the distance to a star; therefore, scientists must rely on indirect methods to measure distances in space. Three of these methods are triangulation, parallax, and spectral analysis.

Triangulation is the method used by surveyors to measure the distance to an inaccessible place (e.g., across a river). If a surveyor wishes to

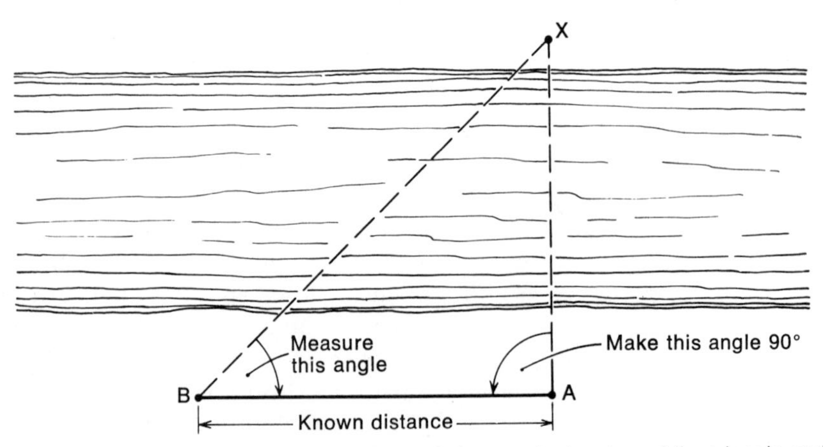

Figure 15–9 Measure the angle at B. Then make a scale drawing of the triangle and measure the distance from A to X on the drawing.

find the distance across a river from point A to X, he or she can carefully measure a line AB and then sight at both points A and B to determine the angle that a straight line to X will make with line AB. Knowing the length AB and the size of angles A and B, the surveyor can calculate the distance to X. (See Figure 15–9.) This method is also used in the calculation of distances to planets. A planet is observed simultaneously from two observatories a known distance apart and the distance to the planet is calculated. Because any base line on earth is very small compared with distances in space, a base line on the earth cannot be used to measure distances to the stars.

Parallax is a method for measuring astronomical distances. Here the base line is usually the distance across the earth's path around the sun (about 300 million km, or 180 million miles). Pictures are taken of the star and the stars in its background at six-month intervals, when the earth is at opposite points in its orbit. The background stars are so far away that they can be assumed to be stationary. By noting the shift of the nearby star against the background stars, the angle of parallax can be measured and the distance to the star calculated. (See Figure 15–10.)

Spectral analysis is a way of determining the intrinsic brightness of stars. Once this is known, distances to very remote stars can be estimated by comparing the apparent brightness of the distant stars with their intrinsic brightness. An analogy can be taken from the way in which the distance to a pair of automobile headlights is judged. We have a concept of the intrinsic brightness of automobile headlights. If the headlights appear bright, they are probably nearby, but if they appear dim they are probably more distant.

In spectral analysis the light from a star is passed through the slit of

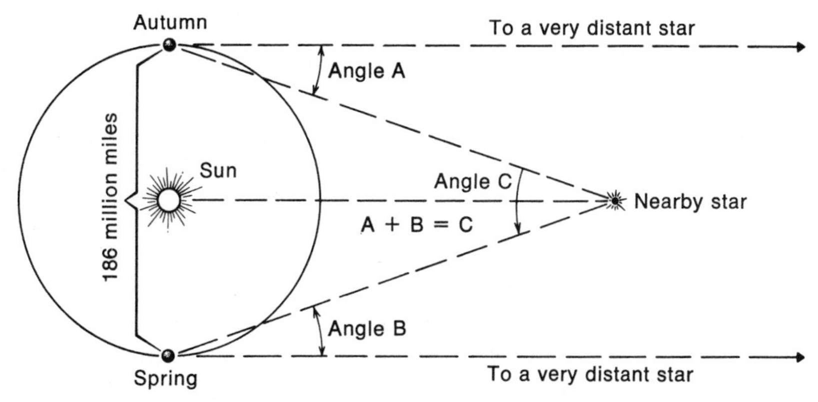

Figure 15–10 The base line is the distance across the earth's orbit. By measuring angle A in the autumn and angle B in the spring, angle C can be calculated. A scale drawing can thus be made and the distance to the nearby star measured.

a spectrograph that has been placed at the foucs of a large telescope. The starlight is dispersed into a spectrum, and by referring the spectrum to a special chart the intrinsic brightness of the star is determined.

After the absolute magnitude (or intrinsic brightness) of the star has been determined, its apparent magnitude is measured on earth with a photometer. The astronomer can then calculate how far away the star must be in order to give the measured apparent magnitude.

Using triangulation to measure the length of the classroom. Using the method described on page 318, measure a base line across the front of the classroom about 3m (10 ft) in length. Next, sight to a spot at the opposite end of the classroom that is perpendicular (at a 90° angle) to one end of the base line. Next, calculate the angle sighted from the other end of the base line to the distant point on the opposite wall. (A drinking straw attached with a pin to the hole at the center of the base of a protractor can be used to measure angles.) Make a scale drawing of the base line. A convenient scale might be 1 cm = 1 m (or 1 in = 1 yd). Measure the angles on the drawing. The point where the side lines of the triangle meet is the point to which the distance is being measured. The distance to the opposite wall can be found by measuring the length of the perpendicular line and multiplying by the scale on the drawing. The children may wish to check their accuracy by comparing their calculations with the other measurements of the room.

Some of the children may wish to use this method to measure the height of a flagpole or the distance to some inaccessible place.

Astronomers use a variety of instruments and techniques to study the universe. Although the stars are very distant from the earth, a great deal has been learned about them. One technique that has been used in studying stars is to investigate the radiation that comes to the earth from the stars, mainly in the form of light. A careful analysis of this radiation can reveal a great deal about the stars from which it emanates. Another important approach to the study of stars is to apply the basic laws that have been developed on the earth to the universe as a whole.

The light received from stars can be analyzed in many important ways. In considering the color of stars, it was pointed out that a nail held in a flame will turn different colors as it becomes hotter. Studying the color of stars can give us clues to its age and heat. Even stars that do not give off light emit some radiation; this too can be studied.

The chemical elements present in a star can be determined by studying the spectrum formed when the starlight is passed through a prism. Each of the chemical elements has a characteristic spectrum. These have been produced and studied in the laboratory. The spectra of stars can be compared with those of various elements to see whether the specific lines that are characteristic of various elements are present in the light from the stars.

Obvious as it may seem, the *telescope* has been an invaluable aid in the study of the universe. When placed on a mountaintop, its view is relatively undisturbed by city lights, smoke, and dust. The largest telescope in the United States is the Hale Telescope on Mount Palomar in California. Constructed in 1948, the Hale Telescope, a *reflecting telescope,* gathers tremendous amounts of light by means of a huge concave mirror (approximately 5m or 200 in. in diameter). This telescope has provided scientists with much information that was relatively unavailable before its construction. The other major type of telescope is the *refracting telescope,* which gathers light by means of a transparent lens. The National Aeronautics and Space Administration (NASA) plans to launch a space telescope in 1983. This instrument will enable scientists to study distant galaxies, some as far as 14 billion light years away.

Another important approach to the study of the universe is *radio astronomy.* In 1931 Karl Jansky, seeking to find a way to eliminate static interference in radio reception, discovered that some of the static was coming from outer space. Here was another source of information about the universe. Radio waves, which are much longer than light waves, are given off by elements, and their reception by radio telescopes makes it possible to map the distribution of gas and dust. Radio astronomy has also made it possible to penetrate clouds of dust and gas to study areas of space that were formerly considered "empty."

Rockets, satellites, space shuttles, space probes, and manned space flights have been tremendous sources of information in the past few decades. The information gathered from such journeys and probes would fill volumes and has perhaps added more to our knowledge of our solar system than all the information gathered from the beginning of time until 1957, when the first satellite was launched. Today's youngsters are living a most exciting time. They can participate in contemporary space explorations through current-events reporting and the careful following of space developments as they occur. The teacher can encourage and foster enthusiasm in these events by alerting children to new and ongoing developments. NASA publications, available from NASA education offices, are excellent classroom references.

Is There Life Anywhere Else Besides Earth?

This is an extremely interesting question, but one that has not yet been answered. All that can be done is to analyze the physical requirements for life and investigate the possibility that these conditions exist elsewhere in the universe.

In order for life as we know it to exist, certain chemical elements must be present. The list of these elements would include oxygen, carbon, hydrogen, nitrogen, calcium, and small amounts of some others. These elements, found in the earth's crust, also seem to be present in the sun and other stars, and probably in the invisible matter in the regions around stars. Thus, the elements needed for life seem to be present elsewhere in the universe.

Life can probably exist only within a certain temperature range. Life exists in hot geysers and the far reaches of the cold Arctic and Antarctic. If temperatures were too hot, though, the complex molecules that make up living matter would be broken up. Because of the high temperatures, life cannot exist too near a star. Since living matter requires energy such as solar energy, living matter cannot exist too far from a source of heat, light, and other forms of energy.

Apparently, living things also need an atmosphere to protect them from deadly radiation from the stars. Atmospheres may be retained by celestial objects by the force of gravitation, but they have to be of a certain size to retain an atmosphere. The necessity for an atmosphere probably rules out smaller celestial objects such as meteors, comets, small satellites of planets, and the smaller planets as possible habitats of life.

After a consideration of the possible hypotheses accounting for the origin of the solar system (see page 312) and the fact that our sun is quite an ordinary star, it is not so farfetched to believe that other planetary systems may have formed in similar ways. Only a small fraction of such planetary systems would have the requirements to sustain life. However, the number of stars in the universe is so large that even a small fraction constitutes a tremendous number. It has been conservatively estimated that there are more than 1 million planetary systems in the known universe that have conditions suitable for life. With so many possibilities, it is difficult to believe that life would arise on only *one* planet in *one* planetary system. The possibility that there is life elsewhere in the universe is very great indeed.

Another matter to consider is that life might exist in a form totally unimagined up to now. However, if we are concerned with intelligent life (i.e., living beings capable of communication), one of the real problems in trying to communicate would be the very great distances involved. If an intelligent being near a star only 1 million light years away were to send a message to us by the quickest means we know (i.e., the speed of light), that message would reach us 1 million years later. Would that intelligent being still be there to receive an answer? Would its habitat still exist? These questions are very thought-provoking and will capture the imagination of many children. Perhaps today's pupils, living in an age of a burgeoning technology, will witness and participate in finding some of the answers to these important questions.

RELATED ACTIVITIES

Sundials, p. 133.
Why do we believe the Earth is round? p. 327.

How can young children explore some effects of heat? p. 391.
Which colors absorb more heat by radiation? p. 398.
How can we demonstrate that light travels in straight lines? p. 400.
Identifying objects from their shadows, p. 401.
Meteors and meteorites, p. 68.
Using relative frames of reference, p. 14.
Seeing things from different frames of reference, p. 53.
Determining our location within the universe, p. 53.
Finding latitude by sighting Polaris, p. 132.
Which object will strike the ground first? p. 64.
How can concepts of great spans of time be developed? p. 342.
How can we obtain and evaluate information? p. 155.

REFERENCES

For Teaching

ATKIN, J. MYRON AND STANLEY P. WYATT, CODIRECTORS. *The University of Illinois Astronomy Program.* New York: Harper & Row, 1969. A series of six volumes designed to help teachers develop experiences in astronomy in the upper elementary school and junior high school.

INGLIS, STUART. *Planets, Stars and Galaxies.* New York: Wiley, 1972. A detailed general astronomy book with much background information.

JASTROW, ROBERT AND MALCOLM H. THOMPSON. *Astronomy: Fundamentals and Frontiers.* New York: Wiley, 1972. A book with much astronomical information for experienced teachers.

ZIM, HERBERT S. AND ROBERT H. BAKER. *Stars: A Guide to the Constellations.* New York: Golden Press, 1975. An accurate guidebook in this useful series of science information for beginning teachers or advanced students.

For Young Children

BERGER, MELVIN. *Stars.* New York: Coward, McCann & Geoghan, 1971. An elementary account of stars and their characteristics. Offers answers to questions that children often ask.

BRANLEY, FRANKLYN M. *What Makes Day and Night?* New York: Harper & Row, 1961. An interesting and accurate book explaining how the rotation of the earth results in day and night. Suggests some student demonstrations.

LLEWELLEN, JOHN. *The True Book of Moon, Sun and Stars.* Chicago: Children's Press, 1960. Basic concepts of the solar system are presented. Set in large type for primary-grade readers.

For Older Children

BRANLEY, FRANKLYN M.

A Book of Stars for You
A Book of Satellites for You

A Book of Planets for You
A Book of Planet Earth for You
A Book of Astronauts for You
A Book of Moon Rockets for You
A Book of the Milky Way Galaxy for You

New York: Harper & Row, A series of interesting and lively accounts of space topics for elementary school children.

KNIGHT, DAVID. *Thirty Two Moons: The Natural Satellites of Our Solar System.* New York: William Morrow, 1974. A detailed account of our solar system's natural satellites. Each planet and its satellites is considered separately. Includes a discussion of possible origins of satellites.

NOURSE, ALAN E. *The Asteroids.* New York: Franklin Watts, 1975. A wealth of information about the asteroid belt, including photographs.

REY, H. A. *Find the Constellations.* Boston: Houghton Mifflin, 1966. Diagrams are used to help children identify major constellations. Sky charts are provided to help locate constellations and the times at which they can be seen.

ZIM, HERBERT S. *The Sun.* New York: William Morrow, 1975. An accurate, well-illustrated book of the sun and its effects on our life on earth.

Children and the Earth on which We Live

CHAPTER 16

"He has his feet on the ground." "They are the salt of the earth." "Solid as the Rock of Gibraltar." These are expressions from common, everyday language that many of our children will have heard. They imply that our earth is solid, never-changing, and that while everything else may be ephemeral, the earth on which we walk will always be there. In this chapter we will discuss ways in which we can learn more about the earth on which we walk—the earth that we so often take for granted.

With elementary school children it is usually best to start with phenomena that are near at hand. In the study of the earth, we can start with the soil, rocks, minerals, and landforms in the local community. In the community through which children walk and ride, they can see examples of the applications of some of the most important concepts of the earth sciences.

Perhaps the most important concept in the earth sciences is that of *uniformitarianism,* the concept that our knowledge of the geological processes now at work can be used to explain how all of the geological features of the earth have been formed. We assume that the same physical principles that work today have operated in the past. Actually, uniformitarianism is an application to the study of the earth of the broader concept of *universalism,* which states that the principles and processes that we study on the earth probably hold true throughout the universe.

Uniformitarianism is a powerful tool for the study of various geological features. As we shall see, children can use this tool as they study features of their environment. After it has rained, small rivulets can be seen emptying into a mud puddle. These rivulets carry sand and silt. When the rushing water enters a calm puddle, it is slowed and can no longer carry as much sand and silt. These sediments may be deposited at the mouth of the rivulet, forming *deltas* much like the one that continues to be formed at the mouth of the Mississippi River. The processes at work in the mud puddle are much the same as those that have been at work over long periods in lakes and oceans. We can use the concept of uniformitarianism to gain a better understanding of many other features in our environment.

The study of the earth can be an important vehicle for the development of profound concepts of time and change.

The earth is believed to be about 4.6 billion years old. The amount of time involved in the rise and wearing away of mountains may be hundreds of millions of years. It is difficult for anyone to conceive of such great spans of time. However, we can contribute to the development of this concept of long spans of time through the study of the earth. Similarly, many of the materials in our environment are involved in natural cycles. For example, there are a nitrogen cycle, a water cycle, and an oxygen–carbon dioxide cycle. Similarly, the rocks and soils in our environment may be considered to be in various stages of a rock cycle. The rock cycle involves great spans of time. But children can be helped to conceptualize the changes that have taken place in rocks through study of the rock cycle and the cyclical nature of the changes that take place in rocks.

As children study the earth, it is helpful to have globes and a variety of maps available. They can use the globes to develop a clearer concept of the shape of the earth and the distances and directions between different locations on the earth. The usual classroom maps can be useful, as can road maps, which can be obtained from local gasoline stations. The topographical map for the locality of the school is especially useful in studying the local terrain. Index maps for your state can be obtained from the U.S. Geological Survey, Washington, D.C. 20242. From the index map you can determine which topographical maps will be of the greatest use to you.

As in many areas of study, it is often useful to develop a concept of the total phenomenon to be studied. In studying the earth, we can plan activities that will develop the concept of the shape of the earth and the evidence to support this concept.

The earth is shaped very much like a round ball. Our everyday observations and common sense may indicate that the earth is flat. After all, if you live on the prairie or are out in the ocean, the earth may appear to be flat as far as the eye can see. Yet we know that it is shaped very much like a round ball. True, the ball is flattened a little at the poles, but it can be considered to be essentially round. Its average circumference is about 39,680 km (24,800 miles) and its average diameter about 12,656 km (7,910 miles).

Since our common sense and everyday observations tend to indicate that the earth is flat, this provides an excellent opportunity to have children make observations and consider evidence that supports the view that the earth is round like a ball rather than flat.

Why do we believe the earth is round? Wherever and whenever possible, children should be helped to make and interpret the following observations:

1. *The way ships disappear over the horizon.* If your children have access to the shore of an ocean or one of the Great Lakes, they may have an opportunity to watch ships sail toward the horizon. The ships seem to disappear into the water. But with the help of binoculars children can see that the bottom of the ship disappears first and the smokestacks and higher parts of the ship later.

   The ships are sailing over the curved surface of the ocean, so the lower parts of the ship disappear first. Some child might suggest that this only shows that the surface is curved – it might be curved like a cylinder. But when ships are observed in different places and sailing in different directions, the bottom of the ship always disappears before the top.
2. *Circumnavigation of the earth.* It is possible to travel around the earth. Obtain maps and timetables from airlines that have around-the-world service. Have the children trace on the globe the routes of trips around the world.

   While most around-the-world flights are from east to west, some airlines have long flights across polar regions. Have the children trace these routes.
3. *The curved edge of the shadow of the earth.* During an eclipse of the moon, we see the shadow of the earth move across the surface of the moon. From almanacs or the local planetarium, obtain the date of the next lunar eclipse. Have the children observe the eclipse. (You may wish to send a note to their parents alerting them to the opportunity to see a lunar eclipse.) Have them note in particular the edge of the shadow. Is it curved?

Figure 16–1 Hold a ball in a beam of light and compare the edge of the shadow with the edge of the earth as photographed from the moon.

Back in the classroom hold a volleyball or basketball in the beam of light from a flashlight or projector. (See Figure 16–1.) What is the shape of the edge of the shadow of the ball? Does the edge of the shadow resemble the edge of the earth's shadow on the moon? You may wish to place a football, a hat, and other objects in the beam of light to check on the shapes of their shadows. (See Chapter 15 for a review of eclipses and how they occur.)

4. *Pictures of the earth.* We now have pictures of the earth taken from the moon and from earth satellites in space. Figure 16–2 is a picture of the earth taken from the moon. What is the shape of the edge of the earth as seen from space? Does it look like a huge ball?

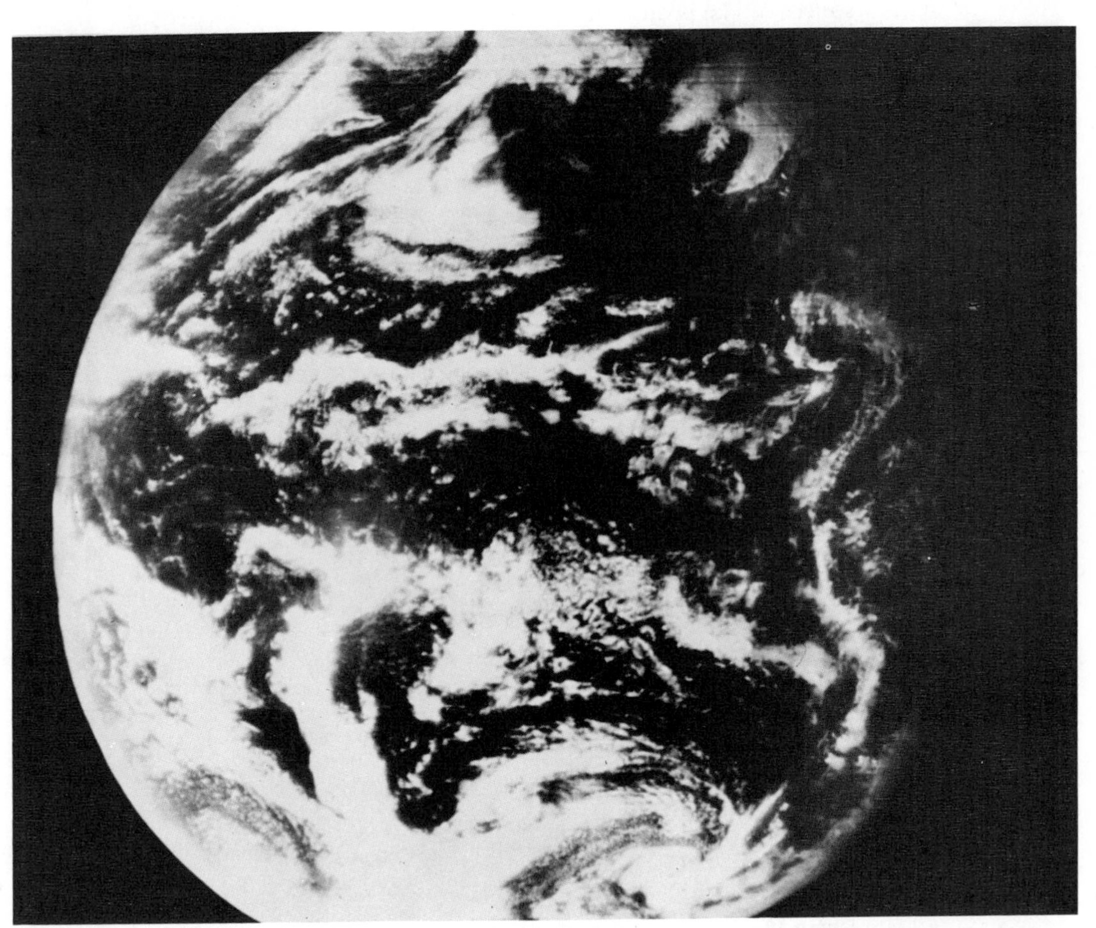

Figure 16–2 A picture of the earth taken from an earth satellite, 40,000 kilometers from the Equator.

SOIL

The soil on which we walk and in which most of the plants in our environment grow is an aspect of the earth with which many children are familiar. Children can begin by studying soils as they can be found in their local neighborhood, and they can bring samples of various kinds of soil for examination in the classroom.

Soils are composed of a variety of materials. The term *soil* refers to the top layers of ground, in which plants can grow. As such, it is one of our most important resources. Most of our food, clothing, and shelter is derived directly or indirectly from plants that grow in the soil. Soil is such an important resource that it is essential that all children, both urban and rural, learn something of the nature of soil and how this resource can be protected.

What is the nature of soil? Obtain samples of soil and spread them on pieces of paper. With the help of magnifying glasses, have the children find answers to the following questions:

What is the color of the soil?
Are the soil particles larger or smaller than a coarse grain of sand?
Do the soil particles have rounded edges or are they jagged and sharp?
Are the soil particles very much alike? If not, what are the differences?
Is there organic matter in the soil sample? If so, what is it like?
Are there living things in the soil? If so, what are they like?

Another way of analyzing soil is to pour some soil into a glass jar full of water. (See Figure 16–3.) Cap the jar and shake the soil-and-water mixture.

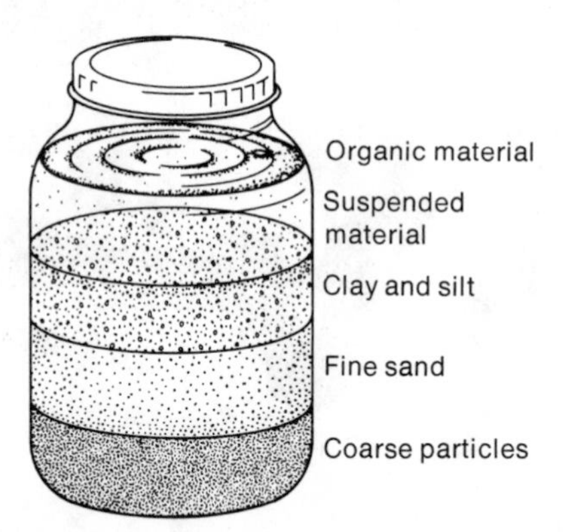

Figure 16–3 When soil is shaken in water and then allowed to settle, the soil particles will tend to settle out in layers. Some material may float to the top and some remain suspended in the water.

Allow the mixture to stand for some time. Does the soil settle to the bottom in layers? If so, how do the layers differ from each other? Is there material floating at the top of the water? What is the nature of this material? What kind of material seems to remain suspended in the water?

In road cuts, stream banks, or freshly dug holes, we can see a cross-section of the soil called a *soil profile*. The top few inches of the soil are called the *topsoil* and are usually darker than the soil below. The topsoil is the soil in which our plants usually grow. It usually contains organic matter that helps fertilize plant growth. The topsoil has been called the lifeline of civilization. It is built up very slowly. Yet a great deal of topsoil can be lost if it lies exposed during a hard rain or windstorm. Most soil conservation practices are aimed at preventing the loss of topsoil.

What evidence of erosion can we find? Searching for evidence of soil erosion can help children become more aware of changes that are taking place in their locality. The following are evidences of erosion by running water or wind. Have the children find as many of them as they can in their neighborhoods.

1. Ditches and gullies.
2. Tree roots that are not covered with soil.
3. Rocks at the surface that were not there some time ago.
4. Soil on sidewalks and hard-surfaced roads.
5. Soil that settles out of water that has been collected from a stream.
6. Ponds and lakes being filled with soil.
7. "Cave-ins" along stream banks.
8. Piles of sand and soil on the lee side of posts and other things that obstruct the wind.

Below the topsoil is the *subsoil*. The subsoil usually is a lighter color than the topsoil and contains little or no organic matter. Usually water will travel both upward and downward through the subsoil. The downward movement of water is called *percolation*. If the subsoil is hard and impervious, the soil will have poor drainage. If it is very porous, little water will be retained in the soil and plants will suffer during periods of little rainfall.

How much water will different soils hold? This activity involves the children in such science process skills as measurement and control of variables and such mathematical skills as addition, subtraction, and division. With rubber bands fasten a piece of cloth over one end of each of two lamp chimneys or pieces of wide-diameter glass tubing. Obtain samples of clay and sand soils. Measure out equal amounts from each of the samples of soil and pour them into the lamp chimneys or glass tubing. Suspend them over glass jars. (See Figure 16–4.)

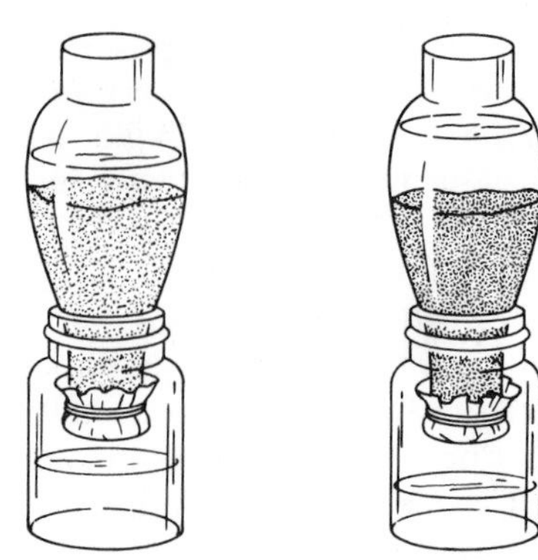

Figure 16–4 Pour equal amounts of water onto equal amounts of soil to find out how much water the soil will hold.

Measure out two volumes of water equal to the volume of soil in each of the lamp chimneys. Pour one volume of water onto the sand and the other onto the clay. Allow the water to percolate down through the soil. When no more water drips from the soil, measure the amount of water in each of the containers. To find the amount of water held in the soil, subtract the volume of water that has passed through the soil from the original amount of water Which type of soil holds the most water?

To find the amount of water held by one unit volume of soil, divide the volume of water held in the soil by the volume of the soil.

Water travels upward through the soil through *capillary action.* If you place a thin piece of glass tubing into water, you can see that the level of the water is slightly higher in the tube than in the surrounding water and that the top surface of the water is curved, with the water near the glass being higher than the water farther away from the glass. This is due to the *force of adhesion* between water and glass. Similarly, there may be a force of adhesion between water and soil particles, and water can rise many feet in soil.

How fast does water travel up through different soils? This activity calls for careful observation, measurement, and control of variables. With rubber bands fasten pieces of cloth over one end of each of two lamp chimneys or pieces of wide-diameter glass tubing. Measure out equal amounts of clay and sand and pour the clay into one lamp chimney and the sand into the other. (See Figure 16–5.)

Lower the two lamp chimneys containing the soil into jars of water so that the water rises to an equal height on the outside of the lamp chimneys. Then watch to see whether the water rises in each of the soils. In which soil does it seem to rise faster? What factor might account for this?

Figure 16–5 In some soils water from below will move upward by capillary action.

Below the subsoil is the *bedrock*. In many places this may be difficult to see because it may be covered with many feet of subsoil and topsoil, but it is often exposed in deep road cuts. In many cases the soil is formed from the bedrock by expansion and contraction due to freezing and heating, the expansion of freezing water, the action of chemicals in ground water, the prying effect of tree roots that grow into cracks in the rock, and the action of wind and running water in places where the bedrock is exposed. Children can compare samples of the bedrock with samples of the adjacent subsoil. Are they the same color? Do they appear to be composed of the same materials? Does it appear that the bedrock was the source of the subsoil?

ROCK

Rocks have been formed in several different ways. Many people believe that all the material at the surface of the earth may have been formed from hot, molten material such as that which comes up from volcanoes. Rocks formed in this way are called *igneous* rocks.

How were igneous rocks formed? The formation of igneous rocks can be demonstrated by heating and cooling chocolate. Heat some chocolate in a pan until it is a liquid. This corresponds to molten rock. Pour some of the liquid chocolate into a pan resting on pieces of ice so that the chocolate cools quickly. Keep the remainder of the chocolate in the hot pan and have it cool very slowly.

Have the children examine the crystals in the candy. In which pan are the crystals larger?

The molten material within the earth is called *magma* and cools very slowly far below the surface of the earth. This igneous rock may contain large crystals. The molten material that flows onto the surface is called *lava* and cools quickly. This rock has very small crystals, or none. Thus, by examining the crystals in igneous rocks we can tell something about how and where the rocks were formed.

How can crystals be grown? Excellent crystals can be grown from any one of such common substances as salt, copper sulfate, or alum. The crystals grow as water evaporates from saturated solutions of these materials.

The solution can be prepared by dissolving as much of the chemical as possible in hot water. This solution usually should be filtered into a clean container to remove foreign substances. When the hot solution cools, some crystals will be formed on the bottom of the container and the solution will be saturated. This is the stock solution that will be used to fill the evaporation dish. The best-formed crystal can be used as a seed crystal.

A ring of vaseline should be smeared around the evaporation dish at the level to which it will be filled. The vaseline will prevent the solution from "creeping" up the sides of the container, evaporating, and leaving a troublesome crust of the chemical.

Some of the stock solution should be poured into the evaporation dish up to the level of the vaseline ring. The seed crystal should be placed in the solution. As water evaporates at the surface of the solution, that part of the solution becomes saturated and somewhat heavier than the rest of the solution. It sinks to the bottom of the dish, and some of the material is deposited on the seed crystal.

The growth of the crystal should be watched carefully. A nearly constant temperature should be maintained. To get symmetrical crystals, the crystal should be turned onto a different face each day. Abnormal growths or "suckers" should be removed from the crystal. Occasionally, fresh stock solution will have to be added to the solution in the evaporation dish. To prevent a shower of crystals at this time, it is usually desirable to add a few drops of water to the evaporation dish before pouring in the stock solution.

With a little care, children can grow large, beautifully shaped crystals. The crystal should be examined every day; it should be turned, and any deformities should be removed. Children may wish to grow large crystals of different substances and compare their shapes.

Most substances change when exposed to the air. To protect crystals when storing them, coat them with a clear nail polish.

Both igneous and sedimentary rocks may change as they are subjected to heat and pressure within the earth. These rocks are called *metamorphic rocks.* Limestone that has been subjected to heat and pressure is changed into marble, sandstone into quartzite, and shale into slate and mica schist. Sometimes we can see layers of rock that have been folded by pressure within the earth. The Appalachian Mountains were formed by the folding of layers of rock. Sedimentary or igneous rocks may be heated when hot, molten material flows between layers and into cracks in rocks. In the zones where this has happened, valuable gems and other minerals are often found. The diamonds in southern Africa were probably formed in this way, as were the wide variety of unique minerals found around Franklin, New Jersey.

Rocks and soil are eroded by wind, water, and ice. The soil particles may be deposited as sediment at the bottoms of lakes and oceans. As the particles accumulate to a considerable depth, tremendous pressure builds up. Sometimes natural cements are also squeezed between the particles. The particles form rocks called *sedimentary rocks*. Sandstone is formed from particles of sand, limestone from lime, and shale from mud. Often the sediment has been laid down in layers, and a great deal can be learned from a study of these layers. For example, a relatively thick layer containing large particles may have been laid down during the rainy season. A thin layer of fine particles may have been laid down when the rivers were low and flowing sluggishly.

How are different layers of sediment deposited under water? Using a large jar full of water and two different kinds of soil, you can show how particles settle out of water. It is more effective to have soils of contrasting colors.

Take a handful of one soil sample and sprinkle it over the water. Watch the particles settle through the water. Do they settle at the same rate?

Take a handful of the other soil sample and sprinkle it over the water. Watch it settle. Alternately sprinkle handfuls from each of the samples into the water.

How are the sediments deposited at the bottom of the jar? Is there any difference in the size of particles deposited at the bottom of a layer compared with those at the top of a layer? Are there particles that remain suspended in the water? If there are, allow the jar to stand for a few days to see whether they eventually settle out. Ask the children if they have seen rocks that appear somewhat like these layers of sediment.

How are folds in rocks formed? The folding of rocks can be demonstrated with layers of different-colored construction paper. Put your hands at opposite ends of the construction paper and push inward on the paper. The paper will tend to fold. Similarly, but under much greater pressure, rocks will tend to fold when subjected to sideward pressure.

Many materials are involved in cycles. There is, for example, the *water cycle*. All surface water—in rain, streams, lakes, and oceans, or as water vapor—is in some phase of this cycle. Sometimes a particular drop of water may pass through the entire cycle in a matter of days. At other times, particularly if the water enters the ground water supply, many years may be required to complete the cycle. There are other cycles involving oxygen, carbon dioxide, and nitrogen. The changes that occur in the materials in the environment often can be interpreted as a part of a cycle of changes.

Rocks may also be considered part of a cycle of changes called the *rock cycle*. This cycle of changes is depicted in Figure 16–6.

The rock cycle involves long spans of time. The original rock material probably came from the mantle in a liquid, molten state. Upon cooling, the molten material hardened to form igneous rock. In some cases this rock may be remelted to form magma. But if the igneous rock is eventually exposed to the work of the elements, weathering takes place and small particles of the rock are broken off and carried away by wind and water. The sediments are eventually deposited, often under the waters of the oceans, and harden to form sedimentary rocks. Sedimentary rocks that are acted upon by heat and pressure or undergo chemical reactions form metamorphic rocks. Under high temperature conditions, the metamorphic rocks may again become magma. Sediments may be formed directly from both sedimentary and metamorphic rocks if they are exposed to the elements.

The rock cycle is not a completely closed cycle. Certainly, material enters the cycle from the mantle. A small amount of material comes from the meteorites that enter the earth's atmosphere. It has been esti-

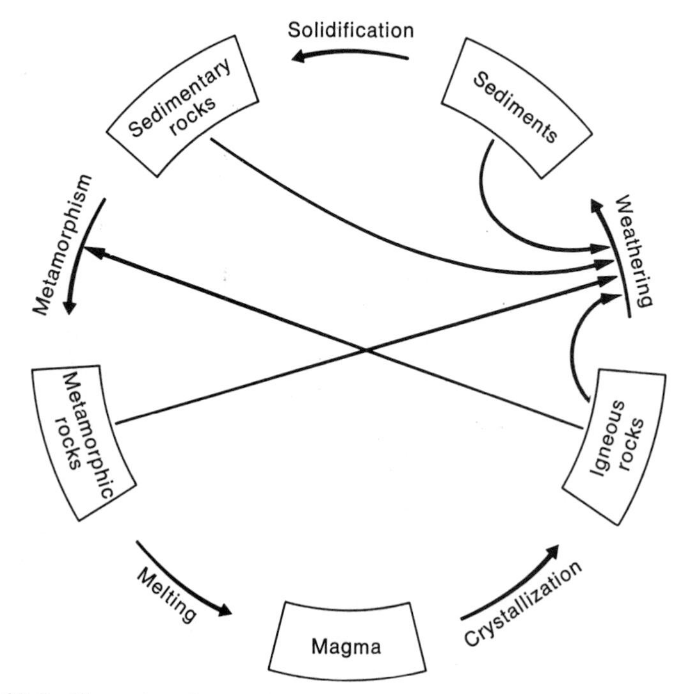

Figure 16–6 The rock cycle.

mated that more than 1 million tons of meteoric dust are deposited on the earth each year. This dust has been found in samples of the ocean bottom in regions where there is very little sedimentation. Also, it may be that some particles of dust from the earth's surface eventually escape from the atmosphere.

The rock cycle is a broad generalization that ties together a large number of discrete facts about rocks and minerals. This is one of the "big ideas" of science. When children examine a rock found in their neighborhood, they can try to analyze it in terms of its place in the rock cycle. By consulting a diagram of the cycle, they can gain some idea of the history of the rock and predict some of the changes that may eventually take place in it.

What are some of the things we can learn from a rock? In this activity children have practice at making inferences from observations. Try to break some samples of rock by placing them in a bag and hitting them with a hammer. (The rock is put in a bag to prevent chips from flying into the face and eyes.) With the help of magnifying glasses, we can examine the fresh surface of the broken rock.

How was the rock formed? If the rock contains grains of sand, lime, or mud, it is a sedimentary rock that was probably formed under water. If the rock contains crystals or is glassy, it is probably an igneous rock. If there is evidence that the rock has been changed by heat and pressure, it is a metamorphic rock.

If it is a sedimentary rock,

Are the grains large or small as compared to the grains in a piece of chalk?

Were the sediments deposited near the mouth of a river or farther out to sea?

Are the sediments of sand, lime, or clay? (A drop of dilute hydrochloric acid dropped on lime will cause it to effervesce. If a rock formed of clayey material is wetted, it will usually have a slight smell of mud.)

Are there layers in the rock? If so, are the layers of equal width? Are they composed of the same materials?

Are there fossils in the rock? If so, what kind of living thing formed the fossil?

If it is an igneous rock,

Does the rock contain crystals?

If it has crystals, are they comparatively large or small?

Are all the crystals composed of the same materials?

Were the rock crystals formed quickly or slowly?

Were all the rocks in the place where this rock was found of a similar nature?

Was there any evidence to indicate whether some of the rocks were younger than the other rocks?

If it is a metamorphic rock,

Are there traces of layers in the rock? If so, how have they been changed?

Are there traces of grains in the rock? If so, how have they been changed?

Are there evidences of fossils that may have been changed almost beyond recognition?

Rocks are composed of minerals. There may be considerable variation within a given type of rock. For example, sandstones may be of different colors, and granites will differ because of the varying proportions of the different minerals. The classification of rocks often involves considerable judgment.

Minerals, however, have a definite chemical composition and have been formed by natural processes. The ice found in glaciers can be considered a mineral. It has a definite chemical composition, H_2O, and it has been formed by natural processes. However, the ice formed in a refrigerator, since it has been made artificially, would not be considered a mineral.

The study of minerals can be an important extension of the description of physical properties, sorting, and classification. (See pages 8–10.) The following are among the physical properties that are used to describe, sort, and classify minerals:

1. Color
2. Hardness—measured on a scale of 1 to 10.
   a. Very soft; easily scratched with a fingernail; greasy (e.g., talc).
   b. Soft; just scratched with a fingernail; not greasy (e.g., gypsum).
   c. Easily scratched with a knife; barely scratched with edge of copper coin (e.g., calcite).
   d. Easily scratched with knife; will not scratch glass (e.g., fluorite).
   e. Can be scratched with knife; scratches glass (e.g., apatite).
   f. Cannot be scratched with knife; scratches glass (e.g., feldspar).
   g. Scratches both knife and glass easily; the hardest common substance (e.g., quartz).
   h. Scratches quartz; uncommon (e.g., topaz).
   i. Scratches topaz; uncommon (e.g., corundum).
   j. Scratches corundum; very rare (e.g., diamond).
3. Streak—the color of the trace that is left when a mineral is scraped on a harder surface such as a rock, a brick, or a porcelain plate. A common streak is that of chalk on a chalkboard.

4. Cleavage—the tendency of a mineral to split in definite directions. Cleavage usually produces smooth surfaces. Different minerals have varying numbers of directions of cleavage.
5. Luster—the way minerals reflect light. Luster is usually referred to as metallic or nonmetallic.

The study of minerals can be a brief but important introduction to chemistry. The following are some of the terms and concepts that are helpful in the study of minerals and will be useful in future experiences in chemistry:

1. *Element.* An element is a substance that cannot be divided into different substances by ordinary means. All materials are composed of one or more of the ninety-two elements that occur in nature. Carbon is an example of an element.
2. *Compound.* Two or more elements may combine to form a compound. Oxygen and silicon are two of the commonest elements. They combine to form silicon dioxide, which is sand.
3. *Mixture.* A mixture is composed of two or more substances that have not combined chemically, and is not uniform. We can make a mixture of sand and salt.
4. *Solution.* A uniform mixture of two or more kinds of substances. For example, when salt dissolves in water a salt solution is formed.

Table 16–1 lists some of the minerals that are found in rocks.

Table 16–1 A Key to Minerals

Mineral	*Color*	*Hardness*	*Streak*	*Description*
Apatite	Reddish brown or green	5		6-sided prisms. Glassy luster. Cleavage in only one direction. A source of phosphoric acid.
Beryl	Light green, sometimes white	7.5		6-sided prisms. Greasy luster. Imperfect cleavage in one direction.
*Calcite	White, sometimes shades of pink	3	White or gray	Bubbles when touched with HCl. Dots seen through it appear double. Perfect cleavage in 3 directions.
Copper	Brownish red	2.5	Brownish red	
*Feldspar	White with faint tinges of color	6		World's most abundant mineral. Perfect cleavage in 2 directions at right angles to each other. Imperfect cleavage in third direction.

Table 16.1 – *(cont.)*

Mineral	*Color*	*Hardness*	*Streak*	*Description*
Galena	Lead gray	2.5	Lead gray	Commonest mineral containing lead. Crystals are usually cubes.
Gold	Deep yellow	2.5	Golden yellow	No cleavage. Very heavy.
Graphite	Black or steel gray	1	Steel gray	Greasy feel.
Gypsum	White or colorless	2	White	Chalklike. One perfect cleavage and 2 imperfect.
Halite (salt)	Colorless; masses are white	2.5		Salty taste.
Hematite	Dark red	6	Dark red	A common compound of oxygen and iron.
*Horn-blende	Black or dark green	5.5		Often found in granite. 6-sided crystals.
Limonite	Brown or yellow, may be black	5.5	Yellow–brown	A compound of iron: rust.
Magnetite	Black	6	Black	Magnetic.
Malachite	Green	3.5	Light green	Bubbles form when touched with acid. An important copper ore.
*Mica	Silvery white, gray, pink, green, or black	2–3	White	Easy to identify. Perfect cleavage in one direction. Possible to peel off very thin sheets.
Olivine	Green	6.5		A common mineral to be found in basalts.
Pyrite	Brassy yellow	6	Greenish black	No good cleavage. Sometimes mistaken for gold.
*Quartz	Colorless, may be colored with impurities	7		6-sided crystals. A very common mineral.
Talc	White or pale green	1	White	Greasy feel.

*Often found in local rocks.

What minerals are found in rocks in the local area? Samples of local rocks can be collected and brought to the classroom. The locations where various rocks were collected should be noted.

Usually it is easier to identify minerals in fresh surfaces. A rock can be put into a bag so that pieces of rock will not fly off when it is hit and broken with a hammer. Magnifying glasses are useful in examining rock surfaces.

To help them recognize common rock-forming minerals, children

Rock ______________________

Where was it found? ______________________

How was the rock probably formed? ______________________

Color	Streak	Hardness	Luster	Other Properties	Mineral

What mineral is the largest crystal that you find composed of? ______________________

Figure 16–7 Mineral identification chart.

should have a chance to examine good samples of such common rock-forming minerals as feldspar, quartz, mica (white and black), calcite, and hornblende. (Samples of minerals can be obtained from suppliers like Ward's Natural Science Establishment, Rochester, N.Y.) Guides to minerals (see the list at the end of this chapter) will be helpful.

Locate minerals in the rock samples and check them for the physical properties listed on pages 338–39. Fill in the chart in Figure 16–7. Try to identify the minerals from the list on pages 338–40 or from a guide to minerals.

THE AGE OF THE EARTH

The earth is very old. The earth is believed to be about 4.6 billion years old. This age and the ages of many old rocks are determined through the analysis of the radioactive decay of such chemical elements as uranium, which decay at a constant rate. Uranium, for example, decays into lead at a rate such that one-half of the uranium will have decayed into lead in 4.5 billion years. Thus, uranium is said to have a *half-life* of 4.5 billion years. By carefully analyzing the amount of uranium and the amount of lead in a rock, the approximate age of the rock can be determined. Rocks have been found on the earth that are more than 3.8 billion years old. However, there are strong reasons for believing that the moon and the earth were formed at about the same time. The rocks on the moon have not undergone the kind of weathering and erosion that those on the earth have, and rocks more than 4.66 billion years old have been found on the moon. However, even these rocks are not believed to have been formed until some time after the formation of the earth and the moon. The consensus is that the earth was formed more than 4.5 billion years ago.

The history of the earth has been divided into eras, periods, and epochs. These major divisions of earth history are shown in Table 16–2, which also shows some of the most important geological and biological events that took place during these divisions.

One of the most significant and distinctive events in the history of the earth was the evolution of life. This could occur on the earth because of its atmosphere, the presence of water and the chemical elements that are found in living tissue, a range of temperatures in which life could exist, and other critical factors. Although life may exist elsewhere in the universe, the earth apparently is the only place in the solar system where life has evolved. Life on earth evolved during the Precambrian Era, probably in the oceans. The first life forms were composed of soft, jelly-like materials that seldom, if ever, form fossils. There are fossil records of primitive life forms in Precambrian rocks. But there seems to be little chance that fossil records of the earliest life forms will be found.

Another significant and distinctive event was the appearance of humans. You will note from Table 16–2 that humans first appeared on the scene in the most recent epoch, only a few million years ago. This is a very recent event compared with the age of the earth. The search for traces of early man continues, largely in East Africa. In a sense, children can participate in this exciting search by noting accounts of new finds as they are reported in newspapers and magazines.

The age of the earth and the various eras, periods, and epochs involve great spans of time that are difficult to conceptualize. This is especially true for children, who have lived such a short time. However, our concepts of time can develop and become more sophisticated over a lifetime, and children can begin to develop these concepts while they are still young.

How can concepts of great spans of time be developed? One approach to concept development is through comparison. We often compare spans of time with the length of our lifetime or periods within our lifetime. Children, of course, have lived such a short time that it may even be difficult for them to conceive of the length of time that their parents have lived. Another way of conceptualizing spans of time is to compare various units of time. Historical geologists develop sophisticated concepts of time by comparing units of geological time. They may compare, for example, the Precambrian Era with the Cenozoic Era or the Triassic Period with the Cretaceous Period.

We can give children practice in comparing spans of time. Have them compare lengths of time like the following:

1. The length of the lunch period or recess with the length of the school day.

Table 16–2 A History of the Earth

Geological Unit	*Number of Years Ago (in Millions)*	*Geological Events*	*Biological Events*
CENOZOIC ERA			
Tertiary Period			
Pleistocene Epoch		Glaciers formed great glacial advances.	Age of humans and mammals.
Pliocene Epoch	1	Continual elevation.	Horses and elephants become modern.
Miocene Epoch	13	Mountain formation: Alps, Himalayas, Cascades.	Apes appear.
Oligocene Epoch	25	Tremendous lava flows in the Northwest.	Hoofed animals become established.
Eocene Epoch	36	Land bridge between Asia and America.	Mammals begin to dominate.
Paleocene Epoch	58	Rocky Mountains continue to rise. Active volcanoes in western North America.	Monkeys and whales appear.
MESOZOIC ERA			
Cretaceous Period	63	Rocky Mountains formed. Large chalk deposits.	Flowering plants appear. Dinosaurs near extinction. Deciduous trees. Modern mammals.
Jurassic Period	135	Colorado Plateau. Dinosaur fossils. Sierra Nevadas begin to form.	Birds appear. Many dinosaurs. Mammals appear.
Triassic Period	181	Large red sandstone deposits. Climate probably very dry.	Age of reptiles
	200		
PALEOZOIC ERA			
Permian Period	230	Appalachian folding. Great glaciation. Continents emerge.	Trilobites become extinct. Reduction in number of types of life.
Pennsylvanian Period	280	Large coal deposits formed.	Earliest reptiles. Coal-forming vegetation. Insects appear.
Mississippian Period	310	Mississippi Valley continuously submerged. Indiana limestone formed.	Sharks. Amphibians become well established.
Devonian Period	345	Core of White Mountains, Acadian Mountains formed. Great sand deposits formed by erosion of mountains.	Amphibians appear. Large numbers of brachiopods. First forests.
Silurian Period	405	Niagara Falls rock formations. New York and Ohio salt beds.	Corals and coral reefs. Land plants appear.
Ordovician Period	425	Most of North America under water. Vermont marble formed.	Vertebrates appear. Large numbers of trilobites, brachiopods, cephalopods.
Cambrian Period	500	Fossils become abundant.	Large numbers of trilobites.
PRECAMBRIAN ERA	600	This era encompasses the bulk of geological time. However, there is only a scanty fossil record.	

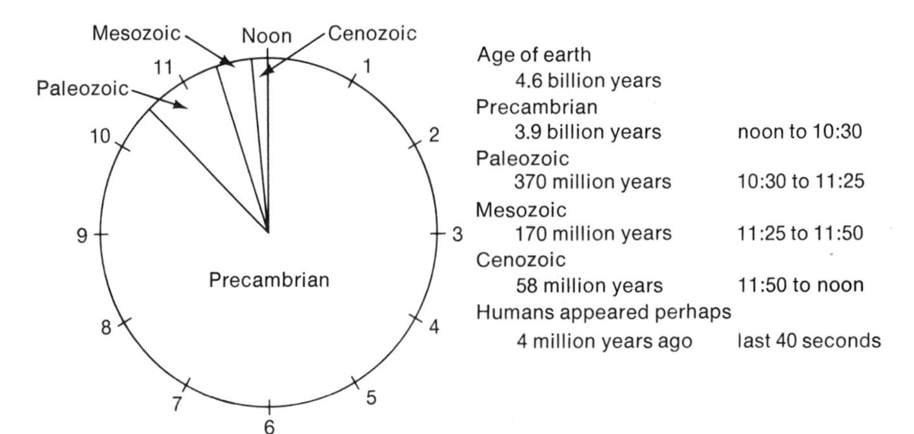

Figure 16–8 The geological clock analogy.

2. The length of the summer vacation with the length of the school year.
3. The length of time since entering school with the length of time before entering school.
4. The length of their lifetime with that of their parents'.
5. The length of lifetime with the time that has passed since their grandparents were born.
6. The length of time since their grandparents were born with the length of time since the U.S. Declaration of Independence (1776).

Another way in which we can help children develop concepts of time and other abstract concepts is through the use of analogies. In the *geological-clock analogy,* the age of the earth and spans of earth history are compared to the face of a clock. The 12 hours shown on the face of a clock represent the time since the earth was formed. The time encompassed by the various Eras are noted on the clock face. For example, if the earth is considered to have been formed at midnight, then the hour and minute hands of the geological clock will have moved to 10:30 to denote the end of the Precambrian Era. The earliest humans lived perhaps 4 million years ago. This is about 40 seconds before noon (now), or at 11:59:20. (See Figure 16–8.)

THE OCEANS

The oceans are an important feature of Planet Earth. If an observer were to view the earth from a space station over the Southern Hemisphere, he or she would see mostly ocean. If children hold a globe so that they look "down" on the South Pole, they, too, will see that the southern half of the earth is mostly ocean. In fact, almost three-fourths of the earth's surface is covered with water. The oceans are such an important feature of the planet that some people have suggested that it should be called the "water planet."

The oceans and other bodies of water have a tremendous effect on conditions throughout the planet. This is because of the unique characteristics of water.

What are some of the characteristics of water? In this investigation children can examine some of the characteristics of one of the earth's most important substances. Place some water in a clear drinking glass and examine it. What is its color? Taste? Smell? Can you see through it?

1. Many kinds of materials are dissolved in water. As children watch the glass of water, they may see bubbles form on the sides of the glass. These are bubbles of air that were dissolved in the water and were released as the water was heated. Fish and other animals that live in water are dependent on this dissolved air. Algae and other water plants get the oxygen and carbon dioxide they need from the dissolved air. Cold water can hold more dissolved air than warm water, and some of the most biologically productive regions of the ocean are in the cold ocean currents such as the Labrador current off the east coast of North America.

   Water can also hold dissolved minerals. Place a little water in a glass jar and allow it to evaporate. The residue left at the bottom is mineral matter that was dissolved in the water. This mineral matter is also used by plants that grow in the water.
2. Water is a solvent. Place equal amounts of water and a liquid such as kerosene in separate glass jars. Put a spoonful of sugar into each of the jars of liquid and stir. In which liquid does the sugar dissolve?

   Life would probably not be possible in the oceans if water were not a good solvent.
3. Water is a heavy liquid. Pour water and kerosene to equal heights in two tall jars, such as graduates or olive jars. Float a new pencil in each of the liquids. (See Figure 16–9.) In which of the liquids does the pencil float higher?

   Because water is one of the heaviest liquids, many solids will float in water. Soil and other materials are eroded by running water. Very heavy ships can float in water. The weight of falling water is converted into electrical energy in hydroelectric plants. Most interesting, it is possible for us to float and swim in water because the density of our bodies is slightly less than that of water.
4. Water changes from a solid to a liquid to a gas at temperatures that can be achieved in the classroom. Place an ice cube in a small amount of water. Use a thermometer to determine the temperature of the water as it melts. Now place the beaker of water on a hot plate and try to determine the temperature at which water boils.

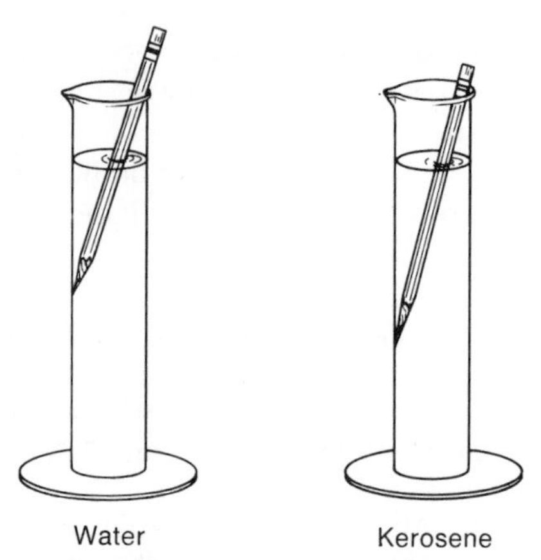

Figure 16–9 Similar pencils float in water and in kerosene.

The fact that water freezes at temperatures that occur frequently means that much water is in the form of snow and ice. Greenland and Antarctica are still mostly covered with ice. During the glacial periods much more of the earth was covered with ice.

Actually, water can change to water vapor at temperatures below the boiling point. The evaporation and subsequent condensation leads to much of our weather, including rain and snow.

5. Water expands when it is heated. Fill a flask or bottle with water and insert a one-holed stopper with a piece of glass tubing in it. (See Figure 16–10.) Wrap a rubber band around the glass tubing level with the top of the water in the tubing. Place the flask on a hot plate and heat the water. Watch the level of the water in the glass tubing. What happens to the level of the water as the water is heated?

   The expansion of water when heated causes warm water to rise and cold water to sink. Many ocean currents are a result of the expansion of water upon heating.

6. Water vapor condenses when cooled. Fill a glass jar with ice. Watch the outside surface of the jar. What happens?

   Because water vapor condenses into a liquid when cooled, clouds form and rain and snow may fall.

7. Ice floats. Place an ice cube in a jar of water. Push the ice cube down to the bottom of the water and release it. What happens to the ice cube?

Water expands when it freezes, and it becomes less dense. In the winter lakes and streams are covered with ice and seldom freeze to the bottom. If bodies of water froze from the bottom up, many would not melt completely during the summer and we would have much colder climates. The expansion

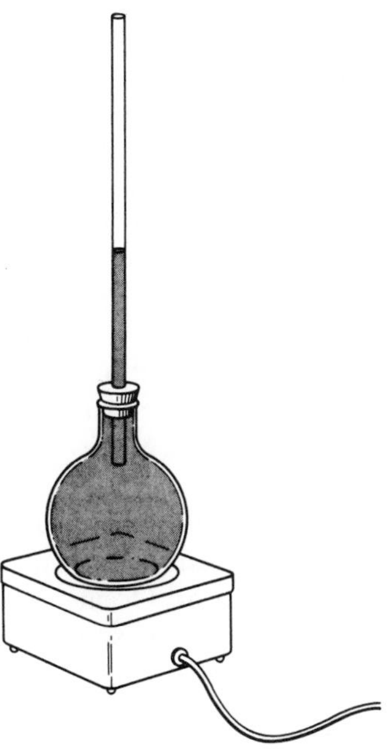

Figure 16–10 What happens to the level of the water when it is warmed?

of water upon freezing can lead to the cracking of water pipes and other containers. (See pages 188–90 for another investigation of the properties of water.)

More heat energy is required to change the temperature of water than other substances such as soil and rock, and this affects the temperature of regions near large bodies of water. About five times as much heat energy is needed to raise the temperature of water a given amount than dry soil. Therefore, large bodies of water, such as the oceans and the Great Lakes, tend to keep nearby regions relatively cool in the summer and extend the period before the first frost in the autumn. This is one of the major reasons that there are large fruit-growing areas near the Great Lakes.

Since water expands when it is heated, ocean water expands when it is heated by the sun. Water near the equator is heated more than that near the poles. This warm water flows in such great ocean currents as the Gulf Stream. The warm Gulf Stream gives Britain and northwestern Europe a comparatively mild climate. Dense cold water tends to sink and flow southward to replace the water that has been heated at the equator. These cold ocean currents tend to chill the nearby land. The cold Labrador Current, for example, gives Labrador and northeastern North America a frigid climate compared to that of the British Isles, which are at about the same latitude. The north–south movements of warm and cold ocean currents are deflected by the spin of the earth. In the Northern Hemisphere, to the observer with his or her back to the current the current will be deflected to the right. In the Southern Hemisphere, the currents are deflected to the left. Children can see these deflections on maps or globes that show the paths of major ocean currents.

Ocean currents are also generated by differences in the salinity of ocean water. Water that has a great deal of salt in it is denser than water with little salt. Since it is heated by relatively direct rays of the sun, water in the Mediterranean Sea evaporates faster than water in the colder Atlantic. The water in the Mediterranean becomes saltier and will tend to sink. At places such as the Straits of Gibraltar, there are deep currents of salty Mediterranean water that flow out into the Atlantic, and surface currents of Atlantic Ocean water that flow into the Mediterranean. Submarines can use these currents to pass through the Straits without using their engines.

How do ocean currents flow? In this investigation children can see how a physical model can be used to learn more about what happens in a physical system. This activity is best done as a demonstration. Since some water may be spilled, make sure there is nothing nearby that can be damaged by water. Two or more glass bottles with fairly wide openings of the same size will be needed. Glass milk or juice bottles are ideal.

Fill one bottle with warm water and the other with cold water. Put a few drops of food coloring in the bottle of warm water. Put a card over the top of the bottle of cold water. Invert this bottle and place it on top of the bottle of warm water. (The card will be held over the opening by air pressure. However, you may wish to hold it on the bottle with your fingers.) Slip the card from between the two bottles so that the water can flow from one bottle to the other. (See Figure 16–11.) Have the children observe carefully what happens where the warm water meets the cold water. In which direction does the warm water flow? The cold water? Have the children explain what happens and relate what they observe to what happens in the oceans.

Repeat the investigation, but this time place the colored warm water on top of the cold water. (The demonstration is especially effective if you have enough bottles so that each demonstration can be left standing for comparison.) Have the children explain what happens when the cold water is on top.

By pressing two bottles together, the top bottle (with warm water) and the bottom bottle (with cold water) can be placed on their sides. Have the children watch as the warm water moves up over the cold water. Usually a very definite dividing line can be seen between the warm and cold water.

This demonstration can be repeated by filling one bottle with fresh water and the other with water to which table salt has been added.

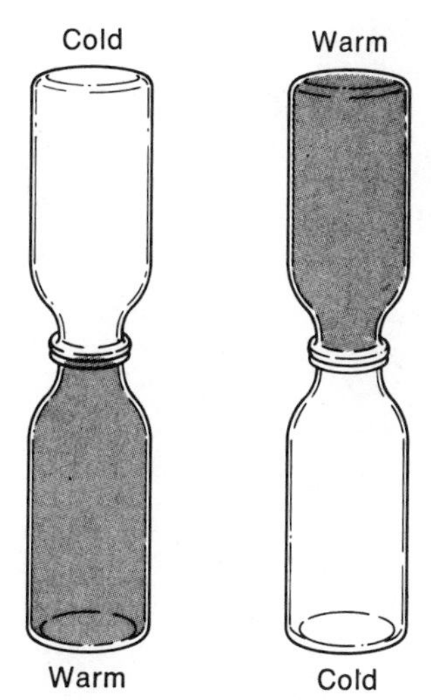

Figure 16–11 Place one bottle on top of the other and remove the card. What happens when warm water meets cold?

If the children have ever visited a beach or an ocean or lake front, they may recall the slapping of the waves against the sand or the roar of huge waves as they beat against beaches and rocks in a heavy storm. During such storms beaches may be completely transformed and huge rocks may be broken off and rolled away. This wearing away of a shoreline is an aspect of *gradation* (i.e., the wearing away of the land). On the other hand, the rivers that flow into lakes and oceans can carry great quantities of sand and soil. Some of this settles to the ocean bottoms, and in some cases huge deltas are formed. Currents may carry sediments along a coastline and eventually deposit them as bars and spits. Many of these phenomena can be seen in mud puddles that form after a rain.

What changes can we see taking place in a mud puddle? During a rain small mud puddles often form on the schoolgrounds or nearby, and in these puddles children can see many of the phenomena that occur when water and land meet. (See Figure 16–12.) (If no mud puddles are accessible, many of the phenomena that are discussed can be simulated in puddles formed in soil in a large tub in the classroom.) Have the children try to see the following:

Waves. Usually a slight breeze will generate small waves, even in a little puddle.

Streams carrying soil. The small streams running into a puddle are often darkened by the soil they carry. How far does the stream seem to flow into the puddle?

Figure 16–12 These children are observing changes in the shoreline of a mud puddle.

Deltas. Sometimes a small stream will deposit soil and build up deltas at its mouth.

Soil in suspension. The puddle looks muddy because of soil particles suspended in the water. Fill a beaker or a glass tumbler with this water. Does any material settle to the bottom of the glass? What is its color? How large are the particles?

Cliffs. Where the wind blows the waves against a shoreline, the shoreline may be eroded at the base of the "hill" and the material above will fall away to form a cliff.

Spits and bars. Where there is a current along the shoreline, the sand and soil may be deposited in the shape of small curved spits or straight bars.

The oceans are a great "sink" for materials that are washed off the land. The water that soaks down through the soil dissolves various chemicals out of the soil, and eventually some of this water seeps into rivulets, rivers, and streams. These, in turn, also erode away particles of soil and sand. The eventual destination of these materials is the oceans. In the oceans some of these materials will be deposited on the bottom to form layer upon layer of sediments. Within these layers may be found fossil remains of some of the organisms that lived in the oceans at the time that the layer was formed. The dissolved materials contribute to the "saltiness" of the oceans. The land surfaces of the earth are slowly being worn and dissolved away and deposited in the oceans.

Over long periods the regions covered by the oceans may be uplifted and become dry land. Much of what is now central North America, for example, was once covered by ocean water. The soil and rocks of these uplifted regions were once the sediment at the bottom of the ocean. Sometimes it is possible to see layers in these rocks that correspond to the different kinds of sediments that may have been deposited from materials washed off the land at different seasons of the year. Also, fossils of organisms that could only have lived in the sea are found in rocks that are thousands of miles from where the oceans are now. But at one time that region was at the bottom of the ocean!

Where did fossilized organisms live? In many communities children will be able to find fossils. If no fossils are to be found locally, sample fossils can be obtained from a scientific supply house.

Have the children examine the fossils. Are the fossils like any plants or animals that they have seen? Are there evidences of shells such as we find in animals like snails and clams? If the fossils are like organisms that live in the water, why do we find them in rocks far from any body of water?

Many scientists now believe that life may have originated in ocean water, possibly in tidal pools at the edges of the sea. The oceans are rich in the chemicals that have been washed off the land, and some of these chemicals may have joined together to form the amino acids; the amino acids may have joined together to form the proteins found in living matter. The oceans provide an environment without great extremes of temperature; the water will provide support for simple organisms; and gases such as oxygen and carbon dioxide are dissolved in ocean water. Many small, simple organisms are found in the oceans, and the first organisms may have originated there.

OUR CHANGING EARTH

We can see evidence of changes that have taken place in the earth. The earth has changed and is changing. A geologist once said that "change is the only thing we can be certain of." We can see hills that at one time must have been higher, beaches and sand bars that once were lower, rivers carrying silt and sand to the oceans, and sediments that are filling in lakes and settling to the bottom of the ocean. These are changes that are taking place now, and we believe that similar changes have been taking place ever since the earth was formed.

Show the children two jigsaw puzzle pieces that fit together. Then have them look at a globe. Are there any land masses that look as if they might once have fit together? Would it be possible to fit Africa and North and South America together? Could Ireland and Britain once have been one land mass? We now believe that there have been gigantic movements of continents and other masses of land. This movement is called *continental drift.* It is believed that great land masses are in a sense floating on plates; hence, the theory used to explain continental drift is called *plate tectonics.*

Continental drift is change on a massive scale, and children can see evidence of it when they examine a globe. The entire surface of the earth is thought to be divided into 6 plates. About 225 million years ago, these plates were close together. The land masses that were to become the continents began to move apart. North and South America moved westward and collided with other plates, and the collision led to the pushing up of the Rocky Mountains and the Andes. Today the separation of continents continues—the continents are separating at the ridges in the ocean at a rate of about 2 to 5 cm (1 to 2 in) per year.

We cannot observe continental drift, but there are many changes for which we can see evidence in our own communities. There are forces that uplift and build land masses. In Hawaii and the Pacific Northwest, we can see how the land has been built up by volcanic eruptions. In these regions and others, the rocks and soil are clearly of volcanic origin. Large areas of North America, such as much of the Middle West, were once covered by ocean waters. In such regions we can see layers of rock formed from the sediments deposited at the bottom of the ocean. About four-fifths of the land surface is underlain by sedimentary rocks. In these shales, limestones, and sandstones, fossils of marine organisms can often be found. The Rocky Mountains were formed largely by the upthrust of rocks as one plate collided with another. The Appalachian Mountains are fold mountains; the folds formed as layers of rock were subjected to sideward pressure. It is sometimes possible to see examples of folded rocks in road cuts and other places where cross-sections of rocks have been exposed. For many regions there are monographs, pamphlets, and books that describe how the hills, mountains, and other landscape features were formed. One way to make one's teaching relevant and interesting is to show children local examples of some of the ways in which landscape features are formed.

As soon as landscape features are raised above the level of the sea, the agents of *gradation* begin the work of wearing away and leveling the earth's surface. Gradation may be seen as a process of "grading" or leveling the surface of the land. Rocks that are exposed at the surface are slowly broken down by wind, water, and changes in temperature to form soil. Wind, running water, and moving ice erode the earth's surface, always moving particles of sand and soil closer to the level of the ocean. If these gradational agents were to continue their work without interruption, all the land would be below sea level and some of the agents would continue to carry the materials from the shallow ocean bottoms into the ocean deeps. But these processes are interrupted by the various forces involved in the rock cycle. (See pages 335–37 for a further discussion of the rock cycle.) The build-up of sediments in one region may actually lead to an upthrust of materials in another area.

Evidences of gradation can be found everywhere, and children should be encouraged to find examples in their neighborhoods. Wind erosion can wear away the softer sections of exposed rock and move sand and soil great distances. Dust storms are dramatic examples of wind erosion on a large scale, in which soil is removed from one area and sometimes deposited in great drifts elsewhere. (In regions where there is wind and snow, children can see some of the striking mechanisms of wind erosion as the wind drifts and then deposits snow.) *Water erosion* takes two major forms: sheet erosion and gully erosion. In sheet erosion, the soil is eroded away quite evenly, exposing rocks and pebbles. In gully erosion, ditches and gullies are carved out as a stream of fast-flowing water carries away sand and soil. *Glaciers* have affected the landscapes of most of northern North America. In some places there are scratches or gouges in rock, pebbles, or boulders that were pushed along by the great masses of ice that once covered northern North America. Children are more likely to see places, such as gravel pits, where soil, pebbles, and sometimes huge boulders are all mixed together. These mixtures of unassorted materials were deposited as the great ice sheets melted. In all communities evidence of some, and sometimes all, of these forms of gradation can be seen.

What evidence of geological processes can be found near the school? Upon achieving scientific literacy, children can participate in certain aspects of the scientific enterprise. (See pages 17–18 for a discussion of scientific literacy.) Part of this enterprise consists of learning to understand and interpret phenomena that occur in our localities. Some of the phenomena in the following list can be found in all localities. Others can be found in some regions and not in others. Children should be encouraged to find as many as possible. When they travel to other localities, they may be able to add to their list of evidences of geological processes.

A. Erosion (see Figure 16–13)

Differential erosion. Soft materials are eroded faster than hard materials. On an exposed rock there may be small ridges of hard materials and small valleys where there were soft materials.

Figure 16–13 Sheet and rill erosion in a cornfield.

Sheet erosion. Sheets of water from rains wash away the soil, leaving rocks, pebbles, and other heavy materials exposed. The soil may be found deposited farther down the hillside.

Gully erosion. Ditches and trenches may be left after heavy rains. Measure the depth of a gully and then see how much it has been changed by the next heavy rainstorm.

Wind erosion. Dust and sand may be seen moving when there are strong winds. Clouds of dust can be seen during dust storms. Rocks and pebbles may be exposed where the wind has blown the sand and soil away.

Glacial scratches and gouges. Scratches and gouges can be found in hard rocks in northern North America. The scratches and gouges indicate the direction the glacier moved.

Stream erosion. Streams can cause the caving in of stream banks. The entire stream valley has been worn away over a long period by the stream.

Ripple marks. Ripple marks under the shallow water of lakes and oceans indicate the movement of sand under the water.

Soil creep. On some steep hillsides the soil will slowly move downhill. Evidences of this soil creep are telephone poles or fence posts that

Figure 16–14 The layers in this rock have been twisted and folded.

are tipped and trees whose trunks are curved to maintain a vertical orientation.

B. Deposition

Layers of sedimentary rocks. In road cuts the bedrock may be exposed. If the rock structure is layered, it is probably sedimentary rock. These rocks were probably formed from sediments deposited at the bottom of the ocean. This means that the region was once covered by the ocean.

Fossils. Sedimentary rocks often contain traces of organisms that lived long ago. Examine the sedimentary rocks for fossils.

Deltas. Deltas are formed as streams flow into lakes or oceans and

Calkins/U.S. Geological Survey

Figure 16–15 Note that the bands of light rock are lower on one side of the fault than on the other.

deposit the sediment they have been carrying. Look for deltas in nearby bodies of water.

Stream deposits. In places where the stream flow is slowed, sand and soil will be deposited. Look for stream deposits at the inside edge of a curve in the stream and in places where the stream may occasionally overflow its banks.

Drifting snow and snowdrifts. We can learn a great deal about wind erosion and deposition by studying how the wind moves and deposits snow. Watch how snow moves across the surface. Note how snow is deposited in drifts in places where the velocity of the wind is slowed.

Mud cracks. When wet mud dries, it often cracks in a characteristic fashion. Sometimes the cracks solidify into hard rock. Have the children see whether they can find any general pattern in the way cracks are formed in mud when it dries.

Glacial deposits. The glaciers carried soil and rocks of many sizes. When the ice melted, these materials were deposited as a mixture of all these materials. Look for glacial deposits in gravel pits and other places where a cross-section of the rock is exposed.

C. Metamorphic Changes

Twisted layers of rock. When layers of rock have been subjected to sideward pressure, the rocks are twisted and folded. (See Figure 16–14.) Sometimes these twists and folds can be seen in road cuts and other places where a cross-section of the bedrock is exposed. Pieces of rock may also have twisted and folded layers.

Faults. Faults are cracks in rock. Huge faults, such as the San Andreas fault in California, are regions of earthquake activity. (See Figure 16–15.) Children can examine cracks in local rocks to see whether there is any evidence of movement of similar sections of rock on each side of the crack.

D. Volcanic Action

Igneous rocks. Igneous rocks have been formed by the cooling of hot, molten material. Many of the rocks to be found in the Hawaiian Islands and the Pacific Northwest are igneous.

Dikes and sills. Dikes and sills are intrusions of igneous rock. When the intrusion is parallel to the layers of adjacent rock, it is called a *sill*. When it cuts through layers of rock, it is called a *dike*.

RELATED ACTIVITIES

How can an eclipse be simulated in the classroom? p. 311.
How far is it? p. 154.

Using triangulation to measure the length of the class-room, p. 320.

Explorations with salt, sand, or soil, p. 110.

Rocks and shells, p. 110.

What are some of the physical characteristics of water? p. 188.

What kinds of materials are there in our rivers and lakes? p. 200.

What is the level of the water? p. 189.

How can we test soil? p. 219.

How does organic material decay in soil? p. 201.

What is the density of a rock? p. 126.

What is the direction of the north magnetic pole? p. 425.

REFERENCES

For Teaching

EARTH SCIENCE CURRICULUM PROJECT. *Investigating the Earth.* Boston: Houghton Mifflin, 1967. An informative guide for background material in earth science.

RHODES, FRANK H. T., HERBERT ZIM AND PAUL R. SHAFFER. *Fossils.* New York: Golden Press, 1962. Information for teachers and older students on the formation and identification of fossils. Well illustrated.

SHELTON, JOHN S. *Geology Illustrated.* San Francisco: W. H. Freeman, 1966. A college textbook with many excellent photographs demonstrating geological phenomena. The illustrations may be useful for elementary school science teaching.

STRAHLER, ARTHUR N. *Physical Geography.* New York: Wiley, 1967. A college textbook on the earth—its shape, climates, weather, and land forms. Provides a wealth of background material for teaching.

ZIM, HERBERT S. AND PAUL R. SHAFFER. *Rocks and Minerals.* New York: Golden Press, 1967. This Golden Nature Guide is a particularly useful reference for any classroom library. Rocks and minerals are described and illustrated.

For Young Children

RAVIELLI, ANTHONY. *The World is Round.* New York: Viking, 1970. A simple treatment of the actual shape of the earth for young readers.

RINKOFF, BARBARA. *Guess What Rocks Do.* New York: Lothrop, Lee & Shepard, 1975. A book on how people and animals have used rocks. Ancient cultures are discussed according to the rocks they used.

SHERMAN, DIANE. *You and the Oceans.* Chicago: Children's Press, 1965. A simple treatment of oceanography.

For Older Children

BRANLEY, FRANKLYN M. *A Book of Planet Earth for You.* New York: Harper & Row, 1975. Earth and its physical characteristics are described. Theories of the nature of the solar system are discussed in an appealing way.

CARSON, RACHEL, *The Sea Around Us.* New York: Golden Press, 1958. A classic edition of this famous work adapted for children.

GALLANT, ROY A. AND CHRISTOPHER J. SCHUBERTH. *Discovering Rocks and Minerals.* Garden City, N.Y.: Natural History Press, 1967. A guide to common rocks and minerals filled with useful information for starting a collection.

KLAITS, BARRIE. *When You Find a Rock: A Field Guide.* New York: Macmillan, 1976. A clear, concise book describing common rocks and where and how to find them.

REUTER, MARGARET. *Earthquakes: Our Restless Planet.* Chicago: Children's Press, 1977. A concise book describing earthquakes and their origins, frequency, and causes.

RITTER, RHODA. *Rocks and Fossils.* New York: Franklin Watts, 1977. A relatively easy-to-read book with a good deal of information on rocks, minerals, and fossils.

WILLIS, WILMA. *Sand and Man.* Chicago: Children's Press, 1973. Sand and its formation and uses are covered in this appealing book for upper elementary school youngsters.

Children and the Air and Atmosphere

CHAPTER 17

"Everybody talks about it, but no one does anything about it." That is an often-heard expression concerning weather. Atmospheric conditions and climate do affect our lives in many ways. We wear certain kinds of clothing because of the weather. We heat, insulate, and air condition our homes because of the weather. In schools, graduation exercises have been suddenly moved because of the weather; class trips have been canceled; and recess may have to be held indoors—all because of adverse weather. But weather need not be always considered a hindrance. Few experiences can be as exhilarating as a crisp autumn morning in northern North America, a gentle breeze on the face, or the first balmy day after a severe winter.

As children become increasingly aware of their environment, they will experience many sensations that they can attribute to weather, climate, and general atmospheric conditions. The eeriness of an approaching storm, a sense of wonder as they see pictures made by clouds, the joy of walking on crunchy leaves or experiencing a first snow—children may have all of these sensations as they "tune into" their physical world. Before they can understand the major factors that govern our weather, though, they must first investigate air and its properties.

The atmosphere that surrounds our planet is relatively thin; half of the materials in our atmosphere are found within 5½ kilometers of the surface of the earth. Although there is no definite boundary line between the atmosphere and interplanetary space, practically all of the atmosphere is within 1,600 kilometers (1000 miles) of the earth's surface.

The air that surrounds the earth is a mixture of gases. The air at sea level is 78 percent nitrogen and 21 percent oxygen. The remaining 1 percent is composed mostly of argon, other rare gases, and water vapor. (See Figure 17–1.) About .03 percent is carbon dioxide. The amount of water vapor in the air varies with climatic conditions. On hot, humid days the air naturally contains more water vapor than on cool, dry days. Our comfort on any particular day is usually affected by the amount of moisture in the air. How often we hear "It's not the heat, it's the humidity!" *Relative humidity* (see pages 378–79) is an indication of the amount of water vapor in the air. If the relative humidity is too high, the perspiration from our bodies will not evaporate readily (since the air is already carrying a great deal of water vapor) and we feel sticky and generally uncomfortable.

Air is a mixture of gases that do not chemically combine with one another and therefore maintain their individual characteristics. Oxygen from the air, for example, is used by most living organisms. As we breathe air, we obtain needed oxygen. Water also contains oxygen (H_2O, the chemical symbol for water, means that it contains two parts hydrogen to one part oxygen) but since it is chemically bound to the hydrogen, the oxygen in water is not in a form that can be used for breathing.

The composition of our atmosphere has changed throughout the history of the earth and is probably still changing very slowly. The com-

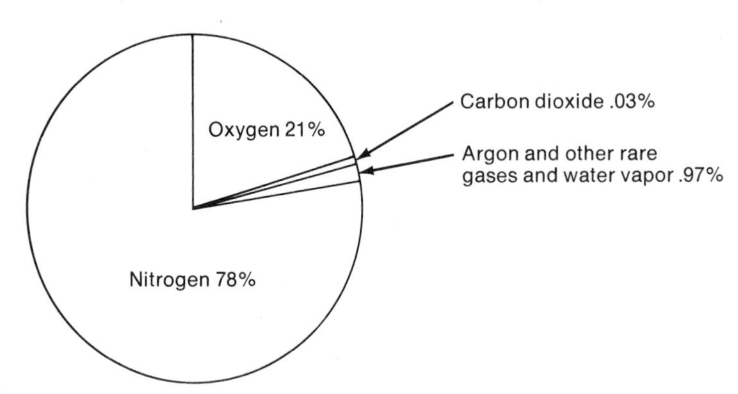

Figure 17–1 Approximate composition of air at sea level.

position of air changes from day to day over small land areas. The air's content of substances such as water vapor, ozone, ammonia, hydrogen sulfide, sulfur dioxide, carbon monoxide, and dust varies over different parts of the earth. Winds mix and thereby alter the proportions of some of these gases. When there is a relatively high concentration of dust and irritating gases in the air, we say that the air is polluted. In some highly industrial cities, the air is so polluted that it is deemed unhealthy and is dangerous, particularly to people with chronic respiratory diseases.

Industrial wastes spewing out of smokestacks, smoke from incinerators, and automobile exhaust fumes all contribute to the concentration of pollutants in the air. Winds usually help disperse undesirable air, but occasionally, in some cities, a situation arises in which a wedge of cool air moves over a layer of warm air, trapping the warm air beneath it. This situation, called *thermal inversion,* can be particularly dangerous if the trapped air is polluted. The undesirable air remains over the land area until weather conditions change.

How can we demonstrate that air is composed of more than one gas? Push a piece of steel wool into the bottom of a large test tube or a narrow olive jar. Make sure the steel wool stays in the bottom of the test tube. (Wash the steel wool first in order to remove any rust retardants.) Invert the test tube in a pan or beaker of water, making sure no water bubbles into it. (See Figure 17–2.) Place another test tube, without steel wool in it, into another pan or beaker containing the same amount of water as the first one. Let the two setups stand for a few days. As oxygen combines with the steel wool to form rust, water will enter the test tube to take the place of the oxygen. How much of the first test tube becomes filled with water? How much water entered the test tube without the steel wool in it? What accounted for the differences? Using this information, can you approximate what percentage of air is oxygen?

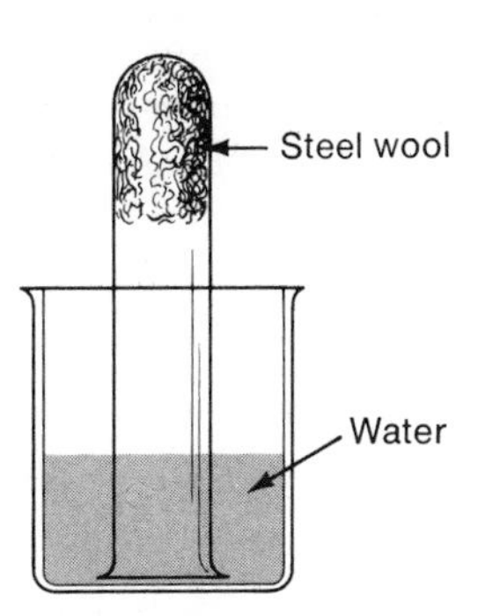

Figure 17–2 Invert test tube containing moist steel wool into a pan or beaker of water.

Demonstrating the presence of water vapor in air. Place some ice in an empty can or a glass beaker. Let the container stand for about fifteen minutes. What appears on the outside of the container? What do you think is the origin of this material?

What is a temperature inversion? A temperature inversion can be simulated in the classroom. Obtain two milk bottles (or similar glass containers), some food coloring (or ink), and a small index card. Fill one of the bottles with cold water and the other with warm water. Place some food coloring (or ink) into the warm water. (It is best to do this experiment in a shallow pan, as a little water is usually spilled.) Place the index card over the mouth of the bottle containing the warm water. Holding the card, tip the bottle upside-down and place it on top of the bottle containing the cold water. Gently pull the index card from between the bottles, keeping the two bottle mouths together. What happens to the water in the two bottles? How does this compare with the usual conditions in the atmosphere? Repeat the experiment, but this time place the bottle with cold water on top of the one with warm water. What happens? How does this experiment demonstrate a temperature inversion?

Air has some unique properties. We cannot see, smell, or taste it, but the characteristics of air are largely responsible for our weather. Air is a gas and behaves much like other gases. Air has no definite shape but will fill its container. An open paper bag cannot be half full of air, because the air will move throughout the bag. Like other gases, air will expand and can be compressed. The air over a burning candle will expand as it is heated.

Air takes up space. One of the evidences that air is real is the fact that it takes up space. Take a clear plastic bag and wave it through the air until it is filled and close it. Can the air in the bag be seen? Push on the side of the bag with a finger. Can the matter inside the bag be felt?

How can we demonstrate that air takes up space? Collect air in a plastic bag and tie the end with a twist-tie. Place the bag on a table top and balance a book on top of the bag. Ask the children what is supporting the book. What substance is between the table top and the book? How does this show that air is real?

Obtain a large glass jug. Mold some modeling clay around the neck of a funnel and press the funnel into the jug so that the joint is airtight. (A funnel in a one-hole stopper may be substituted.) Pour water into the funnel. What happens? How does this demonstrate that the jug was not empty?

Crumple a piece of paper and push it down into the bottom of a clear plastic cup. Turn the tumbler upside-down and push it directly down into a large container of water. Pull the tumbler straight out again and inspect the paper. Did it become wet? Why did this happen? Place a cork on the water and push the tumbler down over the cork. What happens to the cork and the level of water in the tumbler?

Since air is a material substance, it does occupy space. The activities that have been described are ways in which we can detect the presence of air.

Air is all around us. It is often difficult for young children to conceptualize the fact that air permeates our environment. There are few places on the earth where air does not exist. A *vacuum* is a space that is devoid of air. Most vacuums are created by removing air in some way. Children should discuss whether or not various objects in the classroom contain air. For example, boxes, baskets, cans, and bottles all contain air. Does a sponge contain air? The presence of air in an object can be detected by immersing it in a tank of water. If air bubbles appear, then apparently air is present. Children should be helped to realize that even though a door or drawer may be shut, air will still be present in the space that is closed. When asked to identify areas in the school where they thought air *did not* exist, a group of youngsters suggested that air could not exist in a sink cabinet, in their clothes lockers, or in the teacher's desk. When the children collected air in these "unlikely" places, they proved to themselves that air is all around them.

How can children be helped to realize that air is all around them? Give each child in the class a small plastic bag and have the children trap air in the bags by closing them with twist-ties. Ask the children to name places where they think air does not exist and then try to collect air in those spaces. Were there any places where air could not be found?

Air has weight. Hard as it may be for some children to believe, air does indeed have weight. The weight of air, like that of any material, is due to the gravitational attraction between it and the earth. The air in a ball or an automobile tire is much denser than the air in the atmosphere, since it is compressed in a closed container. A cubic meter of air at atmospheric pressure at sea level weighs about 1.2 kg. (2.64 lb). The weight of the air over a surface of one square meter is about 10,356 kg.

Demonstrating that air has weight. Weigh a deflated basketball or volleyball on a sensitive balance. Fill the ball with air and weigh it again. What causes the difference?

Air exerts pressure. Since air does have weight, it is always exerting pressure on the earth and the walls of the containers it occupies. Air exerts pressure in all directions. You might wonder why objects are not crushed by all this air pressure. This does not happen because in most cases the pressure of air on the outside of objects is balanced by the pressure of air from the inside. This may be demonstrated by the following activity:

Demonstrating air pressure on the sides of a container. Obtain a metal can, such as a duplicating-fluid can, and fill it with water. Insert a one-holed rubber stopper with a piece of glass tubing in the hole into the opening of the can. Fill a long (at least 3 m) piece of rubber tubing with water and attach one end to the glass tubing. Hold the can up high in the room or attach it to a high place such as the space over a cabinet, closet, or shelving. (See Figure 17–3.) Allow the other end of the tube to empty into a basin. What happens as the water from the can flows out of the rubber tubing? How can this be explained?

As long as the can was filled with air or water, the pressure (of air) on the outside of the can was balanced by the pressure on the inside. But when some of the water was allowed to run out, there was less counteracting pressure on the inside of the can and the air pressure on the outside crushed it. If the water had been allowed to "gurgle" out of the can through the opening, the water leaving the can would have been replaced by air. The inside and outside pressure would have equalized and the can would not collapse.

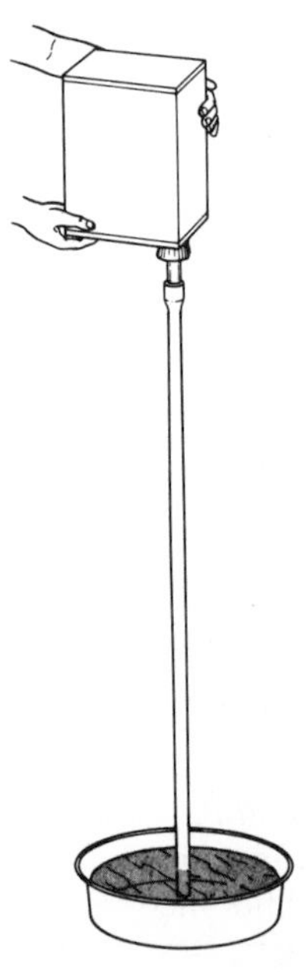

Figure 17–3 As water flows out of the can, the pressure of air on the outside of the can will press in on the can and eventually cause it to collapse.

The pressure of air on the human body is tremendous. But because there is air inside our bodies, we do not collapse. The pressure on the inside counteracts the pressure on the outside.

The pressure of air can be measured with a *barometer.* If a glass tube closed at one end is filled with mercury and inverted so that the open end is immersed in a jar of mercury, the mercury will fall to a certain level. The column of mercury in the glass tube is supported by the outside air pressure. The height of the column will vary with changes in air pressure. This is a *mercury barometer,* one of the most important weather instruments, invented by Torricelli in 1643.

Another type of barometer, the *aneroid barometer,* is basically an evacuated metal box whose sides are sensitive enough to expand and contract with changes in air pressure. These variations are transmitted by a system of levers to a needle that indicates the air pressure. The aneroid barometer has the advantage of being portable and is used in homes and schools.

Air pressure changes as air is heated or cooled. Atmospheric pressure is affected by temperature. If the air in a closed container is heated, the pressure is increased, and if the container's walls are flexible it will expand. This may be easily demonstrated by stretching a rubber balloon over the mouth of a flask. Heat the flask with a candle flame or hot plate. The balloon will expand and become larger. If the flask is placed in cold water, the air contracts and the balloon is deflated. (See Figure 17–4.)

The air in the atmosphere, on the other hand, is not in a closed container, so when it is heated it expands. Thus, there will be less air over a heated portion of the earth's surface, and consequently the air

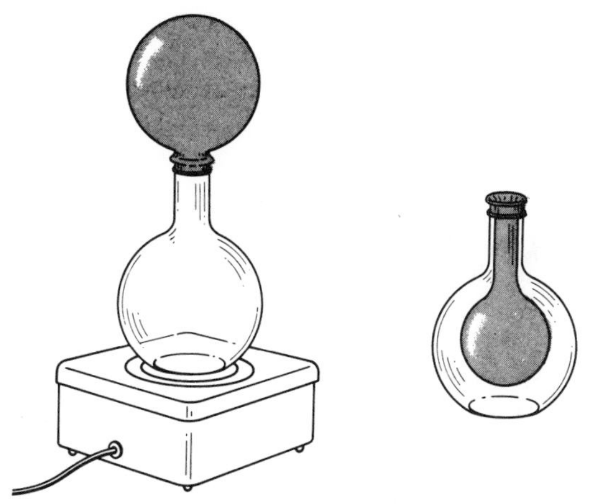

Figure 17–4 As air is heated it expands and the balloon is inflated. When air is cooled it contracts and the balloon is deflated.

pressure will be reduced. A falling barometer usually indicates that a warm mass of air is moving into the region. Conversely, as air is cooled it contracts, becomes denser, and exerts more pressure on the earth's surface. This condition results in increased barometric pressure, usually signaling the arrival of cool, dry air and generally fair weather conditions.

Air pressure changes with altitude. A column of air approximately 1,000 km (600 mi) high stands above most parts of our earth. The height of this column is calculated at sea level. Naturally, at the top of a mountain there is less air pressing down on that part of the earth than there is at sea level. The atmosphere is also thinner at high spots because of the compressibility of air (i.e., air molecules are more densely packed at sea level than they are at higher altitudes). Often, when one is traveling up mountains or in elevators, the air pressure changes noticeably. We swallow or yawn to equalize the inside and outside pressure on our eardrums. Because air pressure decreases predictably as we rise above sea level, altitude can be determined with an aneroid barometer. An aneroid barometer with a dial registering altitude is called an *altimeter.*

THE ATMOSPHERE

The atmosphere consists of a series of distinct layers. The air in the atmosphere is essential for life on earth. Most organisms obtain their oxygen directly from the atmosphere. The atmosphere also serves as a sort of greenhouse, trapping the sun's energy and keeping the temperature in most places within the fairly narrow range that is conducive to life. The atmosphere also shields the earth from X-rays and other harmful radiation that bombards the upper layers of the atmosphere. The millions of small meteors that would otherwise strike the earth's surface each day burn and disintegrate because of the heat generated by friction as they pass through the atmosphere.

The troposphere. The layer of the atmosphere that is in contact with the earth's surface is the *troposphere.* It has an average height of about 13 km (8 mi) and contains most of the mass of the atmosphere. It is in the troposphere that all of our weather occurs. The prefix *tropo* is from the Greek and means "turning" or "turbulence." This turbulence is due primarily to the heating of the lower portions of the atmosphere. Moisture, smoke, and dust particles that originate on the earth do not ordinarily go beyond the troposphere. Also in the troposphere, air temperature decreases with increasing height. The top of the troposphere is called the *tropopause.* Above this "line of demarcation," temperatures cease to diminish with height.

The stratosphere. The portion of the atmosphere from the top of the tropopause to a height of about 80 km (48 mi) is the *stratosphere.* The temperature of the stratosphere is fairly even, the average temperature being approximately −55°C. (−67°F) In the upper parts of the stratosphere, some ultraviolet rays from the sun have enough energy to break apart oxygen molecules, which then recombine to form ozone. This reaction causes a slight increase in temperature, which separates the stratosphere from the next atmospheric layer.

The ozonosphere. The ozonosphere is characterized by the presence of ozone. This layer absorbs most of the ultraviolet rays emitted by the sun. Ultraviolet radiation is harmful in that it can burn skin, blind eyes, and eventually cause death. Were it not for this ozone layer, life on earth as we know it would be impossible.

The ionosphere. Above the ozonosphere is a layer that contains electrically charged particles called *ions,* and this layer is accordingly called the *ionosphere.* It is in this layer that radio waves are bounced back to the earth, making short-wave radio transmission possible. At about 200 km (120 mi) above the earth, the gases in the atmosphere tend to separate according to weight. Between 120 and 1,000 km (72 and 600 mi) above the earth, oxygen appears in the form of single atoms. (Energy from the sun breaks apart the normal oxygen molecule, which consists of two oxygen atoms.) This layer has been studied by means of rockets, balloons, and satellites.

The exosphere. At the 1,000-km (600 mi) level the *exosphere* begins. This layer has been studied by means of satellites. From 1,000 to 2,200 km (600 to 1320 mi) above the earth, the atmosphere is composed almost entirely of the light gas *helium.* Above the 2,200-km (1320 mi) mark, *hydrogen* is the principal gas. At about 10,000 km, (6000 mi) the density of the layer of hydrogen is the same as the density of the gases in space.

WEATHER

Weather is the state or condition of the atmosphere at a given time. As mentioned previously, almost all of our weather occurs in the troposphere. Children can study these conditions directly; they can gather data related to temperature, humidity, atmospheric pressure, winds, and precipitation. Cloud formations are among the best indicators of future weather, as are the speed and direction of winds. As part of their study in this area, children can keep records of cloud and wind conditions to see what kinds of weather they are associated with.

Weather is caused by the interaction of the sun, air, water, and land. The sun warms the earth, but it does not warm all places equally. Since the earth is round, the sun's rays will not strike every spot on its surface at the same angle. Areas near the equator receive the sun's radiation more directly (i.e., rays hit the equator at nearly a right angle) and are therefore warmer than areas to the north or south. As the sun's rays hit a spot at a higher latitude, the same heat energy is spread over a larger surface (because of the tilt and curvature of the earth), and thus less heating of that region takes place. Furthermore, the tilt of the earth's axis causes the earth to receive the sun's most direct rays at different points at different times of the year. (See pages 304–6.) This accounts for our seasons. Also, different earth materials (e.g., rock, soil, sand, water) are differentially affected by the sun's warmth. Day and night also affect how these materials are heated. Some absorb and hold heat, while others are subject to considerable day–night temperature fluctuations. This uneven heating of the earth largely accounts for variations in weather. If all places on the earth were heated equally, the weather would be far less varied and interesting.

As air interacts with differentially heated portions of the earth, it is either warmed or cooled, depending on the surface conditions. Air acts as an insulator, holding in the heat absorbed from the sun. This insulation effect prevents us from experiencing the extreme day–night changes that we would experience if we did not have an atmosphere.

In general, as portions of the earth are heated, the air above them absorbs the heat and expands; the warm air tends to rise, being replaced by cooler air, which is heavier and takes its place. The result is movement or circulation of air.

Air can absorb water vapor. Water enters the atmosphere through the process of *evaporation.* What essentially happens in evaporation is that water molecules bounce out of their liquid surroundings and enter the atmosphere as water vapor. If water is heated, the average speed of its molecules increases and more of them can escape into the air. This is why water evaporates faster in warm regions than in cold ones.

Warm air can hold more water vapor than cooler air. When air is cooled, some of its water vapor condenses and droplets of water form a cloud. Within the troposphere air becomes cooler with increasing altitude. Thus, as air rises it is cooled and can hold less water vapor, so the vapor will condense, form clouds, or fall as rain or other forms of precipitation. This phenomenon may be observed on a cold day, when we see a "cloud of air" as we exhale. The water vapor in our breath is cooled by the outside air, causing it to condense into small droplets.

In sum, the changes that take place in the weather are due to changes in the composition of the air, particularly changes in the amount of water in the air and fluctuating temperatures. These changes can be studied in the following activities.

Are all surface materials affected equally by the sun's rays? Our weather is largely governed by the fact that the earth is warmed by the sun's radiation and then warms the air above it. Obtain six jars or flasks of the same size and fill them as follows:

Jar 1—dry, light sand
Jar 2—wet, light sand
Jar 3—dry, dark soil
Jar 4—wet, dark soil
Jar 5—water
Jar 6—water to which several drops of food coloring have been added

Make sure that approximately the same amount of materal is placed in each jar. Insert a thermometer in each jar and see that all the thermometers have approximately the same reading. Place in bright sunlight or shine a heat lamp (or a projection light) on the jars and let it heat them for five minutes. Have the children predict what changes will occur. What changes actually take place? Do they match the children's predictions? Does fanning (wind) cause changes in temperature readings? Does the angle at which the light strikes the surface matter?

How can heat energy be trapped? Heat energy can be trapped by glass, as it is in a greenhouse, in much the same way that it is trapped by the atmosphere. Place one thermometer in a test tube or a narrow glass jar and another in the open air alongside the test tube. Put them in direct sunlight or shine a projection light on them. Have the children note the temperatures shown by the thermometers every few minutes for half an hour.

Usually the temperature in the open air will rise first, but soon the temperature in the test tube will become warmer. The test tube, like the atmosphere, keeps some of the radiant energy out. This is why the temperature in the test tube rises a little more slowly than that outside. Eventually, however, the test tube traps sufficient heat radiation, which is reradiated by the thermometer and the air in the test tube, thus raising the temperature inside the test tube considerably higher than that outside.

What are the major factors affecting evaporation? Obtain two olive jars (with narrow mouths) and two straight-sided wide-mouthed jars. Fill the jars with identical amounts of water. Tightly cap one olive jar and one wide-

Time Elapsed (in Days)	*Olive Jar Opened*	*Olive Jar Capped*	*Wide-Mouthed Jar Opened*	*Wide-Mouthed Jar Capped*
0				
1				
3				
5				

Figure 17–5 Amount of water in each jar.

mouthed jar, leaving the other two open. Place all four jars on a windowsill. Record the amounts of water left in each jar at the beginning of the experiment and after one, three, and five days. The amounts may be determined by pouring the contents of each jar into a graduated cylinder and then replacing the water in the jar. Records may be kept in a chart like the one in Figure 17–5.

What conclusions may be drawn from the data obtained? What factors influenced the results? How did the size of the jar opening affect the results? What was the significance of capping the two jars? Try some variations to control factors such as heat, wind exposure, and the like.

Wind

Wind is air in motion. Winds result when air moves from a region of high pressure to a region of low pressure. Air heated in the tropics rises by convection, and this results in an area of low pressure. Cool air from high-pressure areas will move in to take the place of the air that has risen, thus causing a wind. In the polar regions, on the other hand, air is cooled and forms a cold mass, which pushes toward the equator. The resulting air currents interact with one another as they circulate over the planet. From the middle latitudes, the *trade winds* blow in an easterly direction toward the equator. Also from the middle latitudes, the prevailing *westerlies* blow in a westerly direction toward the poles. (See Figure 17–6.) The United States is mostly within the belt of the prevailing westerlies, and this accounts for the general movement of winds from west to east in the United States. These winds, caused by the unequal heating of the earth, are called *planetary wind systems.*

Land and sea breezes are also due to the heating of the earth, but here the unequal heating varies according to a day–night cycle. On the seashore the sun heats the land during the day, and it becomes much warmer than the ocean water. The warm air (at lower pressure) over the land is replaced by the cool, heavier air from over the water's surface. This is called a sea breeze. At night, however, the earth cools rapidly and

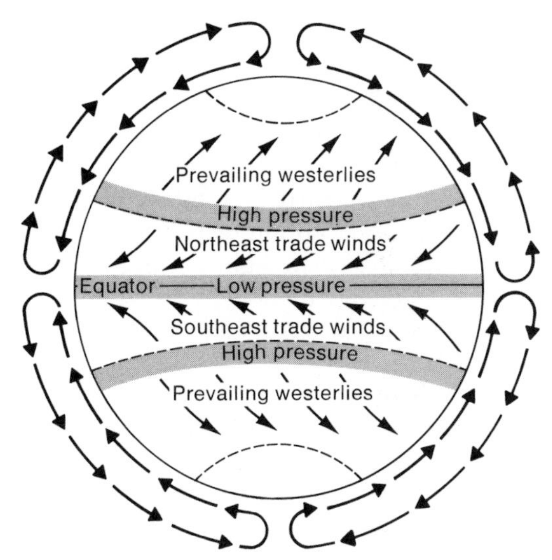

Figure 17–6 The planetary wind belts. Warm equatorial air rises and cold air at the poles descends.

the water (whose temperature fluctuates less from night to day) is warmer than the land. The air will then rise over the ocean and be replaced by the cooler air over the land, causing a land breeze. (See Figure 17–7.)

Air masses and fronts are largely responsible for our daily weather. An air mass is a large body of air that takes on the particular properties of its source. When air stagnates over a given region of the earth, it acquires the temperature and humidity characteristics of the region. Air masses are classified according to source and temperature, as either *tropical* or

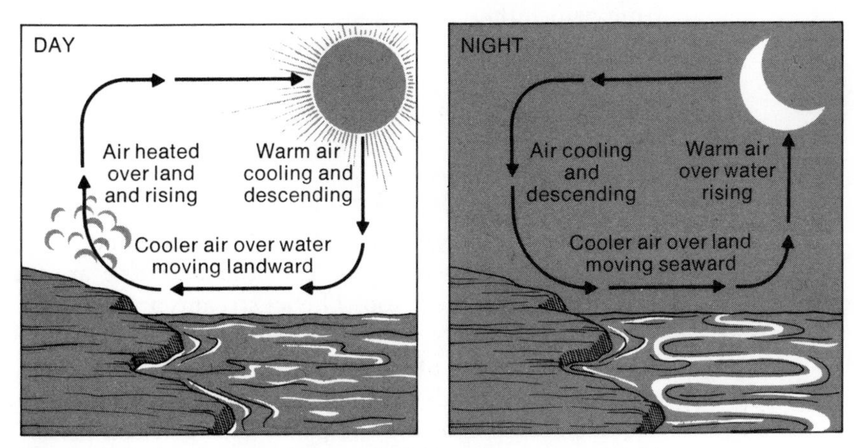

Figure 17–7 Land and sea breezes. (U.S. Dept. Commerce, C.A.A.)

polar. Depending upon whether the source region is over land or water, the air mass is further classified as *continental* (over land) or *maritime* (over water).

A *continental polar air mass* is cold and dry. The air is dense and its pressure is high. The lack of moisture usually means clear skies. The rotation of the earth and other factors tear this air mass away from its source; it can then interact with other air masses. A *maritime tropical air mass* contains warm, light air and is generally associated with low pressures. Since the air mass is warm and originated over water, it usually holds a great deal of water and will cause rain if cooled. (See Figure 17–8.)

When two dissimilar air masses meet, there is usually a sloping boundary between them called a *front.* When cold air advances into warmer air, there is a *cold front.* Similarly, when a mass of warm air moves into a mass of cold air there is a *warm front.* When one front overtakes another front so that there are three air masses in close proximity to each other, an *occluded front* is formed. Tracking the movement of these fronts is a significant factor in weather prediction.

At a cold front cold, dense air will force the warm air upward. As the warm air is pushed up, it is cooled, its vapor condenses, clouds are formed, and some form of precipitation usually occurs. After the cold front passes, the cold air mass arrives and there is a drop in temperature, increased pressure and a general clearing of skies.

At a warm front warm air moves up over the cold air mass. If the warm air contains considerable moisture, clouds will form. Rain will usually begin and become rather steady. As the warm air mass moves in, there is a gradual rise in temperature, a change in wind direction, and gradual clearing.

Usually, in an occluded front one cold front overtakes another cold front, which then wedges under warm air, pushing it up off the ground. When this occurs, the warm front lingers a while and then is followed by the cold front. If a particular air mass remains relatively stationary, the same weather can be expected for several days.

Large air masses may be pulled away from their source locations by the jet streams. The *jet streams* are belts of strong, high-speed winds that flow at altitudes of about 6,000 to 12,000 m. (20,000 to 40,000 ft.) Two jet streams have been found in the Northern Hemisphere, one near the pole and one over the middle latitudes. There are jet streams in the Southern Hemisphere too. The jet streams are believed to be a result of contact between dry, cold, polar air and warm, moist tropical air. High-flying airplanes are often either slowed down or speeded up by the force of the jet streams, depending on the direction in which they are flying. For this reason charting the jet streams and compensating for their effects are of vital concern to airlines.

In weather reports the terms *highs* and *lows* are often used to refer to high and low pressure systems. High pressure is usually associated with fair, dry weather and low pressure with changing and stormy weather. In the United States and Canada, alternate highs and lows (moving from west to east) across the continent are important factors in our daily weather.

Different weather conditions exist ahead of and behind the surface of a front. One side of the surface of a front might contain cold, dry air with strong northeasterly winds. On the other side moist air with southwesterly winds may prevail. The pressure along the front is lower than the pressure at the center of the air masses on either side of the front. Air then moves toward the lower-pressure area along the front and is deflected by the spin of the earth so that it moves around the lows in a counterclockwise direction. This counterclockwise movement of air around a low-pressure area is called a *cyclone*. Clouds and precipitation are usually found in the region of the cyclone. The flow of air around a high-pressure area is in a clockwise direction and is called an *anticyclone*. Cyclones should not be confused with tornadoes, which will be described later.

Wind direction can be used to locate the positions of lows and highs. In the Northern Hemisphere, if you stand with your back to the wind the low-pressure area is to your left and the high-pressure area to your right. This rule is known as *Buys-Ballot's law*.

The origin of cyclones in temperate latitudes is thought to be related to an interaction along a front between warm and cold air masses.

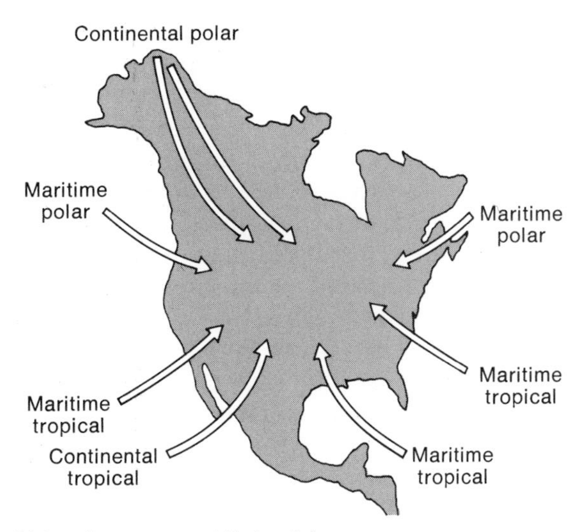

Figure 17–8 Major air masses and their origins.

Warm air moving north from tropical regions is replaced by cold air moving south as a cold front. Often the spinning of the earth causes the air masses to swing into the counterclockwise motion associated with a cyclone.

Precipitation

Water vapor in the air can form clouds and also cause precipitation. When water vapor is cooled and small dust or smoke particles are present, the water vapor will condense into small droplets of water. These droplets form clouds. Often a thick surface layer of air will be cooled. As the vapor in the air condenses into water droplets, a slight motion of the air keeps the droplets suspended. This collection of suspended droplets is called *fog*. At altitudes over about 15 m, (50 ft) this fog would be called a *cloud*. In the mountains clouds often form at the snow line. When air comes into contact with the snow-covered surface, it is cooled and the water vapor condenses into droplets.

There are four general cloud types. *Cirrus* clouds are very high and wispy and are composed of tiny ice crystals. *Cumulus* clouds are rolling and billowy and indicate the presence of air convection currents. Of special interest are the high, anvil-topped *cumulonimbus* or thunderclouds. The anvil tops are formed when rising air currents reach a level of stable air. Cumulonimbus clouds are almost always associated with thunderstorms. (See Figure 17–9.) *Stratus* clouds are sheetlike clouds that indicate very little convective activity in the atmosphere. They usually indicate a chance of rain. *Nimbus* clouds are low, dark rain or snow clouds.

When water vapor is cooled and condensation nuclei are present, the vapor becomes water droplets. Many of the droplets coalesce as a result of air movements and become larger and heavier drops. When the drops become heavy enough so that the force of gravity acting on them overcomes the lifting force of the wind, they will fall as rain.

Snow occurs when the water vapor in air cools so quickly that it condenses directly from its vapor form to ice crystals. If the temperatures are sufficiently cold so that the ice does not melt, the snow will reach the ground. Almost all snowflakes are six sided, but they show a tremendous variety of intricate designs.

Hail occurs when strong updrafts of air cause small ice crystals to be carried upward in a cloud a number of times, forming layer upon layer of ice crystals. The ice will eventually fall as hail. A typical hailstone looks like an onion when it is cut open because of its many layers.

Sleet will fall if raindrops fall through very cold air, in which they freeze and turn into frozen rain.

Special combinations of weather conditions form different kinds of storms. *Hurricanes* are cyclones (see page 373) that occur in tropical

Figure 17–9 Cumulonimbus clouds with anvil top. (Courtesy U. S. Dept. of Commerce, NOAA.)

regions. (See Figure 17–10.) They usually form over oceans and are characterized by very strong winds, rain, and very low pressure. Hurricanes can be extremely destructive and therefore must be carefully monitored by weather officials.

Tornadoes are storms in which strong, intensive winds whirl and twist and cause great damage. The tornado has a *funnel* in which there is very low air pressure. As a tornado strikes, buildings may explode when the higher pressure air inside the building pushes out toward the low pressure above or around it.

Thunderstorms are storms in which the sound of thunder is heard. They are usually small, local storms, only affecting a few square kilometers of surface area. They are caused by convection currents carrying warm, moist air. As the air rises, it is cooled and its condensation produces a cumulus cloud. The upward air currents may be very powerful and reach speeds of over 150 km (90 mi) per hour. Rain, snow, or hail may form in the upper parts of the cloud. As falling rain rushes past the strong updrafts, the friction created produces static electrical energy. If

Figure 17–10 Satellite photograph of Hurricane Agness (Courtesy NOAA).

enough of a charge builds up, it may discharge as lightning from the cloud to the ground or to another cloud. The great heat produced by the flash of lightning causes air to expand at supersonic speeds. This causes a sonic boom called thunder.

Estimating the distance to a thunderstorm. The light from a bolt of lightning reaches our eyes almost instantly. The speed of sound, and therefore the speed of thunder, is approximately one-third of a kilometer (1056 ft) per second. Count the number of seconds between the flash of lightning and the sound of thunder. If you divide this number of seconds by three (this is the same as multiplying by one-third), the result will be the approximate distance of the storm from you in kilometers. How can you tell whether the storm is getting nearer or farther away?

Simulating a cloud. Obtain an empty milk bottle or a thick glass jar. Run hot water into the bottle or jar until it becomes warm all over. Pour out most of the water, leaving the bottle about one-third full. Place an ice cube over the mouth of the bottle or suspend it near the bottle's neck in the toe of a nylon stocking. What happens? How may this be explained?

What are the shapes of snowflakes? Snowflakes have many varied and beautiful shapes. In order to see them, though, they must be caught in a special way so that whey will not melt. Chill a black cloth in a refrigerator or leave it outside in the cold. Spread the cloth across a cold board and catch a few snowflakes. With care the flakes can be kept a little while for study with magnifiers.

Weather analysis and prediction are accomplished by using special instruments, professional expertise, geographical realities, and historical patterns. Meteorologists are trained professionals who study the atmosphere and weather. By studying planetary wind belts (see pages 370–74) and frontal activities, meteorologists can make fairly accurate weather predictions in most parts of the world. In some areas (e.g., the equatorial regions and the trade wind belts), weather prediction is fairly simple, since there are regular daily changes that seldom vary. In the region of the prevailing westerlies, as in the United States, weather conditions are more variable and more subject to unpredictable fluctuations. Seasonal changes also bring obvious variations, as do local geographical conditions such as altitude and proximity to bodies of water.

In North America, because most of the country is affected by the prevailing westerlies, air movements generally pass from west to east. Even cold air masses from the north and warm air masses from the south usually bend and veer in an easterly direction. Thus, if no other information is available, the best way to predict what the weather will be like in New York on Thursday is to find out how things were in Chicago on Tuesday or Wednesday.

Keeping accurate records is an important task of meteorologists. The study of these records may reveal weather trends or patterns that can be helpful in the prediction of weather. Generally, records are kept of temperature, humidity, pressure, wind speed and direction, and precipitation.

Weather instruments are important tools for determining current atmospheric conditions and also aid in forecasting. Thermometers tell us how hot or cold the air is. The bottom of a thermometer has a bulb, usually filled with mercury or alcohol. As the surrounding air is warmed, the liquid expands and rises in the tube. The liquid contracts as air is cooled. Plotting temperature readings in various regions can be helpful in predicting weather.

Wind speed and direction are also important indications of weather. Wind direction is determined in terms of the direction *from which* the wind is blowing. Knowledge of wind direction can help in determining the location of a low-pressure system and the bad weather usually associated with it. Wind speed indicates, to a certain extent, the differences in air pressure between high-pressure and low-pressure regions. The Beaufort scale is used in estimating wind strength. (See Figure 17–11.)

Anemometers are instruments that measure wind speed. An anemometer consists of several round cups attached to the ends of crossbars. The cups catch the wind and the anemometer spins around. Usu-

BEAUFORT SCALE OF WIND FORCE

Beaufort Number	*Wind Effects*	*Miles per Hour*	*Weather Bureau Description*	*Symbol on Weather Map*
0	Calm; smoke rises vertically.	Less than 1	Calm	
1	Direction of wind shown by smoke drift, but not by wind vanes.	1–3	Light air	
2	Wind felt on face; leaves rustle; ordinary vane moved by wind.	4–7	Slight breeze	
3	Leaves and small twigs in constant motion; wind extends light flag.	8–12	Gentle breeze	
4	Raises dust and loose paper; small branches are moved.	13–18	Moderate breeze	
5	Small trees in leaf begin to sway; crested wavelets on inland waters.	19–24	Fresh breeze	
6	Large branches in motion; whistling heard in telegraph wires; umbrellas used with difficulty.	25–31	Strong breeze	
7	Whole trees in motion; inconvenience felt in walking against wind.	32–38	High wind	
8	Breaks twigs off trees; generally impedes progress.	39–46	Gale	
9	Slight structural damage occurs (chimney pots and slate removed).	47–54	Strong gale	
10	Seldom experienced inland, trees uprooted, considerable structural damage occurs.	55–63	Whole gale	
11	Very rarely experienced; accompanied by widespread damage.	64–75	Storm	
12	Great damage.	Above 75	Hurricane	

Figure 17–11 The Beaufort Scale of wind force.

ally the anemometer is connected to an indicator or meter that shows the wind speed. Weather vanes determine the direction from which winds are blowing. The weather vane (or, more accurately, wind vane) is a delicately balanced arrow whose tail is larger than its head. When the wind blows, the larger surface of the tail is pushed away from the wind, causing the head to point into the wind.

Humidity is another important component of weather. The amount of water vapor in the air affects our comfort and also aids in estimating the potential for precipitation. Humidity can be determined in several ways. One way is with a *hair hygrometer*. Human hair stretches when wet and shrinks when it dries. By attaching a pointer to some strands of human hair, calibrations can be made to determine relative humidity. Another, more accurate means of determining humidity is with a *sling psychrometer*. Water can evaporate more rapidly in dry air than in moist air. Temperature also plays a role, since warmer air can hold more moisture than cooler air. In the sling psychrometer there are two thermometers. One is used to measure air temperature; the other is kept moist around its bulb by means of a wet wick. The two thermometers, attached to a bar, are whirled about by a handle. Water evaporates from the wet-bulb thermometer, causing its reading to be lower because the process of evaporation takes up heat. (The evaporation of perspiration from our skin has a cooling effect on our bodies.) Using the readings of the two thermometers, the moisture content of the air can be found on a chart called a *humidity table*. (See Figure 17–12.)

Air pressure is one of the most important aspects of weather. A barometer measures air pressure; its operation was described on page 365. The locations of high and low pressure systems are perhaps the best indicators of impending weather conditions. As mentioned previously, high pressure is generally associated with clear, fair weather and low pressure usually means changing or stormy weather.

A very simple weather instrument is the *rain gauge*. This is nothing more than a straight-sided vessel with calibrated sides, placed outdoors or on top of a building to the record the amount of rainfall. More often, a funnel is placed on top of the rain gauge to increase the area that can catch the precipitation. The ratio of the diameters of the funnel mouth and the collection vessel is determined and compensations are made. For example, if the funnel catches ten times as much water as the collection

Relative Humidity (Percentage) – °C

Temperature of Dry Bulb (°C)	*Depression of the Wet Bulb (°C)*														
	1	*2*	*3*	*4*	*5*	*6*	*7*	*8*	*9*	*10*	*12*	*14*	*16*	*18*	*20*
50	94	89	84	79	74	70	65	61	57	53	46	40	33	28	22
45	94	88	83	78	73	68	63	59	55	51	42	35	28	22	16
40	93	88	82	77	71	65	61	56	52	47	38	31	23	16	10
35	93	87	80	75	68	62	57	52	47	42	33	24	16	8	
30	92	86	78	72	65	59	53	47	41	36	26	16	8		
25	91	84	76	69	61	54	47	41	35	29	17	6			
20	90	81	73	64	56	47	40	32	26	18	5				
15	89	79	68	59	49	39	30	21	12	4					
10	87	75	62	51	38	27	17	5							

Figure 17–12 Chart of relative humidity. The depression of the wet bulb is the difference between the wet and dry bulb readings.

Figure 17–13 Weather satellites such as the one shown above and on p. 381 are invaluable aids in modern weather forecasting.

vessel, then a 10-cm column of water means that the true rainfall was only 1 cm.

The U.S. Weather Service, a division of the National Oceanic and Atmospheric Administration, is the official meteorological agency. It maintains hundreds of observation stations and keeps weather records. In addition, many stations study the upper air and temperatures at various levels by means of weather balloons, radiosonde, and other devices. Recently weather satellites such as Nimbus and Tiros have proven invaluable in determining global weather conditions. (See Figures 17–13 and 17–14.) Satellite photographs clearly show cloud cover over large areas and can provide the means for accurately plotting storms, hurricanes, or other unusual disturbances.

The U.S. Weather Stations are part of an international network of more than 14,000 weather stations that are associated with the World Meteorological Organization, an agency of the United Nations.

To be useful, these data must be assembled in a meaningful way and interpreted. The data are assembled on weather maps, with the readings from each station throughout the nation plotted on the map. Figure 17–15 is an example of a weather map for the United States.

Figure 17–14 Weather satellite.

These maps include information about temperature, barometric pressure, wind speed and direction, areas of precipitation, the locations of fronts, and general sky conditions.

Making a water thermometer. Obtain a small soda bottle, some modeling clay, a clear plastic straw (or glass tubing), and food coloring. Fill the bottle about four-fifths full of water and add some food coloring. Dry the lips of

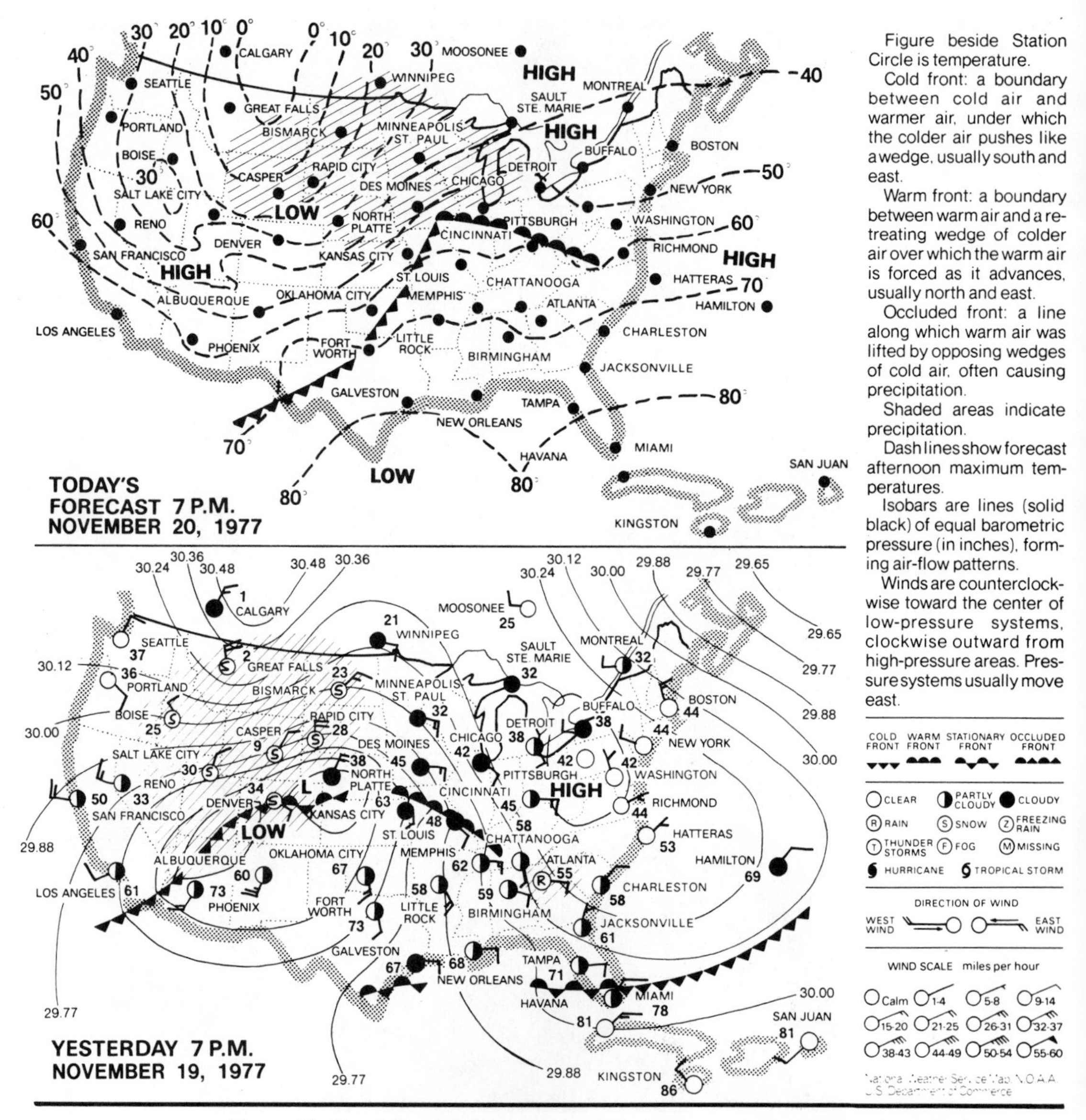

Figure 17–15 Daily weather maps, such as the one reproduced above, can be the basis for many activities. (© 1977 by The New York Times Company, reprinted by permission.)

the bottle and squeeze clay around the rim. Put the straw halfway into the bottle and press the clay around the bottle lips and straw to form a tight seal. (See Figure 17–16.) Note the height of the water in the straw. Tape a card to the straw so that a scale can be made. Place the bottle in the sunlight. Does the height of the water change? In which direction does it move? Put the bottle in a cool place. What happens? Observe and record daily changes.

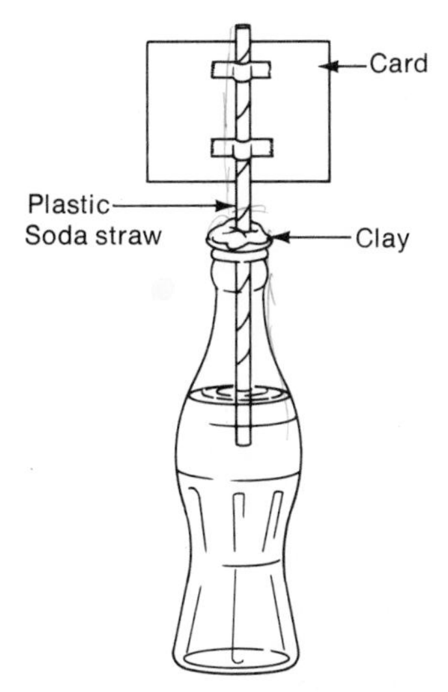

Figure 17–16 A water thermometer made from easily obtainable materials.

Demonstrating the operation of a barometer. It is inadvisable to work with mercury at the elementary school level (because it is a poisonous substance), but the operation of a barometer may be demonstrated with water. Obtain a small, slender soda bottle partially filled with water and invert it in a shallow pan of water. Do not allow water to pour out of the bottle as it is inverted. A strip of paper should be attached to the outside of the bottle to record changes in the water level, which will approximate changes in atmospheric pressure.

How does the wind change? Young children can make some important observations of wind. Have the children turn and face the wind. Hair, coats, scarves, and the like can be seen blowing in the same direction. Have the children point in the direction that the wind is coming from. Little flags can be made in class to be used as wind vanes. With children who can understand direction, north, south, east, and west orientation can be established. If it is possible to sight a flag or a smokestack from the classroom window, daily records can be kept on changes of wind direction. The children can also note whether the winds feel warm or cool. In the springtime, how does a wind from the northwest feel? What kind of weather does it bring? Is this wind different in the winter and fall?

How can relative humidity be determined? A hygrometer can be made using two inexpensive thermometers, a milk container, and a shoelace. The two thermometers must register the same reading when placed together. Cut a 10-cm (4 in) length of a cotton shoelace and slip it over the bulb of one of the thermometers. Tie it with thread above and below the bulb to hold it in place. Attach the thermometers to adjacent sides of the milk container with rubber bands. Cut a small hole (about one square cm) into the milk container just below the thermometer whose bulb is covered with the shoelace. Push the free end of the shoelace through the hole and fill the inside of the container with water to the level of the hole. This will keep the shoelace wet. (See Figure 17–17.) The water around the wet-bulb thermometer evaporates and causes a lowering of its temperature. The wet-bulb thermometer may be fanned with a piece of cardboard. Read the dry and wet-bulb thermometers. Subtract the wet-bulb reading from the dry-bulb reading and refer this difference to the table of relative humidity. (See Figure 17–12.)

How can the amount of rainfall be measured? The simplest way of measuring rainfall when working with young children is to simply use a straight-sided can and mark the units of measurement (e.g., 1, 2, 3, 4, . . . cm) along its side. The amount of rainfall collected should be checked shortly after the rain has stopped, since the rain in the collection vessel will soon evaporate.

A rain gauge can be constructed that will allow the children to accurately measure rainfall even in small amounts. Two straight-sided jars are needed; one should have a wider diameter than the other. A peanut butter jar can serve as the wide-diameter jar and an olive bottle for the narrow-diameter jar.

First the jars should be calibrated. Fasten a thin strip of masking tape along the side of the narrow jar. Fill the wide jar to a height of exactly 1 cm.

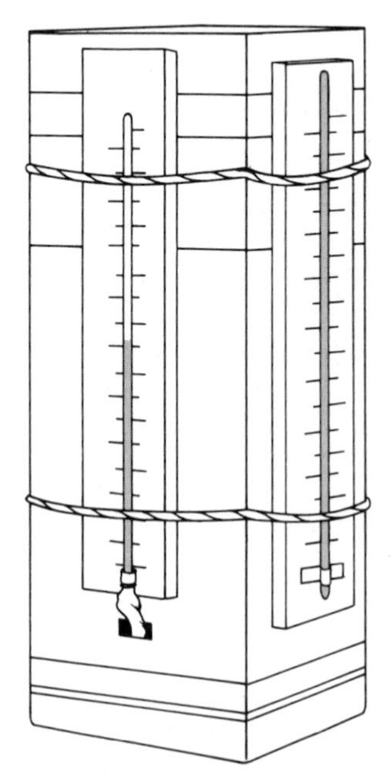

Figure 17–17 A hygrometer, an instrument for the determination of relative humidity.

Pour the water from the wide jar into the narrow jar. Place a mark on the masking tape on the narrow jar that is level with the top of the water. This is the equivalent of 1 cm of rainfall. In a similar way, mark the places on the narrow jar that signify higher amounts of rainfall. A ruler can be used to mark fractions of centimeters.

The wide-mouthed jar should be left outside, away from buildings, to catch rainwater. To prevent tipping, the jar can be placed in a small hole or braced with soil or clay. After a rain the jar can be brought in and the water carefully poured into the narrow, calibrated jar. By sighting the height of the water against the calibrated scale on the narrow jar, the children can determine the amount of water that fell during a rainstorm. Records can be kept for the week, month, and year.

Using weather maps. Many activities can be based on the weather maps (like the one on page 382) that appear daily in most local newspapers. The following questions require accurate map reading:

What was the weather like in your city on the day the map was compiled? What time of day was it?

What was the temperature? What was the barometric pressure?

What was the wind speed and direction?

How far from you was the nearest rainstorm?

Which areas had the highest and lowest temperatures?

What is the high temperature in your city likely to be in the afternoon?

What is the location of major warm, cold, and occluded fronts?

How is today's weather picture different from yesterday's? What do you think the weather will be like tomorrow?

Keeping weather records. Children in the earliest school grades can discuss weather as part of their early morning routine (e.g., at attendance-taking time or with "daily news" or opening exercises). The youngsters may keep daily weather records on calendars on which there is enough room in each box for the placement of a symbol depicting the day's weather. (See pages 111 and 382.) At the end of each month, children may count the number of days that were sunny, rainy, windy, and so forth.

Forecasting the weather. Using thermometers, barometers, rain gauges, wind vanes, wind speed observations, wet- and dry-bulb thermometers, and cloud study, children can obtain a fairly comprehensive description of existing weather. It may be valuable and interesting to have children "publish" a daily weather report.

Children can also make forecasts, giving special attention to the following:

1. Barometric pressure is a good indication of impending weather conditions. A "falling" barometer often indicates the arrival of moist, warm air. When this air meets cold air, there are likely to be clouds and precipitation. A "rising" barometer usually means that fair weather will be moving in.
2. Cloud formations are also helpful. High cirrus clouds moving from the west, for example, are usually associated with cold fronts. Scattered cumulus clouds can mean continued fair weather. Murky nimbus clouds may indicate the arrival of a warm front. A line of cumulus or cumulonimbus clouds may be a squall line with its associated stormy weather.
3. Weather reports can be most helpful in forecasting. With these reports and a knowledge of the general movement of weather in their region, children can often achieve a surprisingly high rate of success in their predictions.

Climate

Climate is the prevailing weather in a region over a long period. Weather is an indication of a prevailing short-term atmospheric condition. After many years of data collection and the observation of weather patterns and trends, an area's climate may be characterized. This is usually done by averaging temperature, rainfall, humidity, and other such factors. If the averages for two places are alike, though, it does not necessarily mean that those two places have the same climate. For example, New York and San Francisco have approximately the same average temperature each year, but New York has many extremes during the year while San Francisco's temperatures are less variable from day to day and month to month. Similarly, Los Angeles may have the same annual temperature as a desert city but will be far less subject to the day–night temperature extremes found in a desert.

It is actually the *distribution* of the weather within a year that is significant in characterizing an area's climate. There are several factors that determine what an area's climate will be like:

1. Generally, the higher the latitude north or south of the equator, the cooler the climate will be.
2. Places in high altitudes are generally associated with a cooler climate than lower areas within the same latitude.
3. Prevailing winds are important in the determination of an area's climate. The westerlies and the trade winds bring certain regular climatic characteristics year after year.
4. Continental (inland) and marine (near the coast) location is another factor that affects climate. In general, continental areas have more temperature extremes than marine areas, the latter's temperatures being moderated by proximity to large bodies of water.
5. Ocean currents also temper a region's climate. Iceland, because of

its northern location, would be expected to have a much more severe climate than it actually does. But its temperatures are rendered milder by the fact that the warm waters of the Gulf Stream pass near the island nation.

6. An area's position with regard to mountain ranges, air masses, and high- and low-pressure systems also affects its climate.

The factors just listed must all be taken into consideration when describing a region's climate. They explain why a particular place's weather or climate may differ from what would be expected from its latitudinal position.

RELATED ACTIVITIES

Air takes up space, p. 44.

Demonstrating that air is a real substance capable of transmitting sound, p. 413.

What is the warmest time of day? p. 299.

Demonstrating the effects of the angle of the sun's rays, p. 306.

How can the presence of water vapor in the breath be demonstrated? p. 274.

What happens to a streak of water on a chalkboard? p. 147.

The case of the candle and the jar, p. 74.

Weather observations, p. 111.

Demonstrating the presence of carbon dioxide in exhaled air, p. 274.

Mixing equal amounts of water at different temperatures, p. 149.

What are some of the characteristics of water? p. 345.

What are some of the physical characteristics of water? p. 188.

What happens when a liquid changes to a gas (evaporates)? p. 190.

How is air heated? p. 191

How can young children explore some effects of heat? p. 391.

How does the operation of a thermometer demonstrate the expansion and contraction of liquids? p. 394.

How can the expansion of gases be demonstrated? p. 394.

How can convection currents be detected? p. 397.

How do ocean currents flow? p. 348.

Observing the effects of heat convection and radiation, p. 398.

How is our daily weather determined by the interaction of the sun, air, and water? p. 41.

Recording and forecasting weather, p. 12.
Explaining the seasons, p. 306.
Predicting the possible consequences of changes in an ecosystem, p. 54.
What kinds of materials pollute the atmosphere? p. 200.

REFERENCES

For Teaching

BAINBRIDGE, J. W. AND R. W. STOCKDALE. *Weather Study—An Approach to Scientific Inquiry.* New York: Barnes and Noble, 1973. A handbook for teachers with many suggestions for activities and investigations.

SPITZ, ARMAND N. *Weather.* New York: Bantam, 1968. A paperback reference for teachers and other adults containing concise and accurate information about weather. Many photographs.

SUTCLIFFE, R. C. *Weather and Climate.* New York: W. W. Norton, 1966. Detailed background information for teachers.

For Young Children

BRANLEY, FRANKLYN. *Air Is All Around You.* New York: Harper & Row, 1962. A book for young children that can stimulate several activities and investigations.

PINE, TILLIE S. AND JOSEPH LEVINE. *Air All Around.* New York: McGraw-Hill, 1960. A book about the nature and characteristics of air and ways in which it can be studied.

SCARRY, RICHARD. *Richard Scarry's Great Big Air Book.* New York: Random House, 1971. A colorful book that provides young children with basic concepts about air and the atmosphere in a cartoon-like format.

For Older Children

BERGER, MELVIN. *The New Air Book.* New York: Harper & Row, 1974. A detailed account of air, atmosphere, weather, and air pollution. Contains a consideration of the chemical properties of air.

BOVA, BEN. *Man Changes the Weather.* Reading, Mass.: Addison-Wesley, 1973. The harmful and beneficial effects of our society on air and weather are considered in this book for children in the upper elementary school grades.

ELLIOTT, SARAH M. *Our Dirty Air.* New York: Simon & Schuster, 1971. An accurate account of air pollution, its effects on weather, and what can be done to alleviate it. Illustrated with photographs.

LEHR, PAUL E. *Storms: Their Origins and Effects.* New York: Golden Press, 1966. A detailed and accurate account of storms and their causes.

TANNENBAUM, BEULAH AND MYRA STILLMAN. *Clean Air.* New York: McGraw-Hill, 1973. This book deals with the major sources of air pollution and concludes with speculations about tomorrow's cities.

Children and Heat, Light, and Sound

CHAPTER 18

There comes a time in each child's life when the child discovers his or her shadow. What a remarkable sensation it is! This patch of darkness follows the child as long as he or she is in the sun's (or moon's) light. As the child runs, so does the shadow. As the child stretches, so does the shadow.

How different would our lives be without light? For one thing, we could not read by conventional methods. Green plants could not grow, for they need light to exist. The children in your class could probably list hundreds of ways in which light affects their daily lives.

The light from the sun brings warmth. Imagine a world with no heat! We could not remain alive. Heat is an important source of energy. Children can speculate about how different the world would be without heat, or whether there could be a world at all. The absence of all heat is hardly imaginable. All matter would be in a solid state and life would be impossible.

Sound is a great aid to communication. We learn to identify events and objects by the sounds they produce. What does the sound of a school bell tell children? How can they tell whether a fire engine is in the vicinity? Sounds can evoke some of the strongest human emotions. The music of a symphony, the clatter of a garbage truck, the pitter-patter of a toddler's steps, or rain on the roof—all convey a message. All signal a feeling.

As children investigate these distinct, yet related phenomena, they will have opportunities to delve into events that touch them daily. The activities that follow should help children realize the importance of these phenomena and the ways in which we depend on them and often take them for granted.

HEAT

Heat is a Form of Energy

Heat is energy associated with motion. It is generated as we rub our hands together, smooth a piece of wood with sandpaper, slide down a rope or pole, and so on. In general, when two materials are rubbed together, heat is produced. The work done (i.e., the energy expended) in rubbing the two materials together is converted into heat, and heat, in turn, may be transformed into other forms of energy.

The production of heat energy cannot be explained without an understanding of the nature and structure of matter. All elements are ultimately composed of atoms, the smallest units of structure within which elements still maintain their identity. Atoms combine to form other substances; for example, two hydrogen atoms combine with one oxygen atom to form the substance water (H_2O). The smallest unit of structure of a substance is the *molecule.* In the case of the substance carbon dioxide (CO_2), a molecule would consist of one carbon atom and two oxygen atoms. If any one of these atoms were to be removed, the molecule would no longer be carbon dioxide; its integrity would be destroyed. Only in the case of the elements themselves can an atom and a molecule be the same thing (i.e., the smallest unit of structure wherein the identity of the material still exists).

Molecules are constantly in motion. In a solid they move in a fairly regular, confined space and their movement is relatively restricted. In a liquid the component molecules move about more freely, and in a gas they move even more freely. As substances are heated, their molecules move faster. The substance expands, since more space is required for the increased movement of the molecules. (See Figure 18–1.) If the substance being heated is a solid, the molecules will eventually move so fast that they will break away from their "assigned places" within the solid and it will melt, forming a liquid. As a liquid is heated, its molecules move faster and faster until they can escape from the main body of the liquid and form a gas. This is called *evaporation.* (Heat can also be removed from a substance; as the movement of its molecules becomes slower, the substance can change from a gas to a liquid or from a liquid to a solid.)

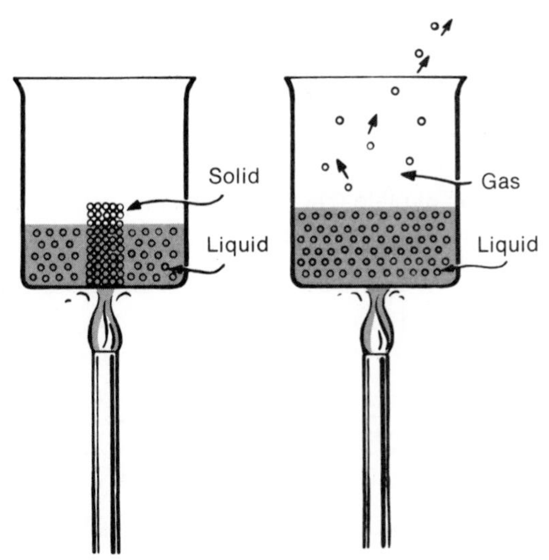

Figure 18–1 As a solid is heated, it changes to a liquid. As a liquid is heated, it changes to a gas. Note the relative density of the molecules in each state.

As hands are rubbed together, the molecules that are exposed in the top layers of the skin are caused to vibrate faster as they are pushed against one another. These surface molecules in turn set the molecules just beneath them to moving about and eventually, when the molecules in several skin layers are affected, the skin becomes warmer because the molecules in it are moving faster.

Similarly, if a nail is placed in a fire, the end in the flame heats up first. Eventually the entire length of the nail will become hot as the molecules closest to the flame are caused to move faster and bump into the adjacent molecules along the shaft of the nail, causing them to move faster until the entire nail is heated.

How can young children explore some effects of heat? The youngest pupils ought to have many experiences with objects that are discernibly hot or cold. Glasses of water can be labeled hot or cold. Daily temperature can be discussed in comparative terms: Is it colder outside than in the classroom? Foods that are usually served hot or cold can be discussed. A thermometer may be used to measure the temperature of various objects or environments and the results compared. Is the temperature in a hole in the soil greater than the temperature in the open air? Do objects become hotter when removed from the shade and placed in the sun?

Young children can also investigate the states of matter by freezing, boiling, or melting water. Appropriate vocabulary can be introduced with each experience. Children might also suggest changes of state as observed in substances other than water.

How can children explore changes in the states of matter? Changes of state should be explored with a variety of materials. Sugar can be melted, poured into a mold, and left to harden again. Butter may be melted and resolidified. Chocolate melts easily and rehardens when cooled.

Children may (with adult supervision) melt candle ends in a container on a hot plate. The liquid wax may then be poured into metal molds, in which it will harden and assume new shapes. Water and other liquids may be boiled until they vaporize. Can the vapors be seen?

Sources of Heat

Fire, friction, electricity, and the sun are major sources of heat. Any substance that can be burned to produce heat is a *fuel.* Fuels assume many forms. Coal, charcoal, and wood are solid fuels; gasoline and fuel oil are liquid fuels; and natural gas is a gaseous fuel. These substances contain carbon and hydrogen and produce considerable amounts of heat as they are burned. The original source of energy in these fuels is the sun. The sun is essential for the growth of trees, which provide wood. Coal is a fossil fuel formed over long periods from buried leaves and other plant material. Petroleum is also a fossil fuel formed from material that was once living, and it is refined to produce fuel oil and gasoline.

Friction, another major source of heat, occurs when one surface rubs against another. Even seemingly smooth surfaces have small irregularities that momentarily interlock as surfaces slide over one another. The molecules of one surface strike the molecules of the other surface, causing them to vibrate more rapidly and heat up. Similarly, if one object is struck by another (e.g., a hammer striking a nail), the surface molecules will move more quickly and heat the substances.

Heat is also produced by electricity as current is passed through wires. In an electric toaster electricity passing through the thin coiled wires causes them to heat up and turn red. Hot plates, room electric heaters, and other appliances operate on the same principle.

The sun is the ultimate source of heat derived from most fuels. Indeed, it is the ultimate source of much of the energy on earth. Even if we do the work to rub one substance against another, producing frictional heat, the work done is a result of the energy applied, and that energy is converted from the foods we eat, either in the form of green plants or as meat from animals that were dependent on plants.

How can children experiment with heat produced by friction? Have the children rub various substances together and touch the surfaces after a minute of such contact. Try objects such as wood, plastic, metals, sandpaper, and so on. Which substances seem to produce the greatest heat? Why?

How can electrical energy be converted into heat energy? Fasten one end of a wire to one post of a dry cell and touch the free end of the wire to the other post. Does the wire begin to feel hot? As soon as it does, remove the wire from the post. The passage of electrical current through the wire generated heat. Most electrical heating appliances, such as toasters and hot plates, generate heat in this way.

Effects of Heat

When most materials are heated, they expand; when most materials are cooled, they contract. Previously it was mentioned that when heat is applied to a substance its molecules move more quickly, collide with other molecules, and cause these other molecules to move faster. This results in the expansion of the substance because more space is required for the greater movement. This is why the liquid in a thermometer rises as it is heated.

Contraction occurs when most objects or substances are cooled. The action of the molecules slows down and the material actually shrinks. Gases also expand and contract as they are heated or cooled. As mentioned in the preceding chapter, air masses expand when they are heated, causing a reduction in air pressure. (See page 368.)

Many examples of accommodations to these phenomena can be

Figure 18–2 Why does heating the screw eye change its diameter?

found. Sidewalks are constructed with large cracks between the blocks of concrete so that the blocks will have room to expand and contract. Bridges, railroad tracks, and highways have expansion joints to prevent them from being destroyed by expansion or contraction.

Water is a notable exception to the general rule that substances expand when heated and contract when cooled. As water is cooled from 4°C (39°F) to 0°C (32°F) (its freezing point), it *expands* slightly. This unique characteristic is due to the particular arrangement of the atoms in the water molecule and accounts for why solid ice floats on liquid water. The expansion of water near the freezing point causes it to be less dense than water that is slightly warmer. Because it is less dense than the surrounding water, it does not sink to the bottom of a lake but floats on its surface.

How can the expansion and contraction of solids be demonstrated? Obtain a large-head screw and a screw eye; the head of the screw should be just a bit too large to fit through the screw eye. Cut a wooden dowel so that you have two 20-cm (8 in) sections. The screw eye should be screwed into the end of one dowel and the screw into the other. (See Figure 18–2.) Demonstrate that the screw head is too large to fit through the screw eye. Heat the screw eye so that it expands enough so that the screw head can be pushed through. Ask the children why they think this has occurred. Other variations may be tried (e.g., cooling the screw eye, heating the screw head, etc.) and the effects noted.

How does the operation of a thermometer demonstrate the expansion and contraction of liquids? Ask the children to explain why the liquid in a thermometer rises in the tube as it is heated. Why does it fall as it is cooled? Refer to the water thermometer activity outlined in the preceding chapter (page 381.) Have the children analyze why it works.

How can the expansion of gases be demonstrated? Place a balloon over the mouth of a flask as explained on page 365. Follow the activity as described. How does this demonstrate the expansion of gases as they are heated? Ask the children to suggest other ways of demonstrating this principle.

Measuring Heat

Thermometers are used to measure temperature. The thermometers that are most likely to be used in an elementary school classroom are basically sealed glass tubes with a liquid in them. The liquid is usually mercury or alcohol. Mercury, which is silver in color, is often difficult for children to see; therefore, thermometers filled with alcohol that has been colored red are more practical for classroom use. Thermometers operate on the principle that materials expand when heated and contract when cooled. At the bottom of the thermometer is the *bulb,* which is a depository for the liquid. This is attached to the glass tube. If the thermometer is immersed in, touched to, or brought to a warmer environment, the liquid in the bulb becomes warmer and expands. Since the tube in the shaft of the thermometer is generally very narrow, small increases in temperature will result in discernable rises in the level of the liquid. Similarly, as the thermometer is brought to a cooler environment, the liquid in the tube will contract and the level of the liquid will fall. Thermometers are *calibrated* when they are manufactured (i.e., definite temperature readings are marked on the thermometer or on the card or metal to which it is attached). This is usually done by immersing the thermometer in a material at a known temperature (e.g., boiling water) and marking it accordingly. The intermediate readings are then etched or marked. When the numbers are etched on the thermometer itself, the thermometer is usually more accurate (but more expensive) than one consisting of a glass tube attached to a card or metal, because the tube can move upward or downward from the markings.

There are other types of thermometers, but they are less common in elementary schools. Many thermostats have thermometers that operate on the principle that all metals do not expand or contract to the same extent when heated or cooled. If two metals are fastened together in a strip, they will expand differentially when heated and cause the strip to bend. If the strip is then attached to a pointer and calibrated, its response to heating or cooling can register accurate readings.

Two temperature scales are widely used in the United States: the *Fahrenheit* scale and the *Celsius* scale. On the Fahrenheit scale, the freezing point of water is 32°F and the boiling point is 212°F. On the Celsius scale, the freezing point of water is 0°C and the boiling point is 100°C. It is easy to see why the Celsius scale is easier to deal with. A third scale, the *Kelvin* scale, is used much less frequently. The degrees in this scale are of the same size as the Celsius scale, but the lowest temperature on the Kelvin scale (0°K) is known as *absolute zero* and is the point at which molecules theoretically no longer move. Absolute zero is approximately −273°C.

The quantity of heat must be measured by other means. Thermometers cannot be used for this purpose, because heat depends on both the quantity of a substance *and* its temperature. A liter of boiling water will register the same temperature as ten liters of boiling water, but ten liters of boiling water has ten times as much heat as a single liter.

The *calorie* is the unit used to measure the quantity of heat. A calorie is operationally defined as the amount of heat required to raise the temperature of one gram of water one degree Celsius. Therefore, the amount of heat required to raise 10 grams of water from 20°C to 28°C is equal to 10 × 8, or 80 calories. The term *Calorie* (with a capital C), as used in measuring the energy value of foods, is equal to 1,000 of these small calories (i.e., it is the energy required to raise 1,000 grams of water one degree Celsius).

How reliable is our sense of temperature? Fill three bowls with water—one with warm water, one with lukewarm water, and one with ice-cold water. Have a child put both hands in the lukewarm water and hold them there for about one minute. Ask the child whether the water seems to be the same temperature for both hands. Next have the child place one hand in the cold water and the other in the warm water, and then have him or her place both hands into the lukewarm water again. Ask the child whether both hands seem to register the same temperature sensation. Check with a thermometer. How reliable *is* our temperature sense?

What is the boiling point and the freezing point of water? Boil water in a pot or beaker on a hot plate. Place a thermometer in the steam immediately above the surface of the boiling water. Leave the thermometer there for three minutes. It should register about 100°C or 212°F. However, the boiling point will be lower at higher elevations. The boiling point is lower in Denver than at sea level. After the thermometer has had a chance to cool for a few moments, place it in a bowl of melting ice. Compare the readings on several thermometers. Do they all indicate the same boiling and freezing points? One of the main reasons for inaccuracy in a mounted thermometer is slippage from the card or metal to which it is attached.

Transfer of Heat

Heat is transferred from one place to another by three means: conduction, convection, and radiation. In *conduction* heat travels through a material. The handle of a pot becomes hot as the pot is warmed because of the conduction of heat throughout the pot. This occurs because the molecules in the metal of the pot are caused to vibrate faster as the pot is heated. Molecules in motion collide with adjacent molecules until the molecular motion (and therefore the heat) spreads throughout the entire material. The same thing occurs when a spoon is placed in a cup of very hot tea or coffee. First the part of the spoon that is immersed in the drink will become hot, but eventually the entire spoon will heat up.

Materials do not conduct heat equally well. Metals are usually good conductors. Wood, hard plastics, and other materials are usually poor conductors. This is why they are often used to make pot and pan handles. A person is not as likely to burn his or her hand when using a pot handle covered with a poor conductor than if the pot's handle is made of metal. Poor conductors are *insulators* and have many domestic and industrial applications (See page 398.)

Ranking materials according to conductivity. Have the children touch a variety of materials that have been placed on a table in the classroom. Ask them to rank the materials from coolest to warmest according to the way they feel. Then have the children place a thermometer on each material, touching the bulb to the surface of each one. Leave the thermometers in place for two minutes and then note the temperatures. Rank the materials again, but this time according to the actual readings. How accurate was the initial ranking performed by touch? What conclusions about the conductivity of wood, cloth, paper, metal, and plastic can be drawn?

Heat is also transfered by *convection.* In this case heat rises as a liquid or gas is warmed. This is not possible in solids because the molecules in solids are relatively fixed and cannot move about as freely as in a liquid or gas. As a liquid or gas is warmed, its molecules tend to move more quickly and the material expands. Since the total number of molecules in the parcel of gas or liquid remains the same, the expansion of the substance results in fewer molecules per unit of volume. Therefore, the heated portion is less dense than its surrounding materials, and rises. Convection takes place on both a small and a large scale. It occurs in a cup of tea, in ocean currents, and in huge air masses. (See page 348 and 371.)

How can convection currents be detected? Have the children check the temperature at various heights in the classroom. Have them graph or chart the temperature gradient from floor to ceiling over the same part of the room. Where is the air the coolest? warmest? How does this demonstrate the effects of convection on the heating of the room? (This investigation is more effective in rooms where the heating devices are in or near the floor.)

The activities described in Chapters 16 and 17 regarding the movements of ocean currents and air masses can also be used to demonstrate the effects of convection.

Radiation is an extremely important means of heat transfer. This is why we feel the heat of the sun. Heat energy radiates outward from the sun in straight lines and in all directions. As we sit before a campfire, the fronts of our bodies are heated by the radiant energy of the fire, but our backs are not heated because of this straight-line transfer. Heat rays are effective in warming a solid object that obstructs its path (e.g., the body in front of the fire) but are far less effective in heating the air because of the low density of molecules in the air.

The amount of heat absorbed by radiation also depends on the color and kind of material absorbing it. Dark colors tend to absorb more heat than light colors. This is one of the reasons that we usually feel cooler if we wear lighter-colored clothing in summer.

Observing the effects of heat convection and radiation. Hold a thermometer about 10 cm (4 in) above a light bulb that has been switched on. Record the thermometer reading. Let the thermometer return to room temperature and then place it 10 cm (4 in) below the lighted bulb. Is there a difference in the temperature recorded above and below the bulb? What could account for this difference? By what means could the thermometer have been heated when it was *below* the bulb?

Which colors absorb more heat by radiation? Obtain six or eight different shades of construction paper. Make certain that one is black and one is white. Put the papers side by side on a sunny windowsill and place a thermometer under each one. Under which color is the temperature highest? lowest?

Bring the children to a parking lot where cars are parked in the sun. Can the same relationship regarding color and radiant heat absorbed be detected? What implications do these results have for our daily living?

Insulators are used to prevent heat transfer from one material to another. Materials that do not conduct heat well are called *insulators.* Wood, asbestos, glass, and ceramic are among the best insulators. Air is also a good insulator. In clothing designed to prevent the loss of body heat, insulating materials are used. Parkas filled with duck or goose down are particularly effective because of the many fine feather vanes that trap air. Wool is also a good insulator in clothing because air is trapped in its fibers.

Insulators are used in cookware handles, ovens, hot-water tanks, and so forth so that we do not burn ourselves with cookware and in order to prevent heat loss from ovens and water heaters. The external walls and attics of houses are also insulated so as to keep the heat in and conserve fuel.

The use of solar energy holds great promise for the future. The sun is the ultimate source of much of the energy on earth. It is the means by which green plants can produce food for animals, and it is responsible for the production of fossil fuels over long periods. These are somewhat indirect uses of the sun's energy. The sun's heat can also be *directly* transformed into other forms of energy. If the sun's rays are "captured" with mirrors and lenses and these devices are brought into contact with water, the water can be heated for household or industrial use. This application is particularly promising in regions that receive a good amount of sunshine.

The sun's heat is also trapped by the "greenhouse effect." Glass walls permit sunlight to penetrate into a room and warm its contents. The contents of the room then reradiate some of this heat energy, but this reradiated energy has longer wavelengths and some of it cannot pass out through the glass walls again; thus, heat energy is trapped in the room. As conservationists, materials developers, and architects act in concert, we can expect to see a proliferation of designs for the use of solar energy in homes. (See page 369 for activities involving the "greenhouse effect.")

LIGHT

Light behaves both as a wave and as a beam of particles. These two important characteristics of light may seem contradictory, but in some instances the nature of light can be explained better if we assume that light is a wave phenomenon, while in other instances it is preferable to assume that it consists of particles. A few simple experiments can be done by children to demonstrate both of these characteristics of light and illustrate the dilemma of trying to explain the nature of light.

If holes are cut in three pieces of cardboard and lined up at the level of a light (see Figure 18–3), the light can be seen through the holes. (You can find out whether the holes are lined up in a straight line by passing a string through them and holding it taut. The string should not touch the cards but only pass through the holes.) This experiment demonstrates that *light travels in straight lines.* This is an important characteristic of light, but one that will be true whether light is a wave or a particle phenomenon.

If the beams of two flashlights are aimed so that they cross each other, the beams do not seem to affect each other in any way. In a some-

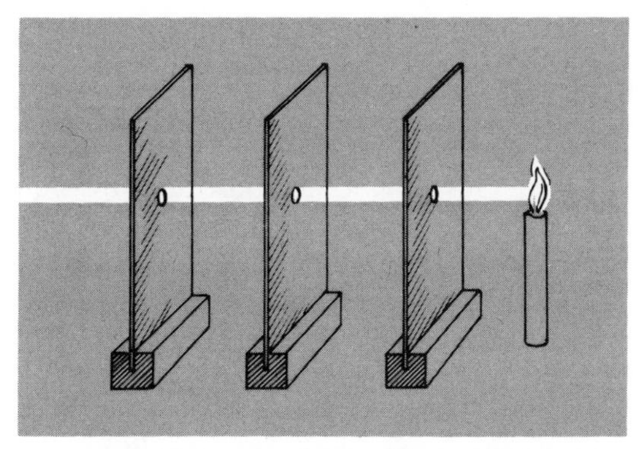

Figure 18–3 Light travels in straight lines. Can the candle flame be seen if the position of one of the cards is changed?

what similar way, if two pebbles are dropped into still water, the waves that are formed cross each other with no apparent effect. This observation supports a wave theory of light.

If a beam of light from a flashlight or projector is aimed through a magnifying glass, the beam of light can be concentrated on a sheet of paper held a few centimeters away. Similarly, sunlight can be concentrated on a piece of dry paper with a magnifying glass to start a fire. This bending and concentrating of light is an example of *refraction* and can be best explained by considering light to be a wave phenomenon.

When a light meter or a photoelectric cell is held in the light, a small electric current is generated that causes the needle on the light meter to deflect. This *photoelectric effect* can best be explained by assuming that light consists of a stream of particles and that these particles knock off electrons in the photoelectric material. The small electric current is the flow of these electrons.

Light is now considered to be both a wave and a particle phenomenon. It has been suggested that light could be streams of small particles traveling in waves. This is not the kind of explanation that is looked upon with favor in science—a simpler explanation would be preferred. But in order to explain various interactions associated with light it is still necessary to consider light as primarily a wave phenomenon under certain conditions and a particle phenomenon under other conditions.

How can we demonstrate that light travels in straight lines? In addition to observing the demonstration described earlier, children can show that light travels in straight lines by bending a length of garden hose or other flexible tube. Ask the children to try to see an object or light through the tube. In what position is the tube when light can be seen? It would be impossible to see light if the tube were not in a straight line.

Reflection

Light makes objects visible. Only when light is given off by objects or reflected by objects can they be seen. A beam of light, by itself, cannot be seen. This can be demonstrated by shining a projector beam onto a screen a short distance away. Ask the children what they see between the projector lens and the screen. The beam of light is visible only if there is dust or smoke in the air. Clap two chalkboard erasers into the beam of light. Is it more visible now? The tiny particles of chalk make the light beam more visible as light is bounced off of them.

Almost everything we see is seen by reflected light. These objects are *illuminated;* that is, they are lit by another source. The walls, floors, desks, chairs, doors, books, and so on in a classroom are all illuminated if

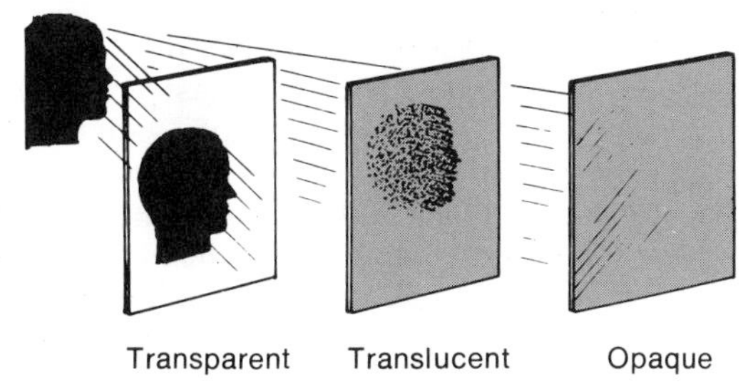

Figure 18–4 Materials are transparent, translucent, or opaque.

we can see them. Objects that emit light are called *luminous* objects. Fire, flashlights, street lamps, "fireflies," extremely hot materials that glow, as well as the sun and the stars, are all luminous. They are sources of light.

We have seen that light travels in straight lines. Because of this property, certain materials in the path of light can block it. Such materials are said to be *opaque,* and we cannot see through them. Metal, wood, and thick paper are examples of opaque materials. Materials that are opaque form shadows as light is cast upon them. Objects can often be identified by looking at their shadows, another evidence that light travels in straight lines. On a sunny day, if we are standing in the shade an object that is blocking the sun can be identified.

Some light passes through some materials, but not enough for us to see objects clearly. These materials are *translucent.* Waxed paper, frosted glass, thin paper, and cloth are all translucent. Materials through which we can see objects clearly are *transparent.* Clear glass and plastic and cellophane are examples of transparent materials. (See Figure 18–4.)

Identifying objects from their shadows. Have the children face a sunlit wall or a screen onto which a projector beam is shined. Place familiar objects (such as a pencil, scissors, a stapler, etc.) in the beam of light or sunlight. Ask the children to guess what the object is from its shadow. Is there a relationship between the shape of the object and its shadow? Children can further investigate shadows by trying to make them appear larger and smaller and by casting shadows from different angles by turning the objects.

What things seem to emit light? Have the children generate lists of things that emit light. The list may be longer than they expected it to be. Make sure the children can distinguish between objects that are truly luminous and those that reflect light (e.g., the moon actually reflects light, as do roadside reflectors and reflective paints). Children can categorize luminous materials according to whether their luminosity is due to "natural" sunlight, batteries, electricity, bioluminescence, heat, radioactivity, chemicals, and so forth.

How do different materials affect the transmission of light? Collect a variety of materials that can be tested for light transmission (e.g., glass, clear plastic, tissue paper, waxed paper, various fabrics, cellophane, typing paper, and sheets of metal, wood, plastic, and rubber). Have the children test each material by shining a flashlight into it and then classifying each one as to whether it is transparent, translucent, or opaque. After the children have tested a few objects, provide them with more objects to test, but this time have them *predict* the results before they test them.

White light is a mixture of many colors. This may be demonstrated by cutting a thin slit in a piece of dark opaque paper or cardboard and taping it onto a sunny window. Let the thin beam of light entering the slit pass through a triangular prism. This should produce a spectrum on one of the walls or the ceiling of the classroom. The spectrum should consist of the colors of the rainbow—red, orange, yellow, green, blue, indigo, violet. Do the colors appear in a special order? If the spectrum of colors can be focused onto and passed through a second prism, the colors should recombine to form white light. (See Figure 18–5.)

Color is actually a property of light and not of the object viewed. This may be explained by using grass as an example. White light from the sun bathes a field or patch of grass. Light of all colors is thus shined on the grass, but only the green portion of that light is reflected. (See Figure 18–6.) All of the other colors in the white light are absorbed by the grass. This causes the grass to appear green. Objects that appear white reflect most of the light that shines on them and absorb very little of it. As colors are reflected (and not absorbed) from objects, they stimulate nerve endings in our eyes. The stimuli are transmitted through nerves to the brain, where they are interpreted as color.

Each color we see is transmitted by light of a specific wavelength. Previously it was mentioned that light sometimes behaves like a wave. Waves have certain characteristics, one of which is the distance from the crest of one wave to the crest of the next. This distance is called a *wavelength.* Each color has a specific wavelength. Visible light is a form of electromagnetic radiation and represents only a small portion of the whole electromagnetic spectrum. This tremendous spectrum of waves includes X-rays, radio, television, and radar waves, and others. Some wavelengths are over 3 km; others are as short as a millionth of a millimeter. The wave-

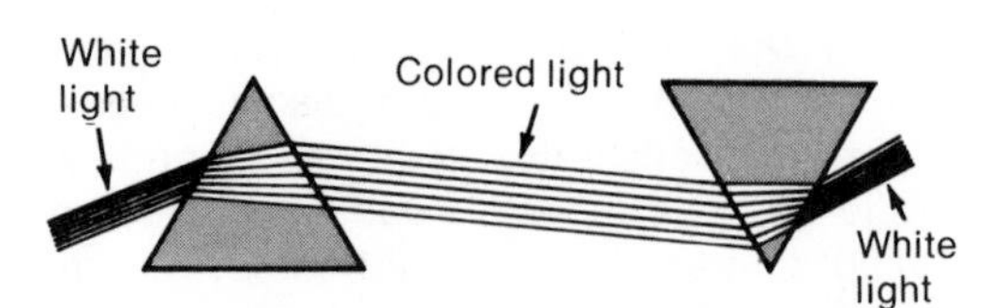

Figure 18–5 The colors in the spectrum may be combined to form white light.

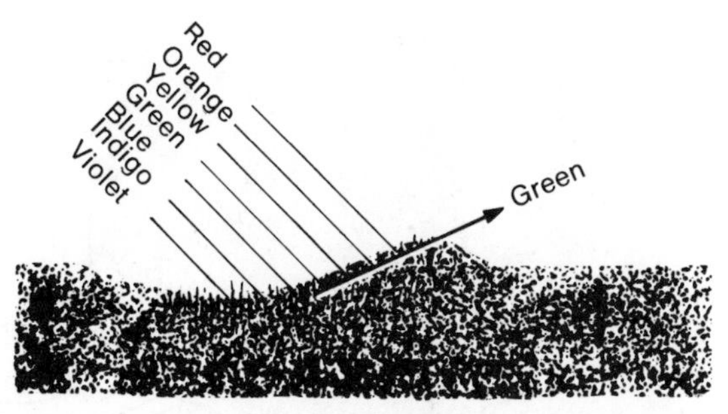

Figure 18–6 Grass appears green because it reflects green light.

length of visible light is between approximately 1/12,000 and 1/24,000 cm. The wavelengths in the electromagnetic spectrum are so precise and reliable that one of them is actually used to define the length of a meter.

The immediate neighbors of the visible spectrum are the infrareds and ultraviolets. As the terms imply, *infrareds* are beyond the red end of the visible spectrum and *ultraviolets* are beyond the violets. Ultraviolet rays are found in sunlight and are the rays that produce sunburn and tanning. They are considered hazardous in excess, yet small amounts produce vitamin D in our bodies. Infrared rays are the radiant heat waves that warm the earth and are responsible for our sensation of warmth.

Light is reflected by mirrors. When a rubber ball is thrown against a wall at an angle, it bounces off at about the same angle. When light strikes a flat, polished surface such as glass, it reflects or bounces in much the same way that a ball would bounce. (See Figure 18–7.) If a ball were thrown against a rough surface, we might not be able to accurately predict the angle at which it will bounce back. Clear images are reflected from very smooth surfaces. However, from most surfaces, such as the

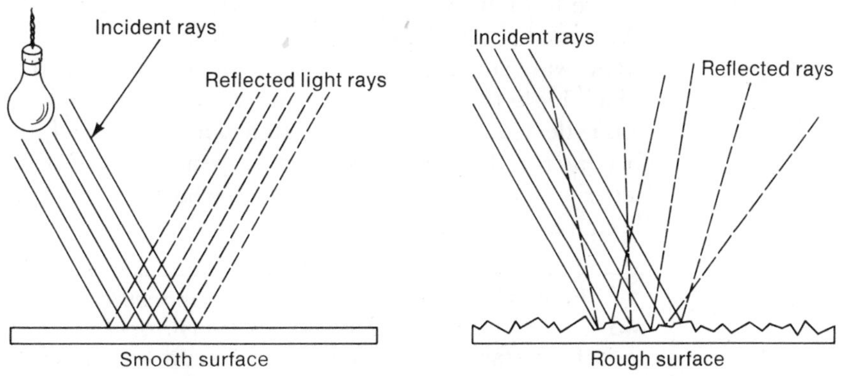

Figure 18–7 Smooth surfaces are good reflectors and rough surfaces are poor reflectors of light.

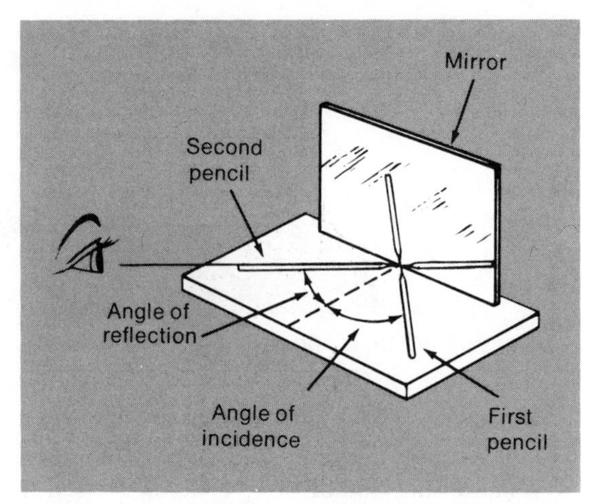

Figure 18–8 The angle of incidence equals the angle of reflection.

surface of this page, light is reflected in such a diffuse manner that we cannot see images.

Physicists have been able to describe the predictable response when light strikes a very smooth surface at a specific angle. This is called the *law of reflection,* and it states that *the angle of incidence equals the angle of reflection.* (See Figure 18–8.) The light ray approaching the mirror is called the *incident* ray and its reflection is called the *reflected* ray. The angles of incidence and reflection are the angle formed between the approaching and reflected rays, respectively, and an imaginary line perpendicular to the plane of the mirror.

There are three types of common mirrors. *Concave mirrors* are like sections of the insides of hollow spheres. The mirrors in large astronomical telescopes are concave, and they gather light from distant sources and reflect it back to a point. The reflectors in automobile headlights are also concave. They are shaped so that the light from the bulbs is reflected outward in parallel rays. *Convex mirrors* are shaped like the outside of a ball. Light that strikes convex mirrors is spread out over a wider area as it is reflected. Because a large area can be surveyed by the use of a convex mirror they are sometimes suspended in stores and elevators for security reasons. In the *plane mirror,* which has a flat surface, light is reflected evenly.

How can the Law of Reflection be demonstrated? Obtain a plane mirror, two pencils, and some cardboard. Fix the mirror onto the cardboard with clay so that it is standing on its long edge. Next draw a broken line perpendicular to the mirror. (You can check this by turning the mirror until the reflection of the broken line is in line with the real broken line.) Next place a pencil with its

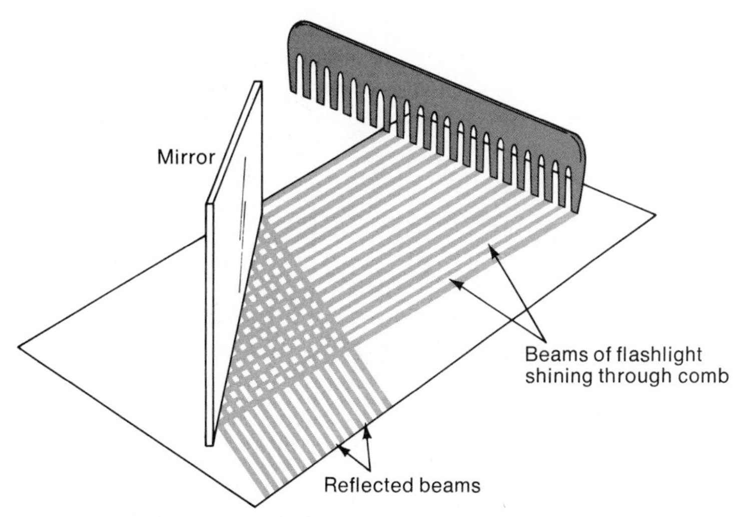

Figure 18–9 As the mirror is turned, what changes occur in the angles of the incident and reflected light beams?

point touching the mirror where it comes into contact with the broken line. Place a second pencil in a similar position on the other side of the broken line so that it is in line with the reflection of the first pencil. With a protractor, check the angle that each pencil makes with the broken line. Does this activity confirm the law of reflection? (See Figure 18–8.) When working with younger children, a hair comb may be held at a distance from the mirror and a flashlight shined through its teeth onto the mirror. (See Figure 18–9.) Make sure that the light rays passing through the comb strike the mirror at an angle. The approaching and reflected beams will shine on the cardboard surface. Note that the beams of light strike the mirror and are reflected at the same angle. Turn the mirror slightly and notice how the angles change.

On what surfaces can we see reflections? Have the children list all of the surfaces on which they have seen a reflection. Some examples might include water, store windows, spoons, automobiles, metal appliances, and so on. How are these surfaces alike?

Usually the back of a glass mirror is coated with a film of metal (most often silver). Mirrors for classroom use are often made of highly polished metal; the greater safety of this type of mirror is obvious, but it is easier to scratch.

Refraction

Light is bent (refracted) when it passes to a medium with a different density. Place a stick or a pencil in a clear glass of water. The stick will appear

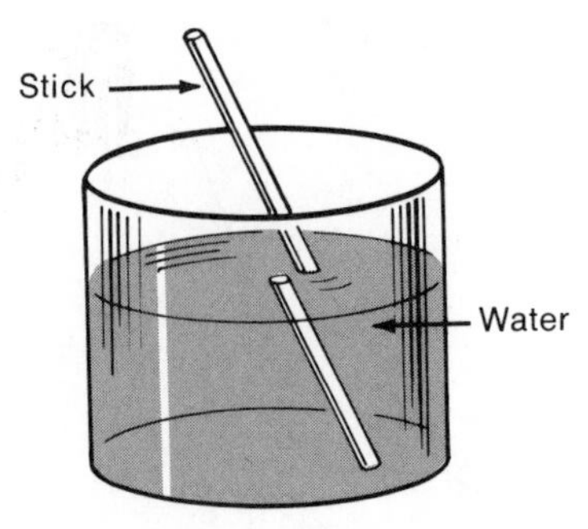

Figure 18–10 The stick appears broken because the light rays are bent by the water.

broken when it is partially under water. (See Figure 18–10.) Place a coin on the bottom of a tin cup and note its position. Slowly add water to the cup. The coin appears to move when the water is added. (See Figure 18–11.) In both of these cases, light rays passed from one medium (air) to another (water). The change of material caused the light to bend because light does not travel at the same speed in all media.

Light travels at a velocity of 300,000 km per second in a vacuum. Its velocity in air is about the same, but in a dense medium such as water its velocity is somewhat less. Therefore, light reflected by the stick and coin in the preceding examples will tend to be bent toward the observer as it leaves the water at an angle and enters the air.

Perhaps the best way to explain refraction to children is to use an analogy. If the right wheels of an automobile rolling down a highway were to strike soft sand, the automobile would tend to swerve to the right. The velocity of the automobile in sand would be reduced com-

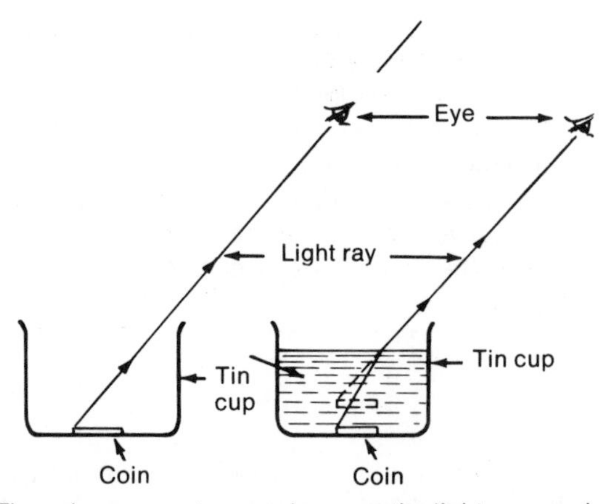

Figure 18–11 The coin appears to move because the light rays are bent by the water.

pared to its velocity on the smooth highway. Similarly, the edge of a ray of light that first strikes a dense medium such as water will tend to be slowed and the entire ray will be bent. Conversely, the edge of a ray that first enters a less dense medium such as air will travel faster than the part of the ray that is still in the water.

When light passes from air into a glass *prism,* the light is bent as it moves from air to glass to air again. The light bends toward the base of the prism. The prism breaks up white light into the colors of the spectrum because each of the colors of the light travel at slightly different velocities. The colors that make up white light are dispersed slightly as they enter and leave the prism.

Many children may notice that the colors of the spectrum of white light are the same as those of the rainbow. Rainbows are formed in very much the same way. As sunlight shines into water droplets left in the air after a rainstorm, the light is spread out. The light is reflected by the other side of the drop and spread out again as it leaves the drop. The refraction and reflection by many drops of water produce the spectrum of the rainbow. Spraying a fine mist from a garden hose into the air on a bright sunny day can have a similar effect.

Two main types of lenses are used in investigating refraction in the upper elementary grades. These are concave and convex lenses. *Concave lenses* are thinner in the middle than they are at the edges. This type of lens is a diverging lens in that it causes light beams entering it to spread out or diverge. (See Figure 18–12.) Normally, if you look at an object through a concave lens it will appear smaller than if you looked at the same object through plane glass. *Convex lenses* are thicker in the middle than they are at the edges. This type is a converging lens in that it causes light beams entering it to converge onto a focal point. (See Figure 18–13.) Common magnifying glasses are convex lenses and cause objects to appear larger than they actually are. Convex lenses are also important parts of cameras, telescopes, and microscopes.

What liquids besides water bend light? Prepare several clear glasses or cups with a pencil in each one (as in Figure 18–10), but fill each glass with a different liquid. Possible choices include light syrup, lemon juice, vinegar, salt water, alcohol, lighter fluid, and ammonia. (Children should not handle any substances which are corrosive or volatile.) Fill each glass to the same

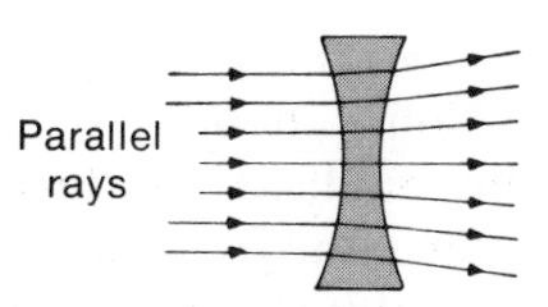

Figure 18–12 A concave lens causes light beams to spread out.

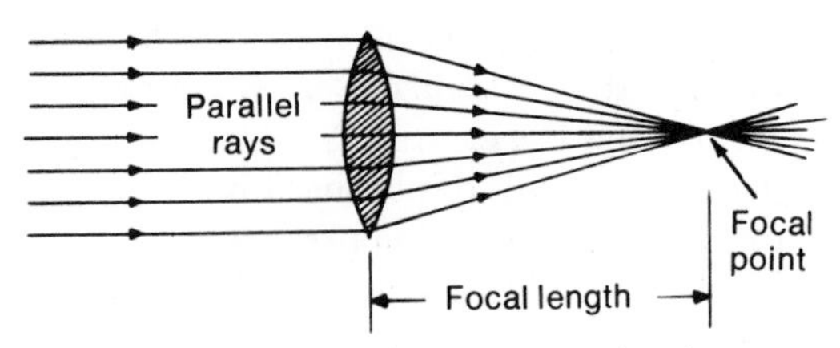

Figure 18–13 A convex lens causes light beams entering it to converge onto a focal point.

level. Have the children observe the pencil in each glass. Does one liquid cause more bending than another? What do you think accounts for this? Ask the children to arrange the cups in serial order according to the degree of bending they cause.

How are lenses used in our everyday lives? Ask children to collect a variety of objects that have lenses in them. The objects might include eyeglasses, flashlights, microscopes, magnifying glasses, and projectors. Have the children discuss the purpose or function of the lens in each case. Can the type of lens be identified easily? (In some instruments or eyeglasses, this may be difficult since a single lens may be a combination of more than one type.) Have the children try to discover what the lenses do to light beams that pass through them. What is the effect of using two lenses in conjunction with one another?

How can a magnifier be made from a drop of water? Using a grease pencil (or china marker), draw a circle about 5 mm in diameter (1/5 in) on a piece of glass. Carefully place a drop of water onto the slide so that it rests inside the circle. Use the drop of water as a magnifier. Why does it work? How does the circle aid in keeping the drop of water intact?

A camera and the eye are alike in many ways. The eye and the camera have parts which serve analogous functions. Children should have access to cameras that they can open and examine as they look for these similarities. (For details see Chapter 14.)

SOUND

Sounds are caused by vibrations. Snap a stretched rubber band. Do you hear a sound? Notice the movement of the rubber band. It moves back and forth quite rapidly. Hold the end of a plastic ruler flatly and firmly on a table top. Snap the free end of the ruler. Is a sound heard? Does the free end of the ruler shake back and forth? Such movements, called *vibrations*, are the basis of most of the sounds we hear.

The majority of the vibrations that account for sounds in our daily lives are less obvious than those just described. The sound we hear when listening to a record player is a result of the rapid vibration of the thin, paper-like material in the speaker. It is hard to see this paper move back and forth, but as one observant kindergartener put it, "It looks fuzzy."

Ask the children in the class to hum while touching their fingertips to their throats. They should feel a vibration. The vibrating materials are the *vocal cords,* two muscular membranes stretched across the *larynx.* The vibrations of the vocal cords as air is forced past them are responsible for our speech sounds.

In order for us to hear sound, the vibrating material must move back and forth at least 16 times per second. Most human ears cannot hear sounds produced by materials vibrating more than 20,000 times per second. The vibrating material responsible for sound may be solid, liquid, or gaseous—as in the case of a vibrating air column.

What sounds can be heard in the immediate environment? Ask the children in the classroom to be silent and listen to the sounds around them. Ask them to list at least three distinct sounds. It may surprise some youngsters to suddenly become aware of the many sounds in their midst. Take a "listening walk" in the school building or outdoors. The children should be encouraged to describe the sounds they hear and judge whether a particular sound is high or low, loud or soft, pleasant or annoying, coming from indoors or outdoors, and so on.

What is vibrating to produce commonly heard sounds? Ask the children to try to identify the vibrating substance producing sounds that they often hear. Can they feel the vibration? As a drum is tapped, what is vibrating? (Placing some talcum powder or chalk dust on the drum membrane will help demonstrate the vibration.)

Ask the children to look at and touch the vibrating strings and wood in a piano or guitar. Ask them to identify the vibrating material of other musical instruments. In some cases more than one material is vibrating.

Experiment with a tuning fork. What is moving? Can the vibrating tines cause other objects to vibrate?

Sound Waves

Sound travels in a wave motion. As a rubber band is snapped and caused to vibrate, its back-and-forth movement pushes against the air near it for an instant, causing the particles of air to compress slightly. As the rubber band moves back, it releases its push on the air and the air particles become slightly rarefied. A series of *compressions* and *rarefactions* make up a sound wave. Air is an elastic substance; therefore, as some

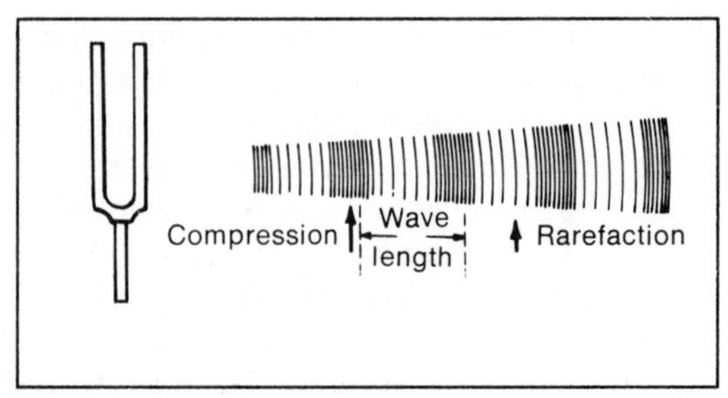

Figure 18–14 A vibrating material causes alternate compressions and rarefactions of the air molecules around it.

particles are alternately pushed together and released, this motion will spread to the surrounding particles. (See Figure 18–14.)

When a pebble is dropped into a pool of water, waves are set up and spread out in all directions. The same is true of sound waves; they spread out in all directions from their source. The waves in a pool become larger as they move outward from their source after the pebble has been dropped in; the same thing occurs with sound waves.

As a sound wave advances through the air, the air particles do not move along with the wave but, rather, tend to stay in the same general location as they are pushed together and released. This is another important characteristic of sound waves and may be demonstrated by placing a small cork in a tub of water and dropping a pebble into the water. The waves will move in all directions away from the pebble, but the cork (representing a particle of air) does not move out with the wave; instead, it moves up and down. (See Figure 18–15.) Air particles behave in the same way.

Sounds produced in the air are carried through the air in this wavelike motion, and the vibrating air strikes a part of the ear (the eardrum), which is also set into motion as the alternate compressions and rarefactions strike it. As the vibration of the eardrum is transmitted by small bones and later by nerve impulses to our brain, we can interpret the sounds we hear. (See page 284.)

Helping young children understand wave motion. A *wave* is essentially a repeating pattern. Young children can learn about wave motion by being involved in the production of repeating patterns. Initially, shape blocks may be used. Have a child establish a pattern (e.g., triangle–square–diamond) and then repeat the sequence several times. (See Figure 18–16.) Practice in such activity may take several forms. For example, children may repeat letter or number patterns, clapped patterns, piano note patterns, and so on. Color sequences may also be repeated. A wave may be drawn on a chalkboard and the children asked to identify the repeating pattern. How is a pendulum's

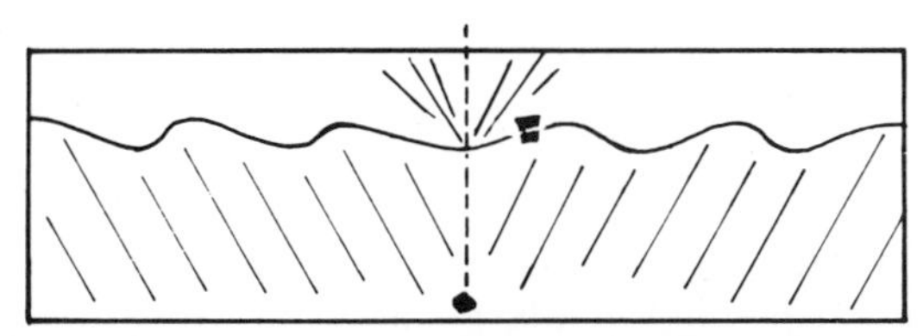

Figure 18–15 As a pebble is dropped into a tub of water and waves result, the cork floating on the water does not move out with the wave.

swinging or a metronome's movements a repeating pattern? Such primary experiences are important if children are to form a concept of the wavelike motion associated with sound.

Producing sound wave models. Models of sound waves can be set up in a variety of ways. A clear glass baking dish half filled with water should be placed on a table top. First have a pupil drop a small pebble into the dish and observe the ripples. A vibrating tuning fork may be touched to the water's surface and the effects noted. This demonstration will be even more dramatic if the baking dish is placed on an overhead projector; the ripples will be projected onto the screen. Instead of water, a loose-consistency gelatin may also be used. The wavelike motion will be even more pronounced.

A large coiled spring ("Slinky") may be used to demonstrate compression and rarefaction. Have one pupil hold one end of the spring and another the other end. One child should pull several coils of the spring toward him or her and quickly release them. The alternate compressions and rarefactions that occur are analogous to what occurs when sound causes waves in the air.

Figure 18–16 A child creates a repeating pattern using blocks.

How Sound Travels

Sound needs a medium in which to travel if it is to be heard. Sounds cannot be heard in a vacuum. In a common demonstration a ringing bell is suspended in a large jar from which the air is slowly evacuated. When most of the air has been removed, the bell can no longer be heard. Most of the time sound travels in the air. It was noted previously that air is an elastic substance whose particles are compressed and released when sound is produced.

Other substances also provide a medium in which sound can travel. Two stones clapped together underwater can be heard by a diver. A rap on a long wooden or iron fence can be heard by a person with his or her ear on the fence at a considerable distance from the person making the sound. Some materials are good transmitters of sound (e.g., wood, iron, water), while others are poor transmitters of sound (e.g., wool, rubber, and other soft materials). Such poor sound carriers are often used for sound proofing, since they absorb sound.

Sound does not travel at the same speed in all materials. In air, sound travels at the rate of approximately 333 m/sec. (1050 ft/sec). In water, it travels at almost four times this rate. Sound travels about fifteen times faster through iron as it does through air. American Indians put their ears to the ground in order to hear the sound produced by distant horses. The hoofbeats of the horses could be heard through the ground before they could be heard in the air. During a thunderstorm the flash of lightning is seen before the sound of thunder is heard. Both events occur simultaneously, but the lightning is seen first because light travels at a much faster rate than sound.

Demonstrating that air is a real substance capable of transmitting sound. The activities described in the preceding chapter providing evidence that air is real should be reviewed or repeated. Only when children believe that air is real and that it is all around them can they imagine that air is a carrier of sound. Children should be helped to realize that there is indeed "something" between a ringing bell and their ears. The development of such concepts takes place gradually. Children can begin by being involved in primary experiences with the phenomena under investigation.

What substances besides air conduct sound? Several activities can be used to demonstrate that a variety of substances carry sound. Ask the youngsters to place an ear flatly on a large table. Stand the stem of a vibrating tuning fork on the table top. (See Figure 18–17.) What do the children hear? What does this demonstrate?

Making a string telephone is a popular activity. Obtain two small frozen-juice cans and remove one lid from each can. Punch a small hole in the center of the intact end of each can. Thread a string into the holes from the outside of the can. Tie each end of the string in several knots to prevent the string from being pulled out of the cans as it is pulled taut. Have two children each take a can and pull the cans apart until there is tension on the string. One child places a can over his or her mouth and talks while the other child listens, holding the other can to his or her ear. What is transmitting the sound?

To experience how string and the bones of the skull transmit sound, obtain a meter of string and tie the center of it around the stem of a metal

Figure 18–17 Their ears to the table top, the children hear the effects of the vibrating tuning fork.

spoon. Have a pupil hold the free ends of the string in his or her ears. Have the child bend over so that the spoon hangs freely and ask another child to tap the spoon with a pencil or another spoon. What is the sound like? What accounts for this?

The transmission of sound by water can be demonstrated by filling a sturdy balloon with water and tying it at the top. Have one child place the water-filled balloon against his or her ear. Another child should hold a watch to the other side of the balloon. Can the watch be heard? Remove the balloon, keeping the watch at the same distance. Which is a better conductor of sound, air or water?

Characteristics of Sound

Loudness, pitch, and quality are three important characteristics of sound. The vibrations that account for sound are not simple. There are three important characteristics of sound, and each is associated with a distinct aspect of sound waves. The three main properties of a sound wave are wavelength, amplitude, and frequency. (See Figure 18–18.) The *wavelength* is simply the distance from the crest of one wave to the crest of the next. The *amplitude* is the height of the wave crest above or below the base line. The *frequency* is the number of waves that pass a designated point within a given time.

The *loudness,* or intensity, of a particular sound is determined by the amplitude of the sound wave. This is associated with the amount of vibration of matter. A piano key that has been struck very hard causes a string to move back and forth farther than if the key had been struck lightly. Therefore, the air particles are also caused to move back and forth a greater distance. This makes the eardrum move back and forth a greater distance. The total effect is that the sound is louder.

The intensity (loudness) of a particular sound also depends on how far away we are from the site of sound production. The closer we are to a sound, the louder it will be. As a sound wave moves out from its source, its amplitude decreases and the sound becomes less loud. (See Figure 18–19.)

The *pitch* of a sound is determined largely by the frequency of the

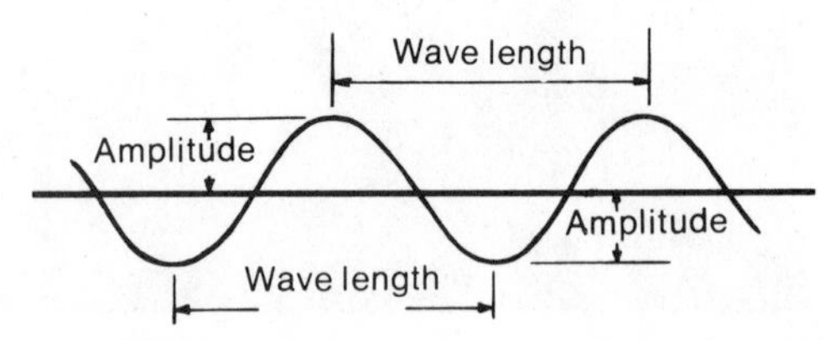

Figure 18–18 Wavelength and amplitude are two important characteristics of waves.

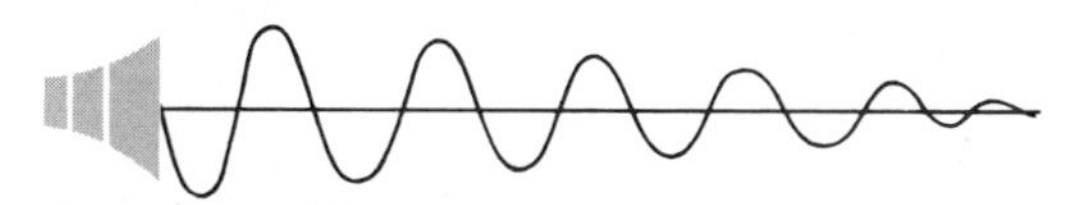

Figure 18–19 As a sound wave moves out from its source, its amplitude decreases and the sound becomes softer. The frequency remains the same.

sound wave. A high-frequency sound (i.e., one in which relatively many complete waves pass a single point in a given time) is a high-pitched sound. Similarly, a low-frequency wave is associated with a low-pitched sound. (See Figure 18–20.) As mentioned earlier, most humans can hear sounds with frequencies between about 16 and 20,000 vibrations per second.

In stringed instruments the pitch of a sound may be determined in several ways. First, the length of the string is an important factor. Longer strings produce lower-pitched sounds than shorter strings of the same thickness held at the same tension. Second, a string that is subjected to a great deal of tension will produce a higher pitch sound than one subjected to less tension, provided that they are the same length and thickness. Third, a thick string will tend to produce a lower-pitched sound than a thinner string of the same length held at the same tension.

The *quality* of a sound is another important characteristic. A "C" produced on a piano, an oboe, and a violin may all have the same frequency and loudness (amplitude), but there is something distinct about each sound. The quality is different. The nature of sound waves is not usually simply a matter of variations in amplitude, wavelength, and pitch. Most musical sounds are a result of a combination of waves. These sounds are due to the vibrations of several frequencies at the same time. A string on a violin, for example, vibrates not only as a whole string but also as though it were composed of two or more parts, each with a different but related frequency. These associated sound waves, in addition to the *fundamental* or main one, are called *overtones*. It is the number and strength of these overtones that primarily accounts for the quality of sound. The kinds of overtones are determined in part by the material

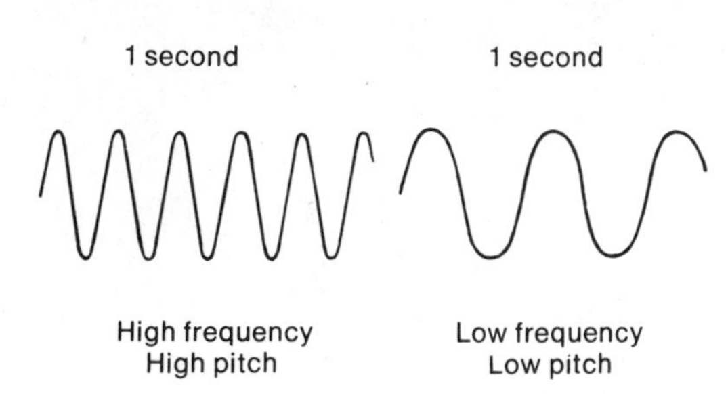

Figure 18–20 The pitch of a sound is determined by the frequency of the soundwave.

and shape of the instrument being played as well as the skill of the musician.

Using a monochord to demonstrate characteristics of sound. A *monochord* is simply a string stretched across the top of an open box. A small desk drawer may be used as the box. A guitar string serves quite well as the string, as will heavy fishing line. Attach one end of the string to the knob of the drawer and pull the free end over the top of the drawer. (See Figure 18–21.)

First have a child pull the free end over the open box and down along the side opposite the point where the string is attached. Ask the child to try to maintain constant tension on the string. Ask another child to pluck the string several times, trying to vary the loudness. Point out the obvious vibration of the string. How can the loudness of the string's sound be controlled? Does the string seem to vibrate as a whole or in parts?

Next the children should investigate various ways of altering the pitch of the string. In order to vary the string's length, obtain a strip of wood longer than the width of the box. (A triangular drafting ruler is perfect for this purpose because of its relatively sharp edge.) With one child holding the string at constant tension, another child can place the wood strip (or ruler) over the sides of the box, yet under the string, so that its effective vibrating length can be altered. Does the pitch seem higher when the string is lengthened or shortened? Can a simple tune be played? How does a guitar player vary the length of a guitar string?

Figure 18–21 Children discover ways to alter the pitch produced by a monochord string.

Pitch may also be varied by changing the tension on the string. Have the string cover the length of the box and have one child vary the tension on the string while another child plucks it. Is the string's pitch higher when more tension is applied or when less tension is applied? The changes of tension may be directly observed by attaching a spring or dial face scale to the free end of the string. As more tension is applied to the string, the number registered on the scale will increase. This will help demonstrate the direct relationship between pitch and tension.

Finally, guitar strings of different thicknesses may be tried. Given the same length and tension (kept constant by using a spring scale), which string produces a higher sound, a thick one or a thin one?

How can various materials be used to make musical sounds? Children can investigate many materials for producing musical sounds. Several soda pop bottles can be cleaned and filled with various amounts of water. As children blow across the top of a bottle, a sound will be produced. What is the vibrating substance in this case?

Drinking glasses may be filled with various amounts of water and tapped with a spoon or pencil. By adjusting the amount of water, the glasses may be "tuned" and familiar jingles can be played. What is vibrating in this case?

In a recorder or "flutophone" the pitch is changed by covering or uncovering holes along the instrument's tube. When all the holes are covered, the air in the entire tube vibrates. As fingers are removed, the length of the vibrating column of air is changed. (See Figure 18–22.) Have children identify the length of a vibrating air column in a recorder.

Plastic drinking straws may also be used to produce sounds. Give each child several straws and have them flatten the first 5 cm (2 in) of each straw. Next they should snip off the corners of the flattened end with scissors. The opposite end of the straws should be cut off at varying lengths to produce a

Figure 18–22 Children identify the length of the vibrating air column in a recorder.

Figure 18–23 Children can produce simple stringed instruments like the ones pictured above.

set of straws of different lengths. Have each child place a straw in his or her mouth so that the flattened edge just touches the tongue. Have them blow into the straws. Which straws account for higher sounds, the longer ones or the shorter ones? What is vibrating?

How can children make simple musical instruments? Applying the principles discovered in the preceding two activities, children can make their own musical instruments out of glasses, water, spoons, wood, nails, rubber, strings of various thickness, rubber bands, and boxes. Some older children may be able to "tune" their instruments and actually play familiar songs.

A group of youngsters produced the simple stringed instruments shown in Figure 18–23 from classroom scraps and fishing line.

Echoes

Echoes are reflected sounds. In the preceding section the reflection of light by a mirror was discussed. Sound, too, can be reflected as it bounces off a distant wall, cliff, or other obstruction. In order for reflected sound to be perceived as an echo by the human ear, it must be heard at least one-tenth of a second after the original sound. This means that a sound must travel a total distance of 33.3 m (110 ft) (i.e., one-tenth of the distance sound travels in one second). Since the echoed sound travels to a reflecting surface and back again to the person producing the sound, the distance from the person to the reflecting surface must be at least half the distance that sound travels in one-tenth of a second, or half of 33.3 m (110 ft). Thus, a sound must be reflected off a wall or obstruction at least 16.65 m (55 ft) away in order to produce an echo. Sometimes whole phrases are echoed if the sounds travel a considerable distance before bouncing back. Draperies, carpets, wooden baffles, acoustical ceiling tiles, and other sound-absorbing materials are often used in large auditoriums to prevent echoes.

The distance to an object can be measured by echoes. For example, the distance to a cliff or an iceberg may be measured by sounding the ship's foghorn and calculating how long it takes to hear the horn's echo. Knowing the speed of sound and the time between sounding the horn and perceiving its echo enables navigators to calculate the distance to these reflecting surfaces.

Echoes in water are also important sources of information. Some ships are equipped with a mechanism (essentially a vibrator) attached to the ship's hull which sends sound down into the water; echoes from these sounds are bounced back from the sea floor. Sensing devices detect the reflected sounds and the time elapsed between the initial sound and the reflected sound. Knowing the speed of sound in water, one can calculate the distance from the ship to the sea floor. *Sonar,* a sounding device, detects both the time elapsed between the two sounds and the direction of the returning sounds, thus making it possible to create a detailed map of the sea floor, including an outline of underwater objects such as submarines, schools of fish, whales and reefs, and depressions.

The thickness of sea floor sediments can be detected by the *hydrophone,* a sounding device that detects two echoes, one from the ocean or sea floor and the other from the bedrock beneath the bottom sediments. The time elapsed between the two echoes is used in determining the thickness of the sediments.

Bats and other animals make use of reflected sound to guide them as they navigate. It is known that bats emit extremely high-pitched sounds (inaudible to human beings) from their mouths. These *ultrasonic* sounds are echoed from obstacles in the bat's path and the echoes are detected by the bat's ears. An internal mechanism then helps the bat sense the distance and size of these nearby objects, enabling it to navigate difficult passages and detect prey. This unique ability was discovered when researchers, suspecting that bats' ears were involved in their navigation, covered their ears and found that they could not fly well and tended to collide with obstacles. More than 100 years later, when researchers taped the mouths of bats *or* covered their ears, they found that the bats' ability to avoid obstacles was significantly decreased. Thus, the role of the bat's mouth in emitting ultrasonic sounds was discovered.

RELATED ACTIVITIES

Discovering hot and cold, p. 110.
How does heat travel along a metal rod? p. 186.
Transfer of heat, p. 73.
What happens when a liquid changes to a gas (evaporates)? p. 190
How is air heated? p. 191.
Does hot water freeze more quickly than cool water? p. 17.
Are all surface materials affected equally by the sun's rays? p. 369.
Will a dull, dark-colored surface absorb more heat than a shiny surface? p. 67.
Do all objects absorb the sun's heat equally? p. 299.
Demonstrating the relationship between color and heat, p. 315.
What is the warmest time of day? p. 299.
How can heat energy be trapped? p. 369.
What are the major factors affecting evaporation? p. 369.
Making a water thermometer, p. 381.
How do temperature changes affect insects? p. 250.
What is a temperature inversion? p. 362.
How can different forms of energy be generated? p. 186.
How can electrical energy be converted into heat? p. 442.
Explorations with light, p. 109.
How can electrical energy be converted into light? p. 442.
How does the length of shadows change in the course of the day? p. 300.
Examining light from the sun, p. 299.
How is the image of an object inverted? p. 282.
Identifying sounds, p. 80.
Hearing similarities and differences in sounds, p. 83.
Sorting objects by sound, p. 88.
How well do we hear? p. 285.

How do we locate sounds? p. 285.
How can auditory fatigue be demonstrated? p. 285.
Estimating the distance to a thunderstorm, p. 376.

REFERENCES

For Teaching

RAINWATER, CLARENCE. *Light and Color.* New York: Golden Press, 1971. A well-illustrated, very informative book that provides much background information for teachers.

RUBLOWSKY, JOHN. *Light: One Bridge to the Stars.* New York: Basic Books, 1964. A book for older students or teachers describing the nature of light and its use as a tool for gathering astronomical information.

WEART, SPENCER R. *Light: A Key to the Universe.* New York: Coward-McCann, 1968. Light, its use in astronomy, and a brief history of developments in the field are covered in this book.

For Young Children

ALEXENBERG, MELVIN L. *Light and Sound.* Englewood Cliffs, N.J.: Prentice-Hall, 1969. An appealing book suggesting many experiments for children in light and sound.

BALESTRINO, PHILIP. *Hot as an Ice Cube.* New York: Harper & Row, 1971. An appealing book about heat, with provocative ideas for inquiry.

JACOBSON, WILLARD J., CECILIA J. LAUBY AND RICHARD D. KONICEK. *The Sounds You Hear.* New York: American Book, 1969. A book about familiar sounds, including investigations. Part of a series.

KOHN, BERNICE. *Light.* New York: Coward-McCann, 1965. A simple book about light and its importance in children's lives.

SIMON, SEYMOUR. *Let's-Try-It-Out Light and Dark.* New York: McGraw-Hill Book Company, 1970. A book about light and shadows for young children including many experiments which children may conduct at home.

For Older Children

ADLER, IRVING. *The Story of Light.* Irvington-on-Hudson, N.Y.: Harvey House, 1971. Background information about light and how knowledge about it was acquired. Includes many investigations.

CATHERALL, E. A. AND P. N. HOLT. *Working with Light.* Chicago: Albert Whitman, 1969. Includes many experiments that youngsters can undertake at school or at home.

——. *Working with Sounds.* Chicago: Albert Whitman, 1969. A book of experiments and information about the production and transmission of sound.

COBB, VICKI. *Heat: A First Book.* New York: Franklin Watts, 1973. This book describes the various properties of heat and the role of heat in our daily lives.

FREEMAN, IRA M. *Sound and Ultrasonics.* New York: Random House, 1968. Contains much background information for upper elementary school children about sounds that we can and cannot hear.

JACOBSON, WILLARD J., CECILIA J. LAUBY AND RICHARD D. KONICEK, *Light and Heat.* New York: American Book, 1968. A book of information and activities dealing with light and heat.

LISS, HOWARD. *Heat.* New York: Coward-McCann, 1965. A clear book about the nature of heat using everyday examples. Includes investigations and a simple explanation of molecular theory.

SOOTIN, HARRY. *Experiments with Heat.* New York: W. W. Norton, 1964. Includes many experiments demonstrating the nature of heat. Deals with the nature of matter and molecules.

———. *Science Experiments with Sound.* New York: W. W. Norton, 1964. An advanced book about sound and sound experiments. These activities should be reserved for the most highly motivated pupils.

WINDLE, ERIC. *Sounds You Cannot Hear.* Englewood Cliffs, N.J.: Prentice-Hall, 1963. An interesting book about ultrasound. Includes treatment of radar, animal communication, bat navigation, and future prospects in the field of ultrasound.

Magnetism and Electricity

CHAPTER 19

If you pass a magnet over a paper clip, the paper clip will be attracted to the magnet. If you rub a rubber rod or comb with a piece of wool or plastic and pass it over small pieces of paper, the paper will be attracted to the rod or comb. In both of these instances, there is no visible physical connection between the objects that attract and those that are being attracted. It is "action-at-a-distance." We can see the evidence of interaction, but we cannot see the actual mechanism by which the attraction occurs. Usually these "action-at-a-distance" phenomena are intriguing to children. Through experiences like those described in this chapter, they can develop mental models that can be used to explain these phenomena.

Models are "pictures" or representations that help us explain observations. *Physical models* are physical representations, often on a different scale, of objects and observations. Model airplanes, for example, are representations of much larger airplanes. Atomic models may be constructed of small sticks and balls and serve as "pictures" that help us explain the nature of matter. *Theoretical models* are "mental pictures" that help us interpret phenomena. Children may have a mental picture of what the inside of a black box is like; scientists may have a mental picture of how inherited physical traits are transmitted from one generation to the next. While theoretical models are mental pictures, they have to be expressed in some way if they are to be communicated. Scientists often express theoretical models in mathematical terms. In the elementary school we usually communicate theoretical models through words, diagrams, or pictures or by developing a physical model. In the study of magnetism and electricity, children can have practice in developing models to help explain the results of observation and experimentation.

As in almost all areas of elementary school science, children should have extensive experience with concrete materials as they study electricity and magnetism. Play is not a waste of time! Children should have chances to play with magnets, generate electric charges, experiment with electrical circuits, and use electricity in a variety of ways. The mental models they develop should be based on these firsthand experiences. Their models can be tested through other experiences with concrete materials.

A *dry cell* is the recommended source of electric current. Do not experiment with electricity from electrical sockets in the room. Always use electric current from dry cells such as those used in flashlights or the larger no. 6 cell.

MAGNETISM

Materials can be classified as magnetic or nonmagnetic. Actually, many materials have magnetic properties. However, children can operationally define materials that are attracted to a known magnet as *magnetic;* materials that are not attracted to a magnet can be operationally defined as *nonmagnetic.* Most of the materials in our environment that would be operationally classified as magnetic have iron or steel in them.

What materials are magnetic and nonmagnetic? Children may have had a variety of experiences in classifying materials on the basis of their physical properties. In this investigation they classify materials on the basis of a property that is not so obvious.

Collect a variety of objects such as buttons, paper clips, coins, nails, bolts, nuts, and pencils. Bring a magnet next to each of them. Which are attracted to a magnet? Which are not? Are some parts of some objects attracted while other parts are not? Of what materials do objects that are attracted to a magnet seem to be made?

Magnets have poles where the magnetic effect is concentrated. If a magnet is laid upon a number of paper clips spread across a table and then lifted, the paper clips will seem to be concentrated at certain points, usually two, on the magnet. These places where the magnetic effect is concentrated are called *magnetic poles.* Magnets have two poles. However, some

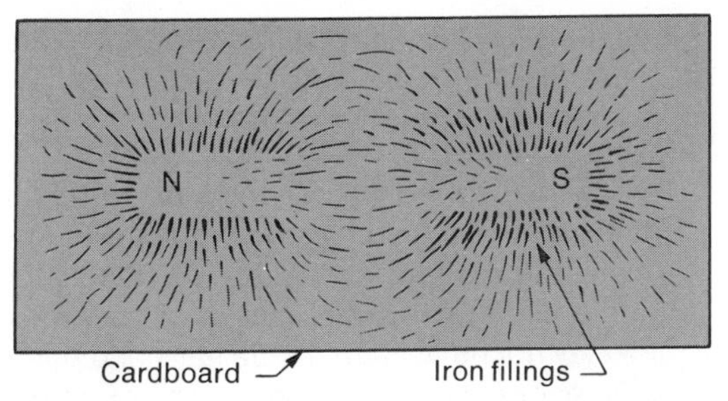

Figure 19–1 The locations of magnetic poles will be indicated when iron filings are sprinkled onto a cardboard laid on top of a magnet.

"magnets" are actually composed of more than one magnet and therefore seem to have more than two poles.

Where are the poles of a magnet located? Take a small compass and move it along the side of a magnet. Watch the compass needle. It will tend to point to one of the poles. Does the same end of the compass needle always point toward a pole?

Place a piece of cardboard over a magnet and then sprinkle iron filings onto the cardboard. (See Figure 19–1) Tap the cardboard. How do the iron filings line up? Where are the poles of the magnet? (Try not to have the children bring iron filings into direct contact with the magnets. They're hard to clean off.)

If a bar magnet is suspended so that it can swing freely, one pole of the magnet will tend to be oriented toward the north. This pole is arbitrarily defined as the *north pole.* The other end of the bar magnet, which is oriented toward the south, is the *south pole.*

Like poles tend to repel and unlike poles attract—the basic law of magnets. If a north pole is brought near the north pole of a magnet that is free to move, these poles will tend to repel each other. Similarly, a south pole

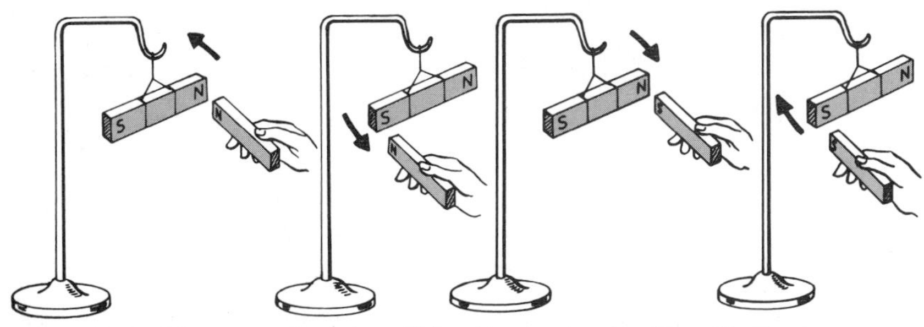

Figure 19–2 Like magnetic poles will tend to repel and unlike will attract.

will tend to repel another south pole. However, a north pole and a south pole will tend to attract each other. (See Figure 19–2.)

How can we test for magnets? It is important that children recognize that sometimes the seemingly obvious test for a property is not a test at all. This is the case in the test for a magnet.

If an object attracts magnetic materials such as nails and paper clips, it still may not be a magnet. The nails and paper clips may be magnets! The test for a magnet is whether some part of an object repels one pole of a known magnet.

Move a pole of a magnet along the side of a nail. Then bring one end of the nail near one end of a compass needle and then the other end of the needle. If one end of the compass needle is repelled by the the nail, the nail is a magnet.

Move a compass along the side of an iron or steel bar. If one end of the compass needle is repelled by some section of the bar, the bar is a magnet. (See Figure 19–3.) Many iron and steel objects have become magnetized by standing in the earth's magnetic field. Have the children move compasses along the sides of various iron and steel objects in the classroom and school. Often such objects as chair and table legs, ring stands, radiators, and flagpoles have been magnetized, and parts of them will repel one end of a compass needle.

In addition to providing a way to test for magnets, the fact that unlike magnetic poles attract and like poles repel is used in many useful devices such as compasses and electric motors. As we will see, this "basic law of magnets" is also similar to the "basic law of electrostatics."

The earth is a magnet. If a known bar magnet is suspended so that it is free to turn, it will line up in a north–south direction. This occurs be-

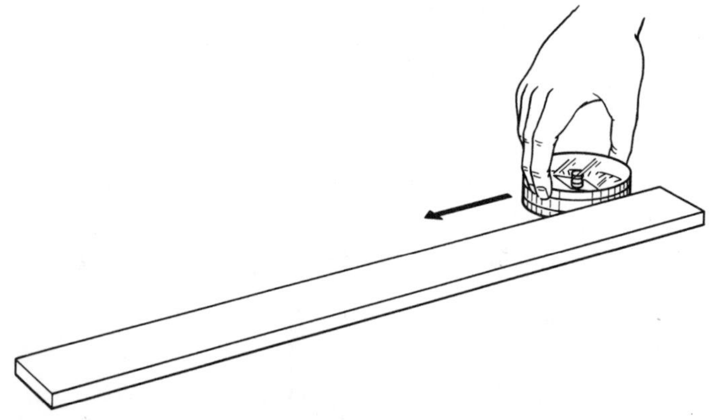

Figure 19–3 Is one end of the compass needle repelled by some section of the bar?

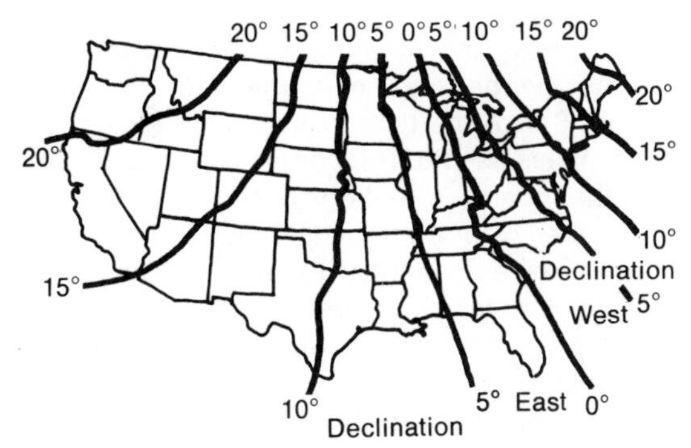

Figure 19–4 Magnetic declination for places in the United States. For example, compasses in Texas will point about 10° East of geographic North and compasses in New York about 10° West.

cause the earth is a magnet, and one pole of a magnet will tend to be attracted to the north magnetic pole and the other pole to the south magnetic pole. Compasses are essentially small magnets that are free to turn.

What is the direction of the north magnetic pole? With a simple compass children can study an important feature of the earth's magnetic field: the direction of the magnetic poles. Place a compass in a position where it is far from any magnet or magnetic material. In the northern hemisphere one end of the compass needle will point toward the north magnetic pole. Try another position. Does the compass needle point in the same direction?

The north magnetic pole is about 1,600 km (1,000 mi) from the north geographic pole. Therefore, in many places the compass needle may point several degrees away from geographic north. This angle is known as the *magnetic declination.* Children can find the direction of geographic north by drawing a line toward Polaris, the north star. (See pages 315–16 for directions for finding Polaris.) What angle does the compass needle make with the direction to geographic north? The magnetic declination for various places in the United States is shown in Figure 19–4.

The naming of the poles of magnets is an example of *operational definition.* The pole of a freely swinging magnet that tends to be oriented toward the north is operationally defined as a "north pole." All the poles that it repels are north poles, and those that it attracts are south poles.

Magnetic materials can be magnetized. If an iron or steel nail is stroked in one direction with a pole of a strong magnet, one end of the nail will repel one end of the compass needle. The nail has been magnetized! Similarly, strong magnets can be used to magnetize other magnetic materials and to strengthen weak magnets.

How can a magnet be made from iron filings? Fill a test tube or some other slender glass container with iron filings and put a stopper in the open end. Shake the tube of iron filings and bring one end of the tube next to a compass needle. The compass needle will be attracted to the iron filings. But if the tube of iron filings is a magnet, one end of the compass needle will be repelled by one end of the tube.

Have the children stroke the side of the tube in one direction with one pole of a strong magnet and describe what happens to the iron filings. Now bring one end of the tube next to a compass needle. Is one end of the needle repelled?

Shake the test tube again. Now is it a magnet?

What happens when a magnet is cut? With pliers or tin snips, cut a long piece of wire from a clothes hanger. Stroke this wire in one direction with one pole of a strong magnet. Move a compass along the wire. Where are the poles located?

Cut the wire in half. Where are the poles in the shorter wires?

Cut the wires again and again. Do the wires continue to be magnets?

A modern theory of magnetism. A theory of magnetism must explain the observations made in the preceding activities.

First, some materials such as iron and steel are magnetic while others are not. The magnetic properties of materials depend on the spinning of electrons in atoms. If there are an equal number of electrons spinning in each direction, the magnetic effects cancel each other out. But in the iron atom there are four more electrons spinning in one direction than in the other. Therefore, iron and materials such as steel that are largely iron are magnetic materials.

However, objects made of iron or steel may not be magnets. As we have seen, they can be magnetized by being stroked in one direction with a strong magnet or by being placed in a magnetic field (e.g., the earth's magnetic field) for a long time.

It is believed that magnetic materials such as iron are made up of many very small magnets called *domains.* In an unmagnetized piece of iron, these domains are lined up in a helter-skelter fashion so that their magnetic effects cancel each other out. If a test tube is filled with iron filings and tested to see whether it is a magnet, the result is usually negative. But if the test tube is stroked with a magnet, it will become magnetized. The iron filings, which are small magnets, become lined up so that their north poles lie in one direction and their south poles in another direction. In a similar way, if the domains in a piece of iron or steel can be lined up so that the north poles tend to be in one direction and the south poles in the other, the object becomes a magnet.

A magnetic substance can be magnetized by being stroked with another magnet or stored in a magnetic field, or by an electric current. In each case, the magnetic domains become lined up. A magnet can be demagnetized by heating or pounding. This tends to rearrange the domains again into a helter-skelter arrangement. In the iron filing magnet, the filings were rearranged by stroking.

If magnetized materials are cut, the pieces will continue to be magnets because the domains are still lined up.

The explanation of the earth's magnetic field involves the relationship between magnetism and electricity. Electric charges in motion set up magnetic fields. It is believed that much of the core of the earth is molten liquid and that it is in motion. The liquid in motion carries electric charges that contribute to the buildup of the earth's magnetic field.

ELECTRICITY

The effects of electric charges resemble magnetic effects. The discussions and activities that follow involve the study of electric charges. In the past some of these activities were difficult to do on damp, humid days. Now it is recommended that plastics and synthetic textiles such as nylon be used. They are more effective for generating static electric charges than the traditional materials. The plastic knives and spoons used on picnics and in lunchrooms are excellent. These can be rubbed with pieces of plastic from garment bags and with pieces of wool.

In many of the experiments, an object such as a plastic knife should be suspended so that it is free to turn. This can be done by placing the object in a folded piece of paper that is suspended by a string. (See Figure 19–5.)

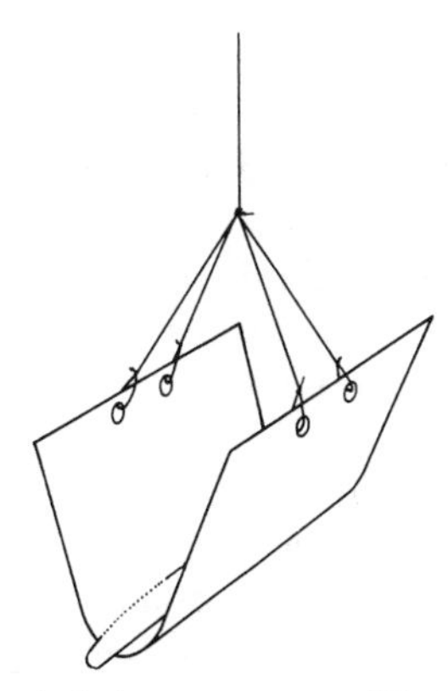

Figure 19–5 A holder for plastic knives or spoons. It is free to turn so that interactions between charges can be studied.

An electrically charged object tends to attract an uncharged object. Sometimes when hair is combed the hair seems to stick to the comb and a slight crackling noise may be heard. Garments made of wool or certain synthetic fibers seem to collect dust and lint. If a nylon garment is removed in the dark, flashes of light may be seen. All of these are electrostatic phenomena.

Electrostatics, as the word implies, refers to electric charges that tend to remain stationary. A great deal can be learned about electricity by studying electrostatic charges. Consider the following experiment that children can do.

Electrostatic phenomena. Tear some very small pieces of paper and place them on the table. Rub a rubber rod or comb with a piece of plastic and bring the rod next to the pieces of paper. The pieces of paper will probably be attracted to the rubber rod. Some pieces of paper may seem to be alternately attracted and repelled by the rubber rod. Try attracting other materials with the rod or comb. Try attracting materials with the piece of plastic. Try rubbing the rubber rod or comb with other materials such as wool and paper.

The explanation that is suggested is that when a rubber rod or comb is rubbed with plastic, some electrons are transferred from the rubber to the plastic, leaving the rubber with a positive charge. When the positively charged object was brought near small pieces of paper, the portions of the paper that were nearest the rod became negatively charged and were attracted.

As with magnetic poles, electrostatic charges are defined operationally. The charge that is generated on rubber when it is rubbed with wool is arbitrarily defined as a negative charge (−). As we will see, the charges generated, whether negative (−) or positive (+), can be determined by their reactions to the defined charge.

Objects with like electric charges tend to be repelled and those with unlike charges attracted—the basic law of electrostatics. The basic law of electrostatics is very similar to the basic law of magnetism, but they should not be confused; there are important differences. Only magnetic materials such as iron and steel can be magnetized. Electrostatic charges are best generated on materials such as rubber, plastic, nylon, and wool. However, this basic law can be used to determine the kind of electric charge that has been generated.

What happens when objects that have been rubbed with the same materials are brought near each other? Rub a plastic knife or spoon with wool and place it in a suspended paper holder so that it is free to turn. Rub another plastic knife or spoon with wool and bring it near the suspended object. (See Figure 19–6.) What happens?

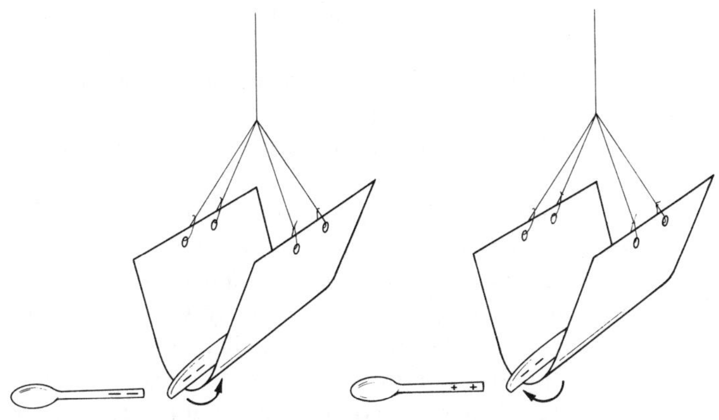

Figure 19–6 What happens when the plastic objects are rubbed with the same materials? Different materials?

Rub a plastic object with plastic and suspend it. Rub another object with plastic and bring it near the suspended object. What happens?

What happens when objects that have been rubbed with different materials are brought near each other? Rub a plastic object with wool and place it in the suspended paper holder. Rub another plastic object with plastic and bring it near the suspended object. What happens?

Rub a plastic object with plastic and place it in the suspended paper holder. What happens when another plastic object that has been rubbed with wool is brought near it?

Try the experiment with a variety of materials. Under what conditions is the suspended plastic object attracted? repelled?

According to our model, when rubber or plastic is rubbed with wool a negative (–) charge is generated in the object. We believe that there is a transfer of negatively charged electrons from the wool to the object, making the object negatively charged. When two such objects are brought near each other, they tend to repel each other.

When rubber or plastic are rubbed with plastic, we believe that electrons are transferred from the objects onto the plastic, leaving the objects positively charged (+). We hypothesize that the electrons are held more tightly in the plastic sheet and that therefore the electrons will tend to be rubbed off the objects, leaving them positively charged. (Children can feel that it is possible to hold a rubber rod or a plastic spoon tighter with plastic than with wool.) When a positively charged object is brought near a negatively charged object, there is attraction. However, when a positively charged object is brought near another positively charged object, there is repulsion. We can generalize to the basic law of electrostatics: *Objects with like electric charges tend to be repelled and those with unlike charges attracted.*

Since the charge on a rubber rod or comb rubbed with wool is by definition negative (—), we can find the charge generated on various objects when they are rubbed with a variety of materials. If the rubber rod or comb that has been rubbed with wool is attracted, the other material has a positive charge. If it is repelled, it has a negative charge.

In the generation of static electric charges, we believe that it is the negatively charged electrons that are transferred. An object that receives more electrons will have more electrons than positively charged protons and, hence, will have a negative charge. However, an object from which, on balance, electrons are removed will have more positively charged protons than electrons and be positively charged.

In general, we generate static electric charges with such materials as wool, rubber, plastic, glass, and paper. These kinds of materials are called *insulators,* and in insulators electrons tend to remain stationary and not to flow from one section of the insulator material to another. Actually, children may be able to generate a positive charge on one end of some insulator material and a negative charge on the other. The tendency of the electric charge to remain at the place in the material where it was generated is denoted by the term *static electricity.*

Static electricity was the form of electricity that was studied by the early experimenters. But it has several limitations. The discharge, for example, is almost instantaneous. If you rub some nylon against your skin in the dark, you can see these instantaneous discharges. You can sometimes hear them when you comb newly washed hair in a dry room. For scientific study it became important to have a steady, reliable source of electricity. Later electricity from other sources became very useful.

Generating Electricity

Electricity can be generated chemically. The Italian scientist Volta discovered that electricity could be generated chemically by dipping some paper in a salt solution and placing it between pieces of zinc and copper. Later he made batteries of cells and generated more electricity by placing paper dipped in a salt solution between alternating strips of zinc and copper. This zinc-and-copper "voltaic battery" provided experimenters with electricity that would flow for a considerable time.

How can electricity be generated chemically? Children can duplicate

Volta's early experiment by dipping a piece of paper toweling into a salt solution and placing it between strips of zinc and copper. If a bare portion of one wire is held on the zinc and the bare portion of a second wire on the copper and the two wires are connected to a galvanometer, an electric current will be indicated. (See pages 435–36 for instructions for making a galvanometer.) Children can experiment with other metals, such as aluminum, steel, and lead, to see whether they can be used to generate electricity.

The voltaic cells contain the essentials for the chemical generation of electricity. Two *electrodes* of different materials are needed. The two electrodes should react with the liquid at different rates. In the voltaic cell the zinc reacts faster with the salt water than the copper. An *electrolyte* that will react with the electrodes is needed. In the voltaic cell the salt solution is the electrolyte. Acids and a variety of other liquids are used as electrolytes.

How does the dry cell generate electricity? The dry cell is widely used as a source of electricity for study. (Active dry cells are not really dry; they contain a moist paste.) Dry cells can be cut apart and the key parts identified, and the materials from the dry cell can be used again to generate electricity. (If it is difficult to cut a dry cell apart, a cell can be made using zinc plate, carbon rod, and ammonium chloride.)

Dry cells can be cut apart with a hacksaw. Clamp a dry cell in a vise so that the long side is horizontal. (If a vise or clamps are not available in your classroom, the dry cell can be cut apart in the school shop.) Then saw the dry cell along the long side. You may wish to put some newspaper or a container underneath in order to catch any material that may fall out of the cell. When you have cut about two-thirds of the way through the cell, take it out of the vise and bend the case open.

The children should be able to identify the three essential components of an electric cell. The black carbon rod in the center is one electrode. The outer case of zinc is the other. The black powdery or pasty material is a mixture of ammonium chloride and manganese dioxide that serves as the electrolyte.

Put some of the black material from the dry cell into water in a glass jar. Take the zinc case from the cell and put it into the water. Place a flashlight bulb in a socket and wire one connection of the socket to the zinc plate and the other to the carbon rod. Insert the carbon rod into the electrolyte. Usually enough electricity is generated to light the bulb. (See Figure 19–7.)

Essentially, electric cells generate electricity because one of the electrodes reacts with the electrolyte faster than the other. In the voltaic cell zinc reacts with the salt solution faster than copper. When the zinc reacts with the salt, electrons are released. These electrons can be conducted through a wire to the copper electrode. This flow of electrons is an electric *current*. If a flashlight bulb is connected into the circuit, the

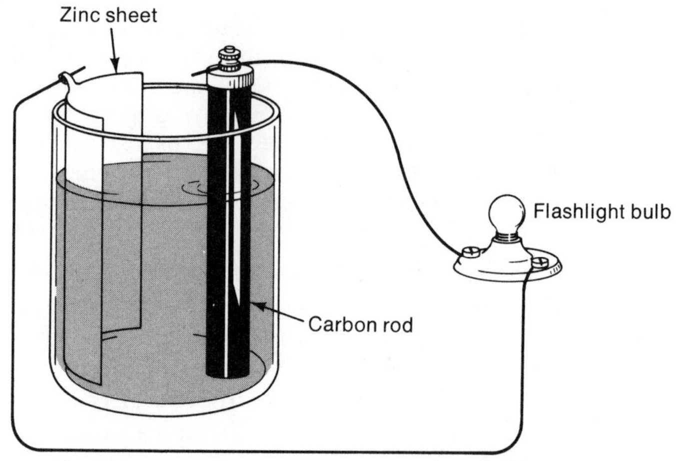

Figure 19–7 An electric cell can be made from the zinc plate, carbon rod, and black powder from a dry cell.

flow of electrons may heat the filament of the bulb to the point at which light is given off. Different kinds of cells may have different electrodes and electrolytes. But more electrons are released at one electrode than at the other. Since all electrons have a negative charge, they repel each other and are forced through the circuit to the other electrode.

If a new "dry" cell is cut apart, the electrolyte will be found to be a wet paste. In fact, one of the ways in which a dry cell may become "dead" is for the paste to dry out. The black, damp paste consists of ammonium chloride and manganese dioxide, which gives off oxygen. The oxygen combines with hydrogen to form water. Without the oxygen, the hydrogen released by the reaction tends to surround the carbon rod and inhibit the reaction. The inner case of the dry cell is usually composed of zinc. The zinc reacts with the electrolyte to release electrons. The electrons flow through an outer circuit through the carbon rod back into the cell. (See Figure 19–8.)

Magnets can be used to detect an electric current. News of Volta's invention of a source of electric current spread quickly among the many experimenters who were studying electricity. Among them was Hans Christian Oersted in Copenhagen. Oersted placed a magnetic compass needle near a wire that was conducting an electric current. He noticed that although there was no visible connection between the wire and the compass needle, when the electric current was turned on the compass needle tended to line up at right angles to the current-bearing wire. As we see, this relationship between electricity and magnetism provides a way of converting mechanical energy into electrical energy and then changing electrical energy back into mechanical energy. It also provides

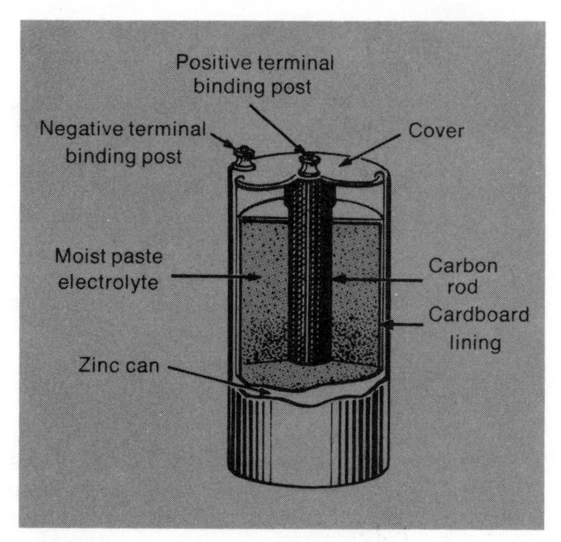

Figure 19–8 The basic components of a dry cell.

us with a very sensitive way of detecting and measuring electric current.

A *galvanometer* is a device for detecting an electric current. Essentially, it consists of magnets and a conductor of electricity. Either the conductor or the magnets must be free to move when an electric current flows through the conductor. Often a needle is connected to the part that moves and indicates the extent of the movement against the background of a dial or scale.

How can we detect an electric current? A very simple, but useful, galvanometer can be made by winding a few turns of wire around a small compass. Have the children touch the ends of the wire to the posts of a dry cell. What happens? This compass galvanometer can be used to detect fairly small electric currents. A more sensitive galvanometer can be made by floating a magnetized needle in water in a glass or plastic dish with a number of turns of wire wound around it.

Wind a number of turns of insulated wire across the top of a shallow glass or plastic dish. A petri dish is ideal. Cover the bottom of the dish with water. Magnetize a thin, light sewing needle by stroking it in the same direction with one end of a magnet. Float the magnetized needle on the surface of the water in the dish. A small cradle of wire can be used to lower the needle gently onto the surface of the water. The needle can float because of the surface tension of the water. It is more likely to float if it is rubbed on your hand so that a little natural oil or grease adheres to the needle. (If difficulty is encountered, the needle can be pushed through a small piece of cork and the cork floated on the water.)

The floating magnetized needle is really a very sensitive compass. In

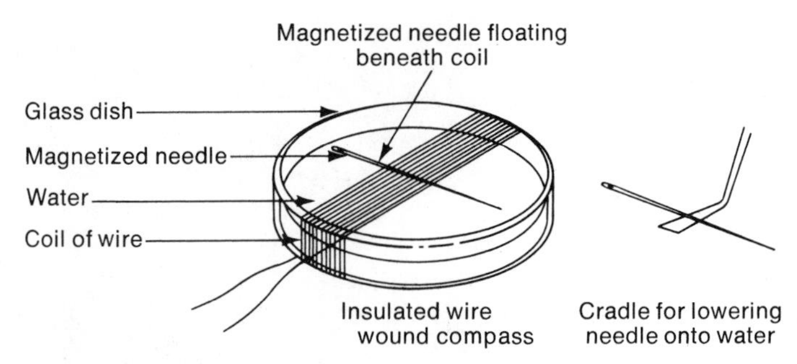

Figure 19–9 Pupil-made compasses.

the absence of other magnets, the needle will line up in a north–south direction. When even a small electric current flows through the coil, it will be indicated by a movement of the needle. (See Figure 19–9.)

The galvanometer is the basic component in the direct-current ammeter and voltmeter. The *ammeter* is used to measure the amount of current flow. Basically, it consists of a galvanometer with a shunt across the connections so that only a small part of the total current flows through the galvanometer. The amount of current flow is measured in *amperes.* The voltmeter is used to measure electrical pressure, and this pressure is expressed in *volts.* In the *voltmeter* there is a resistor that prevents most of the current in a circuit from flowing through the galvanometer. (See Figure 19–10.) An ammeter is always connected in series in a circuit, and a voltmeter is connected in parallel. (See pages 187–88 for a discussion of series and parallel circuits.)

Magnets can be used to generate electricity. Oersted had noticed that a magnetic needle would move if an electric current was sent through a nearby conductor. Michael Faraday, an English experimenter, reasoned that if a magnetic needle moves when a current is sent through a conductor, then something should happen in a conductor if a magnet is moved near it. This was a brilliant insight and has sometimes been called "reverse reasoning." You may wish to call this kind of reasoning to the attention of children and ask them to try to think of other examples. Faraday's insight was creativity of a high order, and part of children's experiences in science should involve some insight into the nature of creative work in science.

Actually, it was ten years before Faraday was able to demonstrate that when a magnet is moved near a conductor an electric current is generated in the conductor. Faraday worked at other experiments during this period but kept coming back to this suspected relationship between magnetism and electricity. Children should also become aware

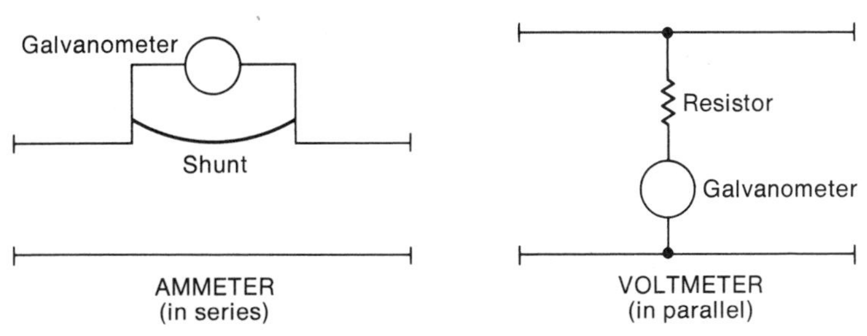

Figure 19–10 An ammeter consists of a galvanometer and shunt and is always connected in a series with a circuit. A voltmeter consists of a galvanometer and resistor and is always connected in parellel with a circuit.

that scientific discoveries do not usually come easily. Usually they involve considerable time and a great deal of effort, but the creative insight is the most precious ingredient.

How can magnets be used to generate electricity? Wind some insulated wire into a coil and connect the ends of the coil to a galvanometer. (The floating magnetic-needle galvanometer described on page 435 can be used.) Then push a pole of a magnet into the coil. (See Figure 19–11.) What happens to the galvanometer? Note what happens to the galvanometer when the magnet is stationary within the coil. Pull the magnetic pole out. What happens to the galvanometer? Repeat by holding the magnet stationary and moving the coil relative to the magnet. Now is an electric current generated?

Three things are needed to generate electricity in this way: a magnetic field, a conductor, and relative motion. In demonstrating the generation of electricity, children will usually use permanent magnets. In electrical generators, however, electromagnets are usually used. Children can wind insulated wire into a coil to serve as the conductor. Actually, the amount of electricity generated will depend on the number of turns of wire that move through the magnetic field. In commercial generators there are many turns of wire in each coil and a number of coils. Children can supply the motion by moving either the coil or the magnet or both. This is another example of relative motion. (See pages 14–16.) The coil and the magnet must be moved *relative* to each other. If the coil and the magnet are held together and moved relative to the surroundings but not to each other, no electricity will be generated. In commercial generators either the coils or the magnets are usually turned inside the other.

A *generator* is essentially a device for converting mechanical energy into electrical energy. The mechanical energy is usually supplied by turbines, which are turned by falling water in hydroelectric plants or by

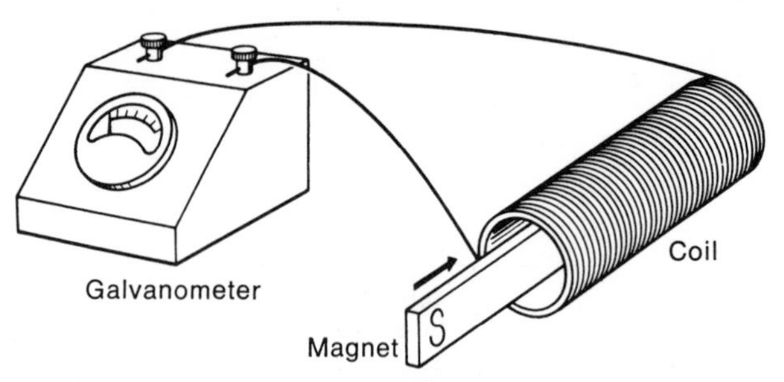

Figure 19–11 When a magnet and coil of insulated wire are moved relative to each other, an electric current is generated.

steam generated by burning coal or oil or from the heat generated by nuclear reactors. Because electricity is such a convenient form of energy, a great deal of energy in other forms is used to generate electricity.

When a magnet is pushed into a coil of wire, the needle of the galvanometer will move in one direction; when the magnet is pulled out, the needle will move in the other direction. This indicates that the electric current alternates its direction of flow. We say it is an *alternating current* (AC). Unless it is altered in some way, the current generated in a generator is an alternating current. The current generated chemically in cells and batteries flows in one direction and is called *direct current* (DC). Direct current is most often used in elementary school science experiments. The current used in homes, schools, and other places in the community is usually alternating current.

Faraday's creative insight led to our modern electrical generating stations. Faraday was never very interested in practical applications of his discoveries. However, when the chancellor of the exchequer visited his laboratory and saw his electrical experiments, he is reported to have said, "Very interesting, Mr. Faraday, but what is the use of it?" There are two versions of Faraday's reply: (1) "What is the use of a newborn babe?" and (2) "Someday, Mr. Chancellor, you will be able to tax it."

Conducting Electricity

Some materials will conduct electric currents while others will not. Children will already have experience in describing objects and materials in terms of color, shape, texture, smell, weight and other properties. They may also have experience in classifying on the basis of whether or not objects and materials have certain properties. Another property that is not as obvious, but is very important, is whether or not a material can conduct an electric current. To conduct an electric current means to transfer the electrical energy from one place to another.

Conductors are materials such as copper, silver, and aluminum through which electric currents will flow. These materials are used in the wires that transfer electrical energy. Nonconductors are materials, such as cotton, linen, wood, glass, and paper, through which an electric current will not flow. Some of these nonconductors are used as insulators to prevent the flow of electricity. For example, insulated copper wire often has cotton wound around it, and wires are hung on glass insulators to prevent or reduce the loss of electricity. There are also semiconductors that conduct electricity to a certain extent. Semiconductors are used in computers, solid-state radio and television sets, and throughout the electronics industry.

What materials will conduct an electric current? Connect one terminal of a flashlight bulb socket with a wire to one post of a dry cell. Connect one end of a wire to the other post of the dry cell and bare the other end of the wire. Connect one end of another wire to the other terminal of the flashlight bulb socket and bare the free end. Test the setup by touching the two free ends of wire together. The flashlight bulb should light.

What materials will conduct electricity? As children suggest various materials that they think will conduct electricity, have a child lay them across the two bared ends of wire to see whether the bulb will light. Place all materials that conduct electricity in a pile and label the pile "Conductors."

What materials will not conduct electricity? Have a child test materials that other children suggest will not conduct electricity. Place all the materials that do not conduct electricity in another pile and label this pile "Nonconductors."

In what other ways do nonconductors seem to be similar?

Electric currents can be conducted through complete circuits. "How can we make the bulb light?" We were in a fourth-grade classroom, and we had placed a flashlight bulb and its socket on a box in the front of the room.

"We can connect it to a battery." "How?" "With wire, of course." The last statement was uttered in the tone that fourth-graders use to indicate that, really, you should know and shouldn't have to ask.

We took a wire and connected it, after the children had instructed us to scrape the insulation off the ends of the wire, to one post of the dry cell and one terminal of the flashlight bulb socket. "The bulb doesn't light!" "You need another wire from the other post of the battery to the bulb." "You need a *complete circuit!*" We suggested that one of the children make this connection. The bulb glowed. The complete circuit provides a path of conductors so that the electrons can flow from the post of the dry cell, through the bulb, and back to the other post of the dry cell. The complete circuit can be compared to a round trip ticket. The elec-

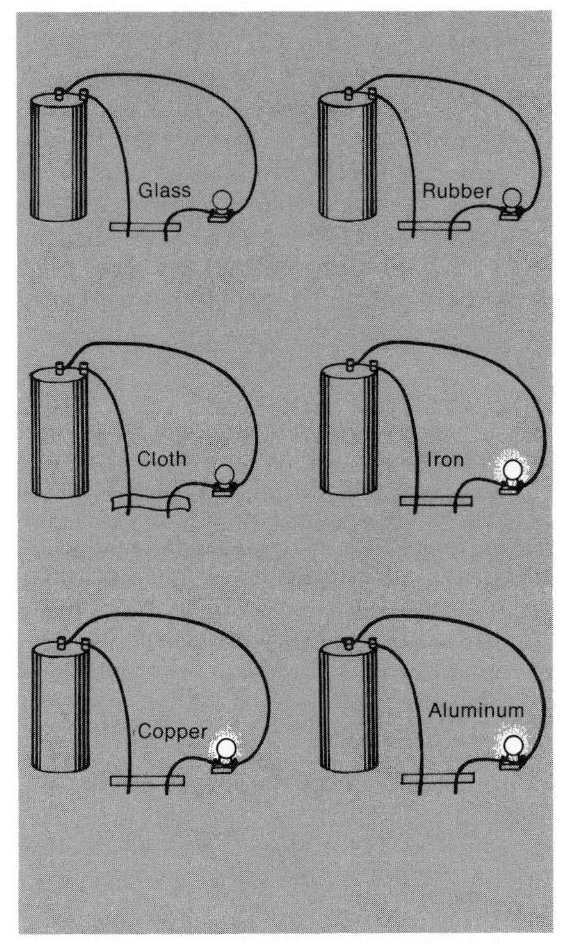

Figure 19–12 A setup for testing whether materials are conductors or nonconductors.

trons do not flow unless there is a pathway of conductors out from the dry cell and back.

Switches are devices for making and breaking complete circuits. A simple switch can be made by arranging the ends of wires so that they have to be pushed together to complete the circuit. It is usually recommended that children incorporate spring switches into their circuits so that the circuit will be automatically broken when it is not in use. This helps preserve dry cells.

"Here are several flashlight bulbs. How can we connect them to the dry cells?" We gave each group of children two dry cells that had a wire connecting the center post of one to the outside post of the other, and three flashlight bulbs in sockets. Most of the groups connected the flashlight bulbs in *series.* One group connected the bulbs in *parallel.*

What are the characteristics of series and parallel circuits?

Connect a flashlight bulb to each of two sets of two dry cells. Do the bulbs shine with equal brightness?

To each of the bulbs connect another bulb. In one case the second bulb should be connected in series and in the other case in parallel. In which circuit do the bulbs shine brighter?

Add a third bulb to the series circuit and to the parallel circuit as shown in Figure 19–13. Compare the brightness.

Unscrew one of the bulbs in the series circuit. What happens to the light from the other bulbs?

Unscrew one of the bulbs in the parallel circuit. What happens to the light from the other bulbs?

Most electrical circuits in homes, schools, and other buildings are parallel circuits. In parallel circuits the amount of electricity to bulbs or appliances is not diminished as more bulbs or appliances are added. Also, if a bulb or appliance is turned off, in a parallel circuit the other bulbs and appliances will continue to operate. This is not the case in a series circuit.

How can a small house be wired? One of the ways in which pupils can apply their knowledge of electrical circuits is to install a flashlight bulb lighting system in a dollhouse (a cardboard box or two with one side cut off will do). They should be encouraged to draw a plan for the most efficient and effective lighting system and then proceed to install the system.

How is electricity conducted in a flashlight? (See Figure 19–14.) The

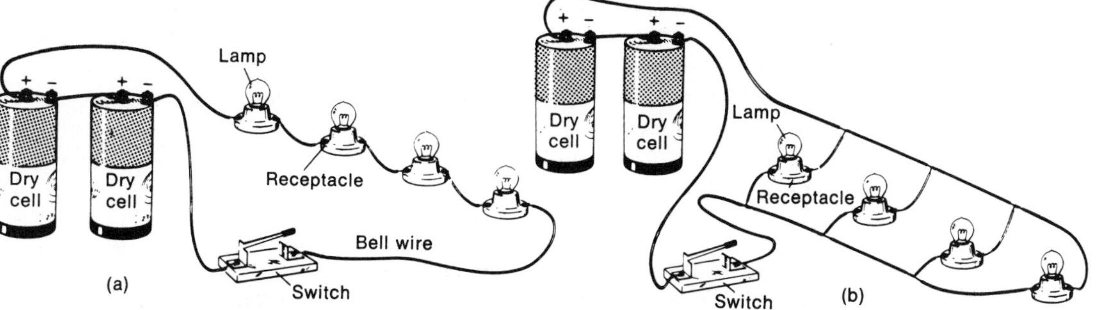

Figure 19–13 In the circuit at the left, the bulbs are connected in series. In the circuit at the right, the bulbs are connected in parallel.

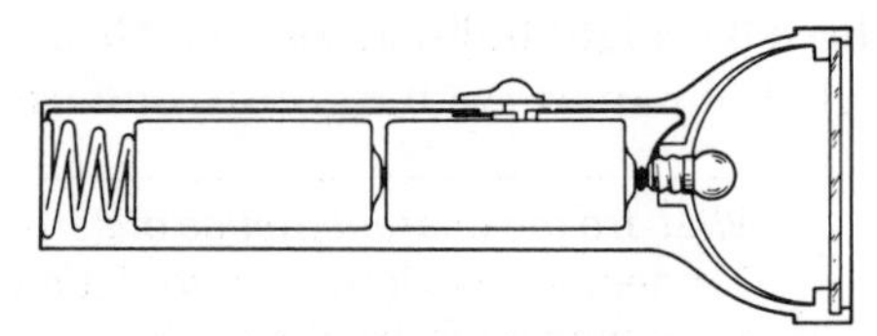

Figure 19–14 Have the children trace the circuit in the flashlight and solve the problems when the bulb does not light.

flashlight is a convenient device for studying electrical circuits. Have the children take a flashlight apart and trace the complete circuit. Have them note how the switch makes and breaks the circuit. After they become familiar with the flashlight circuit, pose problems for them by turning one of the dry cells around, removing the spring at the rear, or placing a nonconductor over the center post of one of the dry cells. See if they can repair the flashlight.

If too much electricity flows through a circuit, the wires may become hot and a fire may be started. This can be prevented by the insertion of a fuse or a circuit breaker into the circuit. In a *fuse,* there is a strip of a metal that will melt and break the circuit when the temperature reaches a certain point. In a *circuit breaker,* when the current reaches a certain point an electromagnet (see page 443) will pull a switch open and break the circuit.

Converting Electricity into Other Forms of Energy

Electrical energy can be converted into other forms of energy. As we have seen, electrical energy can be transferred considerable distances along conductors. Another great advantage of electrical energy is that it can be readily converted into other forms of energy.

How can electrical energy be converted into heat? One end of a wire can be fastened to one post of a dry cell and the other end touched to the other post. Soon the wire will become hot and the connection should be broken. However, the children who do this will have felt some of the heat that was generated from the flow of the electricity through the wire.

Toasters, hot plates, and heaters are among the devices that are designed to convert electricity into heat energy. You may wish to have the children examine such appliances and see the units that actually convert electricity into heat.

How can electrical energy be converted into light? Many heating coils give off light as well as heat. Children may have peered into a toaster, for example, and seen the reddish light that is given off by the heating elements.

If they look at a clear glass light bulb, children can see the filament inside the bulb glow as the electricity flows through it. Note that although they are designed to convert electricity into light, most light bulbs also give off some heat. Children may also see some vibration of the filament in a clear glass bulb as the alternating current surges through it.

An electric light can be simulated by connecting a thin strand of picture wire to the bared ends of two copper wires and then connecting the other ends of the copper wire to the two posts of a dry cell. If the picture wire is very thin, it will usually glow brightly for a short time and then "burn." You may wish to point out that the filaments in light bulbs do not burn in this way because they are not in contact with oxygen.

How can electrical energy be used to generate a magnetic field? How is an electromagnet like and unlike a permanent magnet?

An electromagnet can be made by wrapping a number of turns of insulated wire around a nail or bolt and connecting the wire through a switch to one or more dry cells.

Lower the electromagnet onto paper clips spread over the table. Does the electromagnet appear to have poles?

Hold the ends of the electromagnet next to the ends of a compass needle. Is one end of the compass needle repelled?

Unwind the electromagnet and then wind the wire around in the opposite way. How does this affect the poles of the electromagnet?

Reverse the connections of the wires on the dry cell. How does this affect the polarity of the electromagnet?

Pick up some paper clips with the electromagnet. Then open the switch. Does the electromagnet continue to be a magnet? What happens when the switch is closed again?

Children can move a magnet in and out of a coil of wire and generate electricity. Have them note that they are converting mechanical energy into electrical energy on a much larger scale.

Can electrical energy be converted into mechanical energy? Yes; when we use an electromagnet to attract paper clips, a small amount of electrical energy is converted into a very small amount of mechanical energy. An electric motor is a device for converting electrical energy into mechanical energy on a much larger scale.

How can electrical energy be converted into mechanical energy? Tap, tap, tap a bell with a pencil and then tap, tap, tap it again. This is one way to ring a bell. How else can it be done?

A clapper can be pulled against the bell by an electromagnet. If the circuit is broken, the clapper will spring back, to be pulled against the bell again when the circuit is closed.

An electric bell is an electric motor in which electrical energy is converted into mechanical energy. (See Figure 19–15.) It also involves *automation* (automatic switching). When the armature is attracted to the electromagnets, the clapper strikes the bell. However, the armature is pulled away from the contact so that the circuit is broken. When the

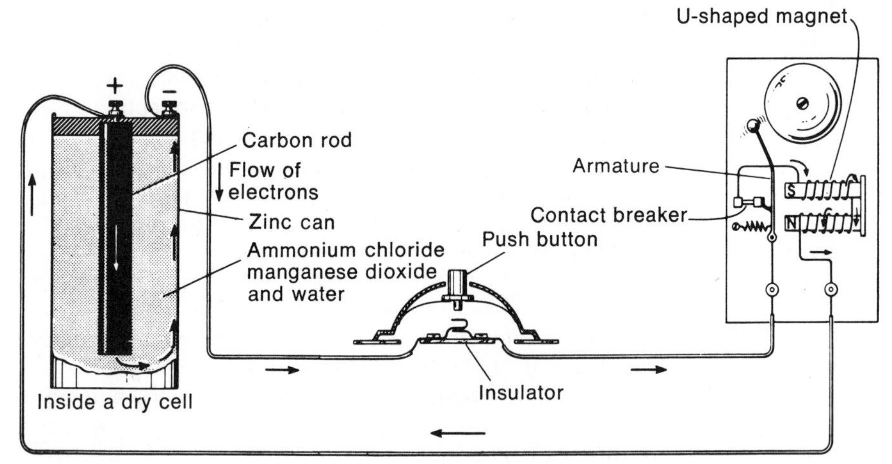

Figure 19–15 The armature is attracted by the electromagnet. The circuit is broken and the armature springs back, closing the circuit again.

armature is released by the electromagnets, it springs back to close the circuit again. This energizes the electromagnets, which again attract the armature. This can be repeated many times a minute.

The Electric Motor

The study and construction of electric motors is almost always a fascinating experience for children. Sometimes we start with an electric bell, which is an electric motor that pulls a clapper against a metal bell. However, children can also build their own motors. As they wind the armature, fasten the bared ends of the armature wire, position the magnets, and arrange the brushes so that they make contact with the ends of the armature wire at the right time, they have a firsthand experience that helps them understand how an electric motor works.

In addition to building an electric motor, children should gain an understanding of the basic principles on which the motor works. Most important, an electric motor is a magnet that is free to turn near or inside one or more other magnets. Almost always, the magnet that turns is an electromagnet.

The electric motor works on the principle that like magnetic poles tend to repel each other and unlike magnetic poles tend to attract each other.

The polarity of electromagnets can be changed by reversing the direction in which the current flows through the electromagnet. This occurs in the electric motor as the electromagnet turns on its axis. By adjusting the position at which the direction of current flow changes, the polarity of the electromagnet can be changed so that the ends of the electromagnet will be alternately repelled and attracted and the electromagnet will turn on its axis.

An electric motor like the one described here has four essential parts. (See Figure 19–16.) The electromagnet that turns is called the *armature.* The magnets near which the armature turns are called *field magnets.* In the motor that can be constructed by children, the field magnets are bar magnets. Fairly strong alnico magnets are recommended. An electrical contact is made with ends of coiled wire in the armature. The contacts are called the *commutator.* In the motor described here, the commutator consists of the bared ends of the wire fastened on the opposite sides of the axis. The sliding contacts are called *brushes,* and they brush against the bared ends of the armature wire.

Building an electric motor. The armature of this motor is an electromagnet that is taped to the bottom of a test tube. The core of the armature can be made of nails, but we have found clothes hanger wire more convenient. With a pair of pliers or tin shears, cut three pieces of wire each about 8 cm (3+ in) long from a wire hanger. The insulated wire for the electromagnet should be about 4 m (13 ft) long. Remove about 8 cm (3+ in) of insulation from each end of the wire. Then, with 10 cm (4 in) of the wire left free at the end, hold the three pieces of wire together and start winding the insulated wire around them. Be sure to *always wind in the same direction.* Wind from the middle of the wires out to one end, back to the other end, and back

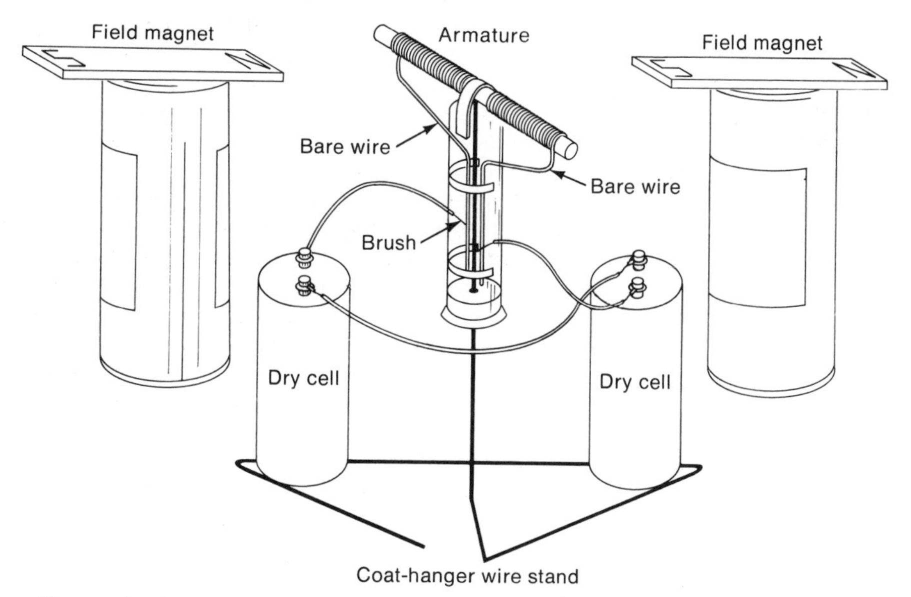

Figure 19–16 A simple electric motor.

to the center. Try to distribute the turns on the wires evenly so that the coil will be well balanced. Leave about 10 cm (4 in) of wire free at the remaining end.

With a piece of tape, fasten the middle of the coil to the rounded bottom of the test tube. This coil will rotate and serve as the armature of the motor.

Stretch the bared ends of the armature coil along opposite sides of the test tube. Fasten the bared ends in place with a thin strip of tape near the coil and another thin strip of tape near the opening of the test tube. The brushes will make sliding contact with the bared ends of the wire.

Cut off the hook part of another wire clothes hanger. Bend the wire as shown in the diagram so that a length of the wire slightly longer than the test tube will be vertical. The test tube will turn on the point of this vertical wire. If the point of the wire is filed so that it is sharp, the friction will be reduced. Some children have stapled the sections of wire to a board to make the motor steadier.

Put the test tube onto the vertical wire so that the inside of the bottom of the test tube rests on the sharp tip of the vertical wire. Spin the test tube to be sure that it can spin freely.

The armature will wobble less if a cork with a small hole in the center is inserted in the opening of the test tube.

Support two alnico bar magnets so that they are at the same height as the rotating coil. These bar magnets are the field magnets of the motor. The magnets can be laid on top of milk cartons or tennis ball cans. The magnets should be opposite each other. The north pole of one magnet should be near the coil and the south pole of the other magnet.

Cut two pieces of insulated wire about 20 cm (8 in) long. Remove about 6 cm (2½ in) of insulation from one end and 2 cm (1 in) from the other end of each wire. These two wires will serve as the brushes.

Connect one bared end of another piece of wire to the center post of one dry cell and the other bared end to the outside post of a second dry cell. Connect the bared end of one wire brush to the outside post of one dry cell and the bared end of the second wire brush to the center post of the second dry cell. Arrange the dry cells so that the bared ends of the brushes will brush against opposite sides of the test tube as it rotates. A small piece of tape can be wound around the wires so that they will be held the right distance apart to continue to make contact with the sides of the test tube.

The brushes should be positioned so that they will make contact at the same time with the bared wire on the sides of the test tube. The brushes should make contact with the wires when the ends of the electromagnet are about one-third of the way past the nearest pole of the field magnets. The exact points of contact between the brushes and the wires on the test tube usually have to be adjusted to get the motor to rotate the fastest. This can be done by moving the dry cells.

Give the armature a spin. Do you see sparking as the brushes make contact with the bared ends of the armature wire? You may wish to adjust the positions of the brushes or of the field magnets to make the motor run more efficiently. When the motor is completed, it should look like the one in Figure 19–16. A number of experiments can be carried out with this or other simple motors:

1. Hold the armature so that the brushes are touching the bare wires on the test tube. Let go. Are the ends of the armature attracted toward the nearest magnet, or are they repelled?
2. Move the magnets until you find the position in which the motor runs fastest. Now stop the motor in the position where both brushes are touching the bared wire on the test tube. In this position how far is it between one end of the armature and the pole of the nearest magnet?
3. Move the magnets into the position in which the motor is running fastest. Remove one of the magnets. Does the motor still turn?
4. Remove one dry cell. Will the motor run on one dry cell?
5. Note the direction in which the motor turns. Looking down from above, is it turning clockwise or counterclockwise? Change the magnets so that a north pole is where the south pole was and a south pole where the north pole was. Now which way does the motor turn?
6. Note the direction in which the motor is turning. Now turn the dry cells around so that the brush from the outside post is where the brush from the center post was and the brush from the center post is where the brush from the outside post was. Now which way does the motor turn?

RELATED ACTIVITIES

Identifying the common property in materials attracted to a magnet, p. 39.

Using magnets, p. 111.

Will a rubber comb rubbed with plastic have the same static electric charge as one rubbed with wool? p. 53.

Ranking materials according to conductivity, p. 397.

How can you make a bulb light? p. 55.

How can electricity be conducted from one place to another? p. 187.

How does a sensing device work? p. 191.

How does a thermostat work? p. 183.

How can electrical energy be converted into heat energy? p. 393.

How can different forms of energy be generated? p. 186.

REFERENCES

For Teaching

ELEMENTARY SCIENCE STUDY. *Batteries and Bulbs.* New York: McGraw-Hill, 1974. A teacher's guide to a series of activities involving electricity. Especially good for descriptions of improvised materials.

GRAF, RUDOLF F. *Safe and Simple Electrical Experiments.* New York: Dover, 1964. A collection of electrical experiments for children. Experiments involving magnetism are included.

For Young Children

BRANLEY, FRANKLYN M. AND ELEANOR K. VAUGHAN. *Mickey's Magnet.* New York: Harper & Row, 1956. Mickey's adventures with magnets, shown in cartoon pictures, should fascinate most children.

JACOBSON, WILLARD J., ET AL. *Magnets.* New York: American Book, 1972. A book for young children that includes some of the basic activities they can undertake with magnets.

For Older Children

FREEMAN, IRA M. *All About Electricity.* New York: Random House, 1957. Background information on electricity.

JACOBSON, WILLARD J., ET AL. *Electricity and Magnetism: Case Histories in Science.* New York: American Book, 1969. A history of the development of our understanding of the relationships between electricity and magnetism.

PODENDORF, ILLA. *True Book of Magnets and Electricity.* Chicago: Children's Press, 1961. A sample of children's activities in electricity and magnetism.

SOOTIN, HARRY. *Experiments with Electric Currents.* New York: W. W. Norton, 1969. Contains complementary sections dealing with "What you should know" and "What to do."

VICTOR, EDWARD. *Exploring and Understanding Magnets and Electromagnets.* Westchester, Ill.: Benefic Press, 1967. An extensive discussion of magnets and electromagnets for older children who wish to go further in this area.

Science Investigations—Continuing the Study of Science

CHAPTER 20

Science investigations are inquiries into questions and problems to which the answers are not known. They are "research" in that children are probing areas that, for them, contain unknowns. In this chapter we discuss several of the important functions of science investigations, describe how several carefully selected investigations can be carried out, and list additional investigations that have been carried out by children and teachers with whom we have worked. Of course, this list is open ended, because some of the best investigations grow out of the questions children ask, the problems they encounter, and the things they are curious about.

Usually science investigations involve questions, problems, and puzzling phenomena that cannot be resolved without the use of science materials in the classroom or laboratory or in the field. Sometimes magnifying glasses or microscopes help children observe the smaller details of an organism, object, or phenomenon. Often, controlled experiments are set up to study the significance of various factors. Systematic observations are made of changes that take place in plants, animals, and other phenomena in the environment. Children should have the experience of using tools and instruments of science as they investigate phenomena.

As in scientific research, there is an important place in science investigations for "going to the literature." Children should learn how to go to books and magazines to discover what others have found out about the phenomenon they are studying and to check their observations and discoveries against those made by others. The science section of a school or public library is an important resource for a good elementary school program; one of its major contributions is to help children in their science investigations. An important by-product is that it gives children opportunities for "purposeful" reading.

Science investigations may deal with questions and problems that grow out of the curiosity of children; they may be suggested by a teacher or parent; or they may be derived from lists of suggested investigations such as those included in this chapter. Some children used magnifying glasses to examine the eyes of a bumblebee that was found dead on the schoolgrounds. They wondered whether other insects had similar eyes. This turned out to be a fascinating investigation. On the other hand, there are times when a teacher may wish to have children acquire a certain skill or develop a concept, and this will lead to a particular investigation. For example, a teacher who wanted children to gain a better understanding of static electricity asked them to investigate the question, "What electric charge is generated when different kinds of materials are rubbed together?" Teachers have found it useful to develop lists of investigations that children can undertake, and a good way to start is with the list in this chapter.

SCIENCE PROCESSES

While carrying out science investigations, children develop skill in using many of the processes that are important in scientific investigation. Among those processes are the following:

Defining Problems. Usually it is important to clarify the problem or the question we want to solve or answer. This is one of the most difficult aspects of research. Children begin to have some experience with problem definition as they think through the nature of an investigation.

Designing Investigations. It is important to plan investigations carefully. What are you going to do? How is it to be done? What information are we going to try to get? What problems may we run into? How can we deal with these difficulties?

Controlled Experiments. Many science investigations undertaken by children take the form of controlled experiments. As discussed in Chapter 8, the thinking processes involved in trying to control all the factors except the variable being tested are believed to be those that contribute to the development of formal, logical thinking. What is the variable to be tested? What variables will have to be controlled?

Observing. Most science investigations involve observing—sometimes with several senses. Often instruments such as magnifying glasses, microscopes, pH paper, and tape recorders are used to extend the range of observation. What should we be able to observe? Of special importance, what kinds of unexpected observations might be made?

Measuring. Measuring makes it possible to increase the precision of our observations and provides a means for recording observations. What measurements should be taken? What tools do we need for these measurements?

Classifying. It is frequently useful to sort objects and observations. Often the classifying is done on the basis of whether or not objects or observations have a certain property. Sometimes children can develop their own systems of classification.

Inferring. It is important that children learn to distinguish between observations and inferences. Sometimes it is necessary to infer beyond the data that have been collected. Such inferences are based on logical deduction and a sense of the probability that an inference may or may not be true.

Formulating Hypotheses. Hypotheses are suggested answers to questions or problems that serve as mental tools in investigations. In fact, many investigations take the form of testing a succession of hypotheses until one is found that withstands the tests. Hypothesis formulation is one of the most creative dimensions of scientific investigation. Through science investigations children can use what they already know to formulate creative hypotheses.

Collecting and Handling Data. Data are collected and handled in a variety of ways. Often tables and graphs are effective ways of finding possible relationships between classes of data. What data should be collected? How should these data be handled?

Interpreting Data. Graphs of various kinds may help children find patterns in data, relationships between variables, and possible trends. How can we find the meanings of the data that we have collected?

Literature Search. Investigations should not be done in a vacuum. Children should learn how to use reference books, trade books, magazines, and other library resources to check their procedures and findings against those of others. Often they will find it helpful to confer with other children, teachers, parents, and other members of the community who may know something about the problems they are investigating.

Relating to Theory. The investigation and its results should be related to broad science concepts. Some of these concepts are listed in the science investigations that follow.

Drawing Conclusions. What reasonable conclusions are warranted by your findings? Here children begin to learn how to make judgments. Investigators are usually cautious in interpreting the data they have collected. On the other hand, excessive caution may prevent investigators from drawing conclusions that are warranted.

In science investigations children use various processes as they are needed. Certainly, there are benefits in having children learn and practice various science processes. But they also have to learn when to use them. One of the reasons for having children undertake science investigations is that they not only have an opportunity to practice various science processes while carrying out the investigations but also use these processes when they are needed and thus have a chance to learn "when" as well as "how."

DEVELOPING SCIENCE CONCEPTS THROUGH INVESTIGATIONS

In addition to developing science process skills, children can develop some of the most basic science concepts through their science investigations. Science concepts are the "big ideas" of science; they are the way we view ourselves and the world in which we live. For example, a group of children investigated the question, "What kinds of plants grow in places where there is a great deal of water and in places where there is little water?" They compared the sedges that grew along the edge of a nearby lake with the cactus plant in their desert terrarium. They could have come out of this inquiry with a catalog of differences between plants that would soon have been forgotten. But the teacher working with them helped them achieve a deeper concept of the fundamental concept of adaptation: "Organisms that live in a particular environment have structures and functions that have made it possible for them to survive in that environment."

We carry science concepts like intellectual building blocks from one experience to the next. For example, once some concept of adaptation has been developed, children can deepen their understanding of this concept in the next experience. If they were to examine the plants that live in the cracks of a sidewalk, they would expect these plants to have adaptations that would help them survive in this harsh environment.

Science concepts are used in defining problems and formulating hypotheses. Concepts can give us clues about a problem. For example, if certain plants are not thriving in a certain environment, children might ask, "Do these plants have the adaptations that make it possible for them to survive under these conditions?" In order to suggest fruitful hypotheses, children have to use what they already know. Among the most useful knowledge they can have are science concepts.

Teachers and other adults who work with children in science can help them develop science concepts as they carry out investigations. They can do this by helping children derive meaning from experience. It is possible to walk down a garden path or sit on a rotting log without seeing or learning much. But an insightful teacher can help children relate their observations to much broader ideas, to "see where they have only looked before." What they "see" can be related to some of the big ideas of science.

SCIENCE INVESTIGATIONS AND EXCEPTIONAL CHILDREN

It could be argued that all children are exceptional in that all children are different, have different interests, and can benefit from a variety of educational experiences. Certainly, all children can benefit from a variety of science experiences, and almost all of the suggestions made in this book can be used with all children but, there are children for whom science experiences are especially important.

Gifted and talented children can benefit from science investigations. They often can see questions and problems in new and different ways. It is especially exciting to see them come up with new and highly imaginative hypotheses as suggested answers to questions and problems. Once these children have learned how to carry out such investigations, they can often work very much on their own. Since our aim in education is to help every child to develop to his or her fullest potential, we need to have a variety of activities that can challenge children. Science investigations are among the ways in which gifted and talented children can be challenged.

Science investigations are also of special importance for children who may not do well in some parts of the school program but have a special interest in science. Paul was in fourth grade when one of us worked with him. Paul was not very verbal, and he was no match for some of the quick, very verbal children in his class in the discussions and quick repartee that form such a significant dimension of many school experiences. In fact, some people suspected that Paul was "a bit slow." But while his ideas might not come fast, there was depth to Paul's thinking. When he was given the chance to digest information and consider all facts, the ideas that he came up with were often of a higher quality than those of children who were glib and quick. Paul did extremely well with science investigations. In fact, Paul's parents felt that this was what made school really worthwhile for Paul. Do you have a Paul in your classroom?

Once started on science investigations, many children, especially exceptional children, find new and challenging directions to pursue. For example, a group of fourth-graders were investigating whether or not it made any difference in germination when seeds were oriented in different ways (up, down, slanted, etc.). They found that it didn't. But they wondered how a full-grown plant would be affected if it were turned upside-down. They found that the leaves fell off and were replaced and the stem of the plant eventually turned and grew upward. Then they wondered whether it was gravity or the light source that had caused the change in the plant. Thus, imaginative children ask questions and these questions lead to further investigations. This is a characteristic of scientific inquiry–science has been said to have "an endless frontier." This characteristic makes scientific investigations especially appropriate for exceptional children.

SOME SCIENCE INVESTIGATIONS

The following investigations have been tried and tested by elementary school teachers. Most of the investigations are illustrative of the use of science processes and can lead to the further development of science concepts. An attempt has been made to select investigations that use materials that are readily available to most children and teachers.

How Does a Ball Bounce?

In science investigations observations are made with the senses. In many investigations we get our information by seeing. In this investigation most of the information is obtained by listening.

For this investigation two balls, such as a tennis ball and a rubber ball, along with a measuring stick or tape are needed.

The investigation is best done by two children working together. One investigator drops the ball from a measured height while the other listens and counts the number of bounces heard.

First have the children practice by closing their eyes and counting the number of times they hear a ball bounce when they drop it.

Have one investigator drop the ball from heights of 0.5, 1, 1.5, and 2 m. In each case the other investigator should count and record the number of bounces heard. The number of bounces should be plotted on a graph like the one in Figure 20–1.

Have them predict how many bounces there will be when the ball is dropped from a height of 2.5 meters. Have them mark their predictions on the graph, together with a question mark. Then have them count the bounces when the ball is dropped from a height of 2.5 m. (They may have to stand on a chair.) How close was the prediction to the number of bounces counted? What might account for any difference that occurred?

Have the children change roles and repeat the investigation using a different kind of ball.

Science processes:

1. Observing and counting by listening.
2. Measuring heights with a meter stick or tape.

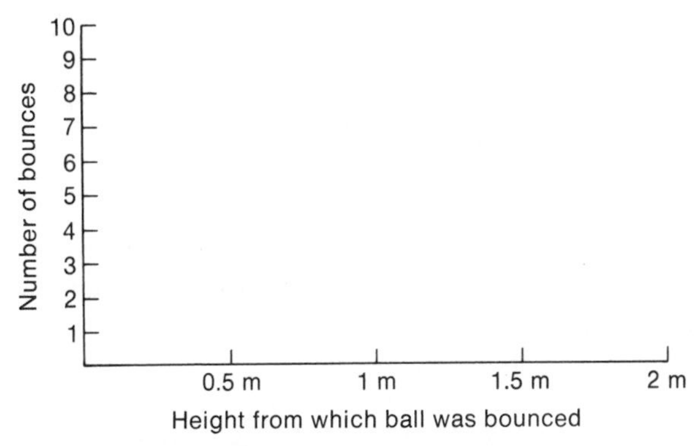

Figure 20–1 Distance dropped and number of bounces.

3. Graphing and extrapolating a line on a graph.
4. Predicting on the basis of data already gathered.

Science concepts:

1. The number of times an elastic ball will bounce is related to the speed or velocity with which it strikes a surface.
2. The speed or velocity of a dropped ball is related to the height from which it is dropped.

Rate of Water Flow

How is the rate at which water flows out of a bottle related to the size of the opening? This is essentially an investigation of the relationships between two factors: size of opening and rate of water flow. The investigation can focus either on how long it takes a given amount of water to flow out of the bottle or on how much water flows in a given amount of time.

Holes with diameters of 1, 2, 3, 4, and 5 mm can be cut in pieces of aluminum foil. The circles can be traced with a pencil and then cut with a knife, different-sized cork borers, or the sharp end of a pencil.

Fill a glass bottle with water and fasten a piece of aluminum foil with a hole in it over the opening of the bottle with a rubber band. Place a finger over the hole and invert the bottle. Using a sweep second hand, measure the amount of time required for 50 cc of water to flow through the hole. (Or, measure the amount of water that flows through the hole in a given time.) Repeat several times and find the average time required for 50 cc of water to flow through a hole of that size.

Repeat with holes of different sizes.

Observe the flow of the water. Is it a steady flow? What factors seem to affect the flow of the water?

Construct a line graph showing the relationship between size of opening and time required for a given amount of water to flow out of the bottle. What is the nature of the curve?

Science processes:

1. Observation of the flow of water.
2. Measurement of time and volume.
3. Study of the relationships between two factors.

Science concepts:

1. The downward flow of a liquid is affected by such factors as gravity, air pressure, and size of opening.
2. The area of a circle varies with the square of its radius.

How Does Popcorn Pop?

Most children will probably have had the experience of eating popcorn. Do they know the source of the popcorn? Do they know how a hard kernel of corn is changed into the white fluffy popcorn that we eat?

Give each child one or more kernels of popcorn. Have them examine the kernels very carefully. In what ways are these kernels different from the fluffy popcorn that we eat? In what ways are they alike? Where do they think the kernel will split when it pops?

Have the children put a small amount of cooking oil into a pan and drop a small number of popcorn kernels into the oil. Have them place the pan on a hot plate and observe the kernels very carefully as they are heated. Do the kernels change? What happens when they pop? Examine the popped corn. Where did the corn kernel split?

Science processes:

1. Careful examination of an object.
2. Prediction of where change may take place in the object.
3. Careful observation of change.

Science concepts:

1. Materials such as popcorn can undergo great changes when heated.
2. Substances such as water expand upon heating. Rapid release of pressure can result in an explosion.

What Really Happened?

Different people observe and remember an event in different ways. The reports of an accident by different observers may differ so much that it may be difficult to believe that they are recalling the same event.

There are many possible approaches to such an investigation. One is to arrange for an "accident" such as the spilling of a liquid or the dropping of a weight. Without prior discussion, have each child write on a sheet of paper what he or she saw. The resulting accounts can be analyzed in terms of questions like the following:

1. What observations are reported in all of the descriptions?
2. What are some things that some people saw and others didn't?
3. Which elements of the descriptions are direct observations and which are inferences?
4. What are some factors that seem to affect what different people observed?
5. Is it possible to reach a consensus on a description of what happened?

Science processes:

1. Distinguishing between observation and inference.
2. Possible effects of location and frame of reference on observation.
3. Possible effects of previous experiences on what is observed.

Science concepts:

1. The observer can be thought of as a participant in an event.

Bread Molds

On what kinds of substances will bread molds grow? If bread is left in a moist, dark place, molds will eventually grow on it. This will happen even though most commercially prepared breads contain the mold inhibitor calcium propionate. Often there will be molds of several different colors. On what other substances will these molds grow?

Moisten pieces of such materials as orange peel, apple peel, wood, paper, potato slice, cloth, leather, plastic, and aluminum foil and put each in a separate shallow jar. Place approximately equal amounts of the same kind of bread mold on each of the substances. Store the jars in a dark place and keep the materials moist.

Periodically examine the various materials. On which materials does the mold begin to grow first? On which is the mold growing best? On which materials does it not grow? How long does the mold continue to grow on various substances?

The investigation may be repeated with different-colored bread molds. Does the color of the bread mold seem to make a difference?

Science processes:

1. Observation of changes.
2. Control of all factors except the variable factor.

Science concepts:

1. Molds are among the plants that do not manufacture their own food but depend on other materials for food.
2. All organisms have certain requirements that must be met if they are to live and grow.

Decay

What happens to dead plants? Important changes take place in dead plants. Eventually dead organisms such as plants become nutrients that can be used by living plants.

Obtain some leaves or grass, two apples, and two bananas. Place one of each of these in a separate plastic bag and twist the top of the bag shut. Bury the other in several centimeters of soil. Be sure to mark where the items are buried so that they can be examined periodically.

Each week look for any changes that may take place in the dead materials in the plastic bags. Record the changes and the date in a notebook.

Every month excavate the dead materials in the soil. Take care not to damage their structure during the excavation. Record the changes and the dates in a notebook.

Science processes:

1. Careful observation of changes.
2. Keeping accurate records of observations.

Science concepts:

1. Dead materials eventually decay (biodegrade) into nutrients that can be used by growing plants.
2. Organisms such as scavengers, molds, fungi, worms, and bacteria help break down materials that were once living.

Materials in Running Water

What materials are there in running water? Fasten a strong string to a glass jar or jug and dangle it in a stream or river. (The most striking results will be obtained if the sample of water is taken when the stream is in flood.)

Allow the water to stand in the jar so that some materials may float to the top and the sediment will settle to the bottom.

Skim off whatever material is floating at the top and examine it with the aid of a magnifying glass. What kind of material does it seem to be?

Inspect the material that has settled to the bottom. Compare the amount of sediment at the bottom with the amount of water.

Pour off the water into another container and examine the sediment with the aid of a magnifying glass. What color is the sediment? Are all the particles the same size? Do the particles appear to be sand, clay, or silt? Pour a little of the river water through a paper filter. Are there small particles of soil to be found on the filter?

Pour a little of the river water into a clean evaporating dish. Place the dish on a hot plate and heat until all the water disappears. Is there any material left at the bottom of the evaporating dish? What is its color? These are the minerals that were dissolved in the water. You may wish to compare the amount of mineral matter in the river water with the amount in an equal volume of tap water.

Science processes:

1. Collecting material from the environment for investigation.
2. Careful observation, including observation with the aid of a magnifying glass.
3. Carrying out the process of evaporation.

Science concepts:

1. Running water often carries considerable quantities of material.
2. When flowing water is slowed, some of the sediment being carried by the water settles out.
3. Materials are often in solution in water. Some of these can be separated out by evaporating the water.

Latitude

What is your latitude? Latitude is a measure of the distance a place is from the equator. Latitude is stated in degrees. The equator is at 0° latitude. The North Pole is at 90° north latitude. The South Pole is at 90° south latitude. All places on earth are at some latitude between the poles and the equator.

Since Polaris, the north star, is almost directly above the North Pole, it is possible to find the latitude of a place north of the equator by sighting on Polaris. The latitude of a place is equal to the angle between a line toward Polaris and a horizontal line.

A device for measuring this angle can be made from a protractor, a soda straw, string, a pin, a weight, and a stake. Fasten a protractor to a stick or pole set in the ground. The protractor should be lined up in

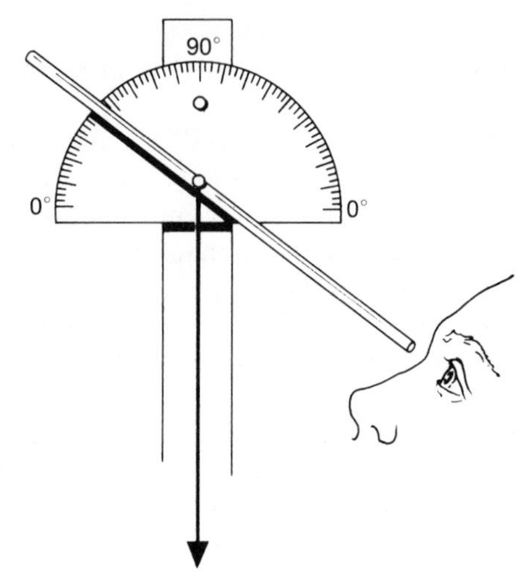

Figure 20–2 A device for measuring the angle between the horizontal and a line of sight to Polaris.

a north-south direction. You can do this by lining up the protractor with the needle of a compass. (See Figure 20 – 2.)

Hang a weight from the hole in the center of the protractor. Hold a ruler or straightedge along the string and adjust the protractor so that the edge of the ruler is at 90°. Stick a pin through the center of a soda straw and the hole in the center of the protractor. It should be possible to move the ends of the straw up and down as it turns on the pin.

On a clear, dark night, locate Polaris by first finding the Big Dipper in the northern sky. Locate the two stars that are farthest from the handle of the Big Dipper. These two stars are the "pointers." Draw an imaginary line through these two stars away from the bottom of the dipper. Extend the line five times the distance between the "pointers." At the end of this line is Polaris.

Turn the soda straw so that you can see Polaris through the straw. Hold the soda straw against the protractor and, with the help of a flashlight, read the number next to the straw. This number indicates your latitude. Children may wish to check their finding with the latitude of their location on a map.

Science processes:

1. Locating a constellation in the sky and using that constellation to locate a star.
2. Measuring angles.

Science concepts:

1. Latitude is a measure of the distance of a location from the equator.
2. Polaris is almost directly above the north pole of the earth.
3. The latitude of a location in the Northern Hemisphere can be found by finding the angle between a line of sight to Polaris and the horizontal.

Friction

On what factors does the force of friction depend? Attach a string to a block of wood. To carry out these experiments, the block of wood should be pulled across a wooden board at a constant velocity. (If the block is pulled at a constant velocity, its inertia can be disregarded.) The amount of force necessary to pull the block can be measured with a spring scale. (See Figure 20–3.) This experiment will also give children practice in handling data as they record the forces that are exerted.

First pull the block along the board several times. Each time note the force that is necessary to pull the block at a constant velocity. Ask the children what figure should be used for the force necessary to pull the block across the board. (They may suggest that the figures be averaged.)

Turn the block of wood over on its side and pull it along the board. How does this affect the force necessary to pull the block?

Now place progressively greater weights on the block of wood and in each case measure the force needed to pull it across the board. Have the children graph the force required as the weight is increased. (See Figure 20–4.) What relationship is there between the force of friction and the force pressing the surfaces together?

Tack a piece of sandpaper to the front edge of the block of wood and then bend the sandpaper underneath the block of wood. Pull the block of wood over this sandpaper surface. Does changing the nature of the surfaces that are in contact change the force of friction?

Put dowels or round pencils under the block of wood. How does this affect the amount of friction?

Science processes:

1. Measuring forces with a spring scale.
2. Maintaining a pull at a constant velocity.
3. Studying the effects of different variables.

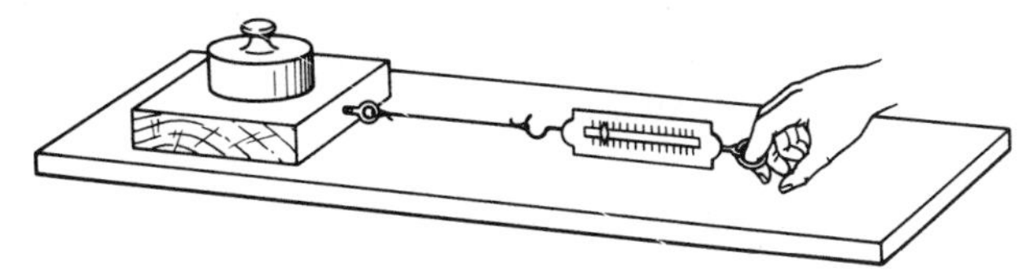

Figure 20–3 Setup for measuring the force of friction.

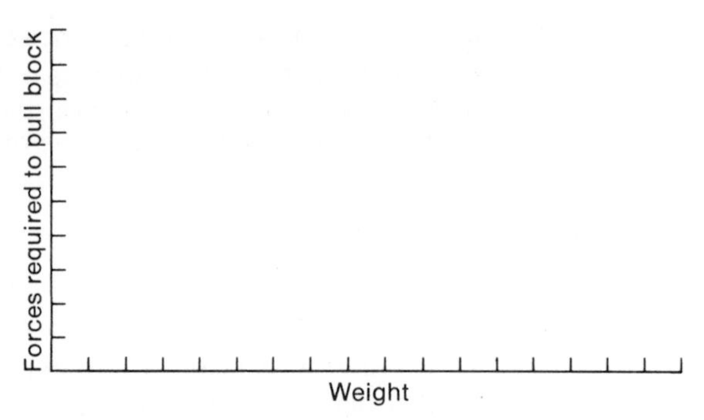

Figure 20–4 Weight and the force of friction.

Science concepts:

1. The force of friction depends on such factors as the types of materials in contact and the force (weight) pressing the materials together.

Designing an Experiment

How can you design an experiment? One of the important skills that children should derive from their experiences in science is the ability to design an experiment. Essentially, an experiment is an experience that is planned in such a way that we can learn from it. Many children are eager to try things, to manipulate, and to engage in various science activities that are planned for them. But eventually they should learn how to plan and organize their experiences so that they are more likely to learn something from them. This, of course, can lead to formal, logical operations. (See Chapter 8.)

The problems in the following list can be dealt with through experiments in which one factor is varied and an attempt is made to control all other factors.

1. Do insects such as houseflies or mosquitoes prefer a dark environment or one where there is considerable light?
2. Will plants grow better in red light than in blue light?
3. How does temperature affect the movement of mealworms?
4. Is the growth of plants affected by a strong magnet?
5. Does a film of oil on top of water affect the rate at which water evaporates?

When the children design their experiments, the following factors should be checked:

1. Can the experiment actually be carried out?
2. Are all factors except the variable to be studied controlled? Can the experimenter say with some assurance that any differences found are due to the variable being studied?
3. What is the hypothesis that is being tested? What are the reasons for suggesting the hypothesis?
4. What are the observations to be made? Will the observations involve counting or measuring?
5. What plans are there for obtaining and recording data?
6. How are the data that are collected going to be handled? Will graphs be used? Are there plans for the interpretation of data?
7. What plans are there for checking the experiment and its results against the work of others?
8. What plans are there for dealing with unexpected results?
9. Are the experiment and its results seen as leading to other possible experiments?

Science processes emphasis:

1. The planning of experiments.

MORE SCIENCE INVESTIGATIONS FOR CHILDREN

1. Are there events that happen periodically in your life? Have the children list events that seem to take place regularly: every day, every week, every month, every year, etc. Do these events have anything in common? Are there differences between those with short periods and those with long periods?
2. How does a hacksaw blade vibrate? Clamp one end of a hacksaw blade to a table. Tape an increasing number of large nails to the free end of the hacksaw blade and study the effect on the number of vibrations in a given length of time.
3. How does a swing swing? Count the number of swings completed by a playground swing in a given length of time. Study the effects of pulling the swing back farther and of having someone sit in the swing.
4. How can we separate the ingredients in a mixture of sand, sugar, and iron filings? The iron filings will be attracted to a magnet. The sugar will dissolve in water. The sand can be filtered out of the water. The dissolved sugar can be removed from solution by boiling off the water.
5. Does food coloring spread faster in cold water or in hot water? Put cold water in one glass container and hot water in another. Drop three drops of food coloring into each container and study the spreading of the color.
6. Through what kinds of solids will heat pass? Get sticks or spoons

made of silver, steel, aluminum, plastic, wood, and glass. Dip one end of each into a cup of hot water. Through which materials can you feel the heat pass?

7. What kinds of materials are given off when a substance, such as a candle, burns? Burn a candle under a glass jar. Drops of water may be seen to form on the sides of the jar. Black soot may form above the candle. If clear limewater is swished around in the jar, it will usually turn milky in color, indicating the presence of carbon dioxide.
8. How high can the jar be raised over a candle before the candle will not go out? If a large glass jar is placed over a burning candle, the flame will eventually go out. Raise the glass jar 2 cm above the table surface. Now does the flame go out?
9. How does a compound pendulum work? A compound pendulum is made by twisting the strings of two adjacent pendulums around opposite ends of a stick or stiff wire so that the stick or wire will be held up by the strings. (See Figure 20–5.) What happens to the other bob when one bob starts swinging? When both bobs are swinging, how does the direction of swing change?
10. How far can a bath towel be pulled down across a towel rack before it begins to slide off? Measure the length of a bath towel. Then hang it on a towel rack and begin to pull one end down. With a measuring stick or tape, measure the length of the short end when the towel finally begins to slide. What is the ratio of the short end to the long? Try other toweling materials. Is the distance that a bath towel can be pulled before sliding dependent on the nature of the material? What happens if the towel is pulled down at an angle?
11. How much water can lichens absorb? Find some lichens on a rock. Determine the area covered by the lichens. (It may be possible to trim the lichen growth into a rectangle.) Add water to the lichens until no more can be absorbed. How much water can be absorbed per square centimeter? How does the ability to absorb water help the lichens survive on barren rock?
12. How can fish in an aquarium be taught to respond to stimuli? Begin by tapping the side of the aquarium before feeding the fish. Eventually, do the fish come to that part of the aquarium before they get their food? Do they appear to see you and come before you tap? Try teaching guppies, goldfish, and perhaps other fish. Which kind of fish learns fastest? Will fish respond to other stimuli, such as a bright light or a sound?
13. In what direction does water tend to swirl when it drains out of a bathtub or washbowl? Allow the water in a washbowl to become completely quiet. Then open the plug to the drain. Does the water begin to swirl? In what direction? If no other factors affect it, it is believed that in the Northern Hemisphere water will swirl in a counterclockwise direction because of the spin of the earth. Does the water actually begin to turn in a counterclockwise direction?
14. How fast do objects fall through different kinds of fluids? Fill a tall graduate or glass jar with water. Hold a small pebble or ball bearing just above the surface of the water and then drop it. With a

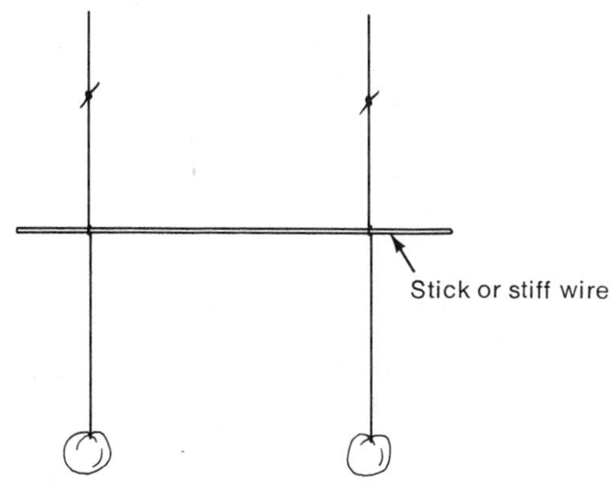

Figure 20–5 A compound pendulum.

stopwatch or a watch with a sweep second hand, determine the time it takes for the object to drop to the bottom. Repeat with equal heights of salad oil, kerosene, detergent, and other liquids. In which liquid does the object fall slowest? fastest?

15. What kind of stimulus is most likely to make a mealworm (or other worm) reverse its direction of movement? When a mealworm is moving in a particular direction, shine a bright light on it. Does it change direction? Try blowing at its head through a drinking straw, tapping it lightly with a straw, putting a drop of water on part of it, and putting a sandpaper surface in front of it. Which, if any, of these stimuli will cause the mealworm to change direction?

16. What roof angle will keep the interior of a box coolest (warmest) under intense sunlight? Cover the open top of a cardboard box with a cardboard roof whose angle can be changed. Place the roof and box in bright sunlight and measure the temperature inside the box. While making certain that all other factors are kept the same, change the angle of the roof and record the temperature inside the box.

17. What roof color will keep the interior of a box coolest (warmest)? Put a roof over the open top of a box. Place the box in sunlight and measure the temperature inside the box. Making certain that all other factors are kept the same, place different-colored paper on the roof of the box and record the temperature inside the box.

18. What objects in the environment are magnets? An object is probably a magnet if some part of it repels one end of a compass needle. Some objects have been magnetized by the earth's magnetic field. Move a compass alongside the metal legs of chairs and tables, scissors, a hammer, pliers, a flagpole, steel fence posts, and other steel or iron objects in the environment. Do any parts of these objects repel one end of a compass needle?

19. How can we make a map? To better understand how maps are made and used, children can make a map of their classroom. Have them measure the length and width of the classroom. For mapping, it may be helpful to use grid paper and choose some scale such as one square = one meter. The children can also locate objects in the room on their map.

A game that is sometimes played with such a map is to point to a location on the map and have the children find the object at that location.

20. What is the widest-diameter glass tubing in which a column of water can be held when the top end is sealed? If one end of a narrow piece of glass tubing is put into the water and a finger is pressed over the other end, air pressure will hold water in the tubing when it is lifted out of the water. The children's ingenuity can be tested as they try to find out ways of holding water in wider-diameter glass tubing.

21. What colored materials are found in the leaves of plants? How do various plants differ with regard to colors? Grind up a few leaves into very small particles. Extract the colored materials from the leaf particles by stirring them around in acetone. Place the colored acetone in a beaker or glass tumbler. Dip one end of a narrow strip of paper toweling or filter paper into the liquid. As the liquid moves up the paper, the different-colored materials will separate in a process called *paper chromatography.*

22. What is the effect of bleaching liquid on different colors of cloth? Immerse different colors of cloth in a bleaching liquid and then leave them in direct sunlight for some time. How are the different colors affected? Does the kind of cloth make a difference?

23. How does the acidity of soil change with depth? Obtain soil acidity testing paper from a garden store. Test samples of soil from different depths at the same location. Try other locations. Are there differences?

24. What is the magnifying power of a lens? Hold a magnifying glass over lined paper. Compare the distance between two lines as viewed through the lens with the distance as viewed outside the lens. How much does the distance appear to increase?

25. How does the length of a shadow change with the time of day and the season of the year? Place a small dowel or stick upright in a place where it will not be disturbed. Measure the length of the shadow at different times of the day. Repeat on days as close as possible to September 22, December 22, March 21, and June 22.

26. What kind of material is the best insulating material? Obtain several glass flasks of the same size. Place the glass flasks inside tin cans that are also of the same size. Fill the space between the can and one flask with crumpled paper, another with cotton cloth, another with wool, and another with commercial insulating materials. Fill the flask with equal amounts of hot water that is at the same starting temperature in each flask. Periodically take the temperature of the water in each flask and record it. With which insulating material does the water cool the slowest?

REFERENCES

For Teaching

BENDER, ALFRED. *41 Science Projects with Electrons.* New York: Sentinel, 1973. The projects in the chapters dealing with electrostatics, magnetism, and electromagnetism can be adapted into science investigations for older children.

BEVERIDGE, W. I. B. *The Art of Scientific Investigation.* New York: W. W. Norton, 1957. A lucid discussion of various aspects of scientific investigation with appropriate examples drawn from the history of science.

GABRIEL, MORDECAI L. AND SEYMOUR FOGEL, EDS. *Great Experiments in Biology.* Englewood Cliffs, N.J.: Prentice-Hall, 1955. A collection of reports of classic experiments in biology. The editors' introductions to the papers show how the reports fit into the context of biology.

NELSON, LESLIE W. AND GEORGE C. LORBEER. *Science Activities for Elementary Children.* Dubuque, Iowa: Wm. C. Brown, 1976. A collection of over 300 elementary science activities. All of them deal with science questions and problems; some of them can be used as science investigations.

RUCHLIS, HY. *Discovering Scientific Method.* New York: Harper & Row, 1963. A collection of puzzle picture problems that can be examined to illustrate various aspects of scientific investigations. Teachers will have to adapt and explain many of the problems before they can be used by children.

TATON, R. *Reason and Chance in Scientific Discovery.* London: Scientific Book Guild, 1960. This translation makes available in the English language this significant book, first published in France. Examples from the history of science are used to illustrate various factors involved in scientific discovery.

VIVIAN, CHARLES. *Science Experiments and Amusements for Children.* New York: Dover, 1963. Seventy-three "experiments" that can be done with easily obtainable materials. Many of the "experiments" are essentially demonstrations of scientific principles.

For Children

BEELER, NELSON F. AND FRANKLYN M. BRANLEY. *Experiments with a Microscope.* New York: Harper & Row, 1957. A description of many investigations that can be carried out by children with the aid of a microscope.

———. *More Experiments in Science.* New York: Harper & Row, 1950. "Why a Balloon Rises" and "Why the Wind Blows" are examples of the investigations described in this book.

BEHNKE, FRANCES L. *What We Find When We Look Under Rocks.* New York: McGraw-Hill, 1971. Look under a rock! There is another world there. This book for young children will help them learn more about what they may find under rocks.

BENDICK, JEANNE. *Measuring.* New York: Franklin Watts, 1971. A series of investigations for children that feature the scientific process of measuring.

COBB, VICKI. *Logic.* New York: Franklin Watts, 1969. Logic, of course, is an important aspect of many scientific investigations. Cobb discusses various aspects of logic in ways that are useful to older children.

GRAY, WILLIAM O. *What We Find When We Look at Molds.* New York: McGraw-Hill, 1970. A series of investigations of molds and how and where they grow.

SIMON, SEYMOUR. *Finding Out with Your Senses.* New York: McGraw-Hill, 1971. A series of investigations for young children that involve many of the senses.

APPENDICES

A Competency-Based Teacher Education Program in Elementary School Science

APPENDIX 1

Many instructors of preservice courses in elementary school science wish, or are asked, to develop a competency-based course. The following is a model one-semester competency-based course called "Science in Childhood Education" that we have used with both preservice and inservice teachers.

PURPOSE OF THE COURSE

The major purpose of the work in "Science in Childhood Education" is to make it possible for teachers and future teachers to master the competencies that are deemed essential for effective work with children in science. Teachers master these competencies through a wide variety of experiences, using science content that is often dealt with in the elementary schools. In these experiences the instructors strive to present model approaches to science that will be helpful in work with children.

APPROACHES TO DEVELOPING SCIENCE EXPERIENCES

The teachers and future teachers are involved in many different approaches to the development of science experiences. Among these are the following:

Laboratory experiments and skill development
Field experiences
Demonstrations
Individual investigations
Cooperative investigations
Reading and critical analysis of elementary science programs and materials
Discussion
Use of audio-visual materials
Use of community and regional science resources
Projects

SCIENCE CONTENT AREAS

The experiences in various approaches to science are developed within the context of content areas that are often included in elementary school science programs. The content areas are selected on the basis of a diagnosis of weaknesses in content background and in response to the interests of the participants. Among the content areas that often provide a context for professional elementary school science experiences are the following:

Heat, light, and sound
The solar system and the universe
Space and space exploration
Electricity and magnetism
Animals
Plants
Ecology
The human body and its care
Air and weather
Water and the hydrosphere
The earth
Science and technology

SCIENCE PROCESSES

In the development of professionalized science experiences, special attention is given to the science processes that are central to many new elementary science programs. These processes include the following:

Observing
Classifying

Communicating
Serial ordering
Using space/time relationships
Predicting
Model formation
Inferring
Measuring
Controlling variables
Interpreting data
Formulating hypotheses
Defining operationally
Experimenting

The work in "Science in Childhood Education" is directly related to work with children. Students use these activities in their work with children in schools, homes, or other agencies. Their science experiences with children are analyzed and discussed.

THE COMPETENCIES

Upon completion of this program, a teacher or future teacher will have the following competencies:

1. Knowledge of some of the basic content essential for work with children in science.

   Program experiences: Many of the science experiences are developed within the context of science content areas. These content areas are chosen as a result of a diagnosis of the weaknesses in content background among the participants. Laboratory experiences, demonstrations, discussion, and reading assignments are used to develop content proficiency.

   Assessment: Participants are expected to answer correctly at least 75 percent of the questions in each of the subcategories of the Elementary School Science Survey. (Single copies of this instrument are available from Prentice-Hall. Other instruments can also be selected and used.)

2. The basic laboratory skills necessary to carry out laboratory experiences with children.

   Program experiences: Participation in planned laboratory experiences in which the emphasis is on the development of laboratory skills and the planning and implementation of laboratory experiences in elementary school classrooms.

Assessment: Laboratory skills practicum and analysis of plans for laboratory-type science experiences.

3. The knowledge and skills needed to maintain plants and animals in elementary school classrooms.

   Program experiences: Participation in the setting up and maintenance of classroom aquariums and terrariums. Each participant also cares for and maintains at least one plant and one animal over an extended period. Study of science resource materials.

   Assessment: Analysis of experiences and problems encountered in maintaining aquariums, terrariums, plants, and animals.

4. The ability to identify and use natural and cultural science resources in a community and region.

   Program experiences: Participants identify at least one question or experience about which they are curious. The importance of modeling curiosity for children is emphasized. Participants use a variety of community resources such as museums, zoos, botanical gardens, and libraries, as well as science articles in newspapers and magazines, to get a better understanding of the phenomena about which they are curious.

   Assessment: Analysis of oral and written reports of investigations of phenomena.

5. The skills necessary to plan and carry out field experiences with children.

   Program experiences: Participation in an intensive field experience in which approaches to field experiences are demonstrated and various possibilities for field experiences in earth and space science and the biological sciences are shown. Usually children are involved in these experiences. Study of resource materials dealing with field experiences.

   Assessment: Discussion and analysis of field experiences and of plans for field experiences with children.

6. The science skills needed to plan and carry out elementary science investigations and develop plans for their incorporation into elementary science programs.

   Program experiences: Participants undertake several elementary science investigations, and in most cases these are carried out with children.

Assessment: Analysis of written and oral reports of investigations and of participant demonstrations of investigations.

7. Knowledge of new approaches to elementary school science.

Program experiences: Through reading, laboratory experiences, examination of program materials, and discussion, participants examine five or six of the most important new elementary school science programs.

Assessment: Participants write critical reviews of each of the programs considered.

8. The ability to formulate a set of criteria and to use these criteria to analyze science program materials available for use in the elementary school.

Program experiences: Using professional materials, participants formulate a set of criteria that are useful in critically analyzing available science program materials. The participants use these criteria to analyze program materials at a grade level of their choosing. The emphasis is on newer program materials.

Assessment: Analysis of oral and written reports of criticisms of elementary school science program materials.

9. The ability to analyze discussions of issues in elementary science education.

Program experiences: Participants read papers written by leaders in elementary school science and discuss them in class sessions.

Assessment: Analysis of participation in class discussions and references to the literature in written assignments.

10. The ability to apply relevant science education research that is deemed applicable to the teaching of science in the elementary school classroom.

Program experiences: Participants read papers written by leaders in science education research and related fields (e.g., developmental psychology) and discuss the applicability of the research findings to the teaching of science. They also view films demonstrating the application of educational research to classroom science teaching and critically analyze the approaches taken in class discussions.

Assessment: Analysis of participation in class discussions and use of research in written assignments.

Elementary School Science Programs

APPENDIX 2

American Association for the Advancement of Science (AAAS). *Science: A Process Approach II*. Xerox.

Atkins, Myron, et al. *The Ginn Science Program*. Ginn.

Berger, Carl, et al. *Modular Activities Program in Science*. Houghton Mifflin.

Biological Sciences Curriculum Study. *Elementary School Sciences Program*. Lippincott.

______. *Me Now* and *Me and My Environment*. Hubbard.

Blecha, Milo K., et al. *The Laidlaw Exploring Science Program*. Laidlaw.

Blough, Glenn O., et al. *Basic Science Program*. Scott, Foresman.

Brandwein, Paul F., et al. *Concepts in Science*. Harcourt Brace Jovanovich.

Brewer, A. C., et al. *Elementary Science: Learning by Investigating*. Rand McNally.

COPES, *Conceptually Oriented Program in Elementary Science*. New York University.

Elementary Science Study (ESS). *Units and Kits*. McGraw-Hill.

Fischer, Abraham S., et al. *Modern Elementary Science*. Holt, Rinehart and Winston.

Holmes, Neal J., et al. *Science: People, Concepts, Processes*. McGraw-Hill.

Jacobson, Willard J., et al. *Investigating in Science*. American Book.

______. *Thinking Ahead in Science*. American Book.

Klopfer, Leo and Audrey Champagne. *Individualized Science*. Imperial International Learning.

Lavatelli, Celia. *An Early Childhood Curriculum*. American Science and Engineering.

MALLINSON, GEORGE, ET AL. *Science Series.* Silver Burdett.

MARSHALL, STANLEY, ET AL. *Curriculum Foundation Series.* Scott, Foresman.

———. *Pre-Primary Science System.* Scott, Foresman.

NAVARRA, JOHN AND JOSEPH ZAFFORONI. *The Young Scientist.* Harper & Row.

NOVAK, JOSEPH D., ET AL. *The World of Science.* Bobbs-Merrill.

OBIS, *Outdoor Biology Instructional Strategies.* Lawrence Hall of Science, University of California.

PALDY, LESTER, ET AL. Science Curriculum Improvement Study, II (SCIS II). American Science and Engineering.

PILTZ, ALBERT, ET AL. *Discovering Science.* Bobbs-Merrill.

ROCKCASTLE, VERNE, ET AL. *Space, Time, Energy, Matter.* Addison-Wesley.

SCHNEIDER, HERMAN AND NINA SCHNEIDER. *Heath Science Series.* D. C. Heath.

SCHOOLS COUNCIL, NUFFIELD FOUNDATION AND SCOTTISH EDUCATION DEPARTMENT. *Science 5/13.* Purnell Educational.

SWARTZ, CLIFFORD E. *Measure and Find Out.* Scott, Foresman.

TANNENBAUM, HAROLD E., ET AL. *Experiences in Science.* McGraw-Hill.

———. *McGraw-Hill Process Science Modules.* McGraw-Hill.

THIER, HERBERT D., ET AL. *Adapting Science Materials for the Blind.* Lawrence Hall of Science, University of California.

———. *Rand McNally SCIIS.* Rand McNally.

THURBER, WALTER. *Exploring Science.* Allyn and Bacon.

THE UNIVERSITY OF ILLINOIS ASTRONOMY PROGRAM. *Six Texts and Guidebooks.* Harper & Row.

Professional Books in Elementary School Science

APPENDIX 3

ALMY, MILLIE. *Logical Thinking in Second Grade.* Teachers College Press.

AMERICAN ASSOCIATION FOR THE ADVANCEMENT OF SCIENCE. *Commentary for Teachers.* Xerox.

______. *Preservice Science Education of Elementary School Teachers.*

ANDERSON, RONALD, ET AL. *Developing Children's Thinking Through Science.* Prentice-Hall.

BLOUGH, GLENN O. AND JULIUS SCHWARTZ. *Elementary School Science and How to Teach It.* Holt, Rinehart and Winston.

BUTTS, DAVID P. *Teaching Science in the Elementary School.* Free Press.

CARIN, ARTHUR AND ROBERT B. SUND. *Teaching Science Through Discovery.* Bobbs-Merrill.

CLELLAN, DOREEN, ET AL. *Exploring Science in the Primary School.* Macmillan.

COMSTOCK, ANNA BOTSFORD. *Handbook of Nature Study.* Comstock.

DEVITO, ALFRED AND GERALD H. KROCKOVER. *Creative Sciencing.* 2 vols. Little, Brown.

ELEMENTARY SCIENCE STUDY. *The ESS Reader.* Education Development Center.

GEGA, PETER. *Science in Elementary Education.* Wiley.

GEORGE, KENNETH D., ET AL. *Elementary School Science: Why and How.* D. C. Heath.

______. *Science Investigations for Elementary School Teachers.* D. C. Heath.

GOLDBERG, LAZER. *Children and Science.* Scribner's.

GOOD, RONALD G. *How Children Learn Science.* Macmillan.

______. *Science—Children.* Wm. C. Brown.

HELGESON, STANLEY L., ET AL., EDS. *Science Education Abstracts and Index for Research in Education 1966–1972.* Education Associates.

HENNESSEY, DAVID E. *Elementary Teacher's Classroom Science Demonstrations and Activities.* Prentice-Hall.

HONE, ELIZABETH B., ET AL. *A Sourcebook for Elementary Science.* Harcourt Brace Jovanovich.

HOPMAN, ANNE B., ED. *Helping Children Learn Science.* National Science Teachers Association.

HUBLER, H. CLARK. *Science for Children.* Random House.

HURD, PAUL D. AND JAMES J. GALLAGHER. *New Directions in Elementary Science Teaching.* Wadsworth.

IVANY, GEORGE. *Today's Science.* Science Research Associates.

JACOBSON, WILLARD J. *Population Education.* Teachers College Press.

———. "USA Post Sputnik Science Curricula." In *Strategies for Curriculum Change: Cases from 13 Nations.* International Textbook.

JACOBSON, WILLARD J. AND ALLEN KONDO. *SCIS Elementary School Science Sourcebook.* University of California.

JACOBSON, WILLARD J. AND HAROLD E. TANNENBAUM. *Modern Elementary School Science.* Teachers College Press.

KARPLUS, ROBERT AND HERBERT D. THIER. *A New Look at Elementary School Science.* Rand McNally.

KUSLAN, LEWIS I. AND A. HARRIS STONE. *Readings in Teaching Children Science.* Wadsworth.

———. *Teaching Children Science: An Inquiry Approach.* Wadsworth.

LANDSDOWN, BRENDA, ET AL. *Teaching Elementary Science Through Discovery and Colloquium.* Harcourt Brace Jovanovich.

LEITMAN, ALLAN. *Science for Deaf Children.* Alexander Graham Bell Association.

LOCKARD, J. DAVID, ED. *Twenty Years of Science and Mathematics Curriculum Development.* University of Maryland.

LOWERY, LAWRENCE F. *The Everyday Science Sourcebook.* Allyn and Bacon.

MATTHEWS, CHARLES C., ET AL. *Student Structured Learning in Science.* Wm. C. Brown.

NELSON, LESLIE W. AND GEORGE C. LORBEER. *Science Activities for Elementary Children.* Wm. C. Brown.

PILTZ, ALBERT AND ROBERT SUND. *Creative Teaching of Science in the Elementary School.* Allyn and Bacon.

RENNER, JOHN W., ET AL. *Teaching Science in the Elementary School.* Harper & Row.

ROWE, MARY. *Teaching Science as Continuous Inquiry.* McGraw-Hill.

SCHMIDT, VICTOR E. AND VERNE ROCKCASTLE. *Teaching Science with Everyday Things.* McGraw-Hill.

SCHOOLS COUNCIL. *With Objectives in Mind.* Macdonald Educational.

SCIS. *SCIS Teacher's Handbook.* University of California.

STONE, A. HARRIS, ET AL. *Experiences for Teaching Children Science.* Wadsworth.

SUND, ROBERT, ET AL. *Elementary Science Teaching Activities.* Bobbs-Merrill.

TANNENBAUM, HAROLD E., ET AL. *Science Education for Elementary School Teachers.* Allyn and Bacon.

THIER, HERBERT D. *Teaching Elementary School Science.* D. C. Heath.

UNESCO. *Sourcebook for Science Teaching.* Unipub.

VICTOR, EDWARD. *Science for the Elementary School.* Macmillan.

VICTOR, EDWARD AND MARJORIE LERNER. *Readings in Elementary School Science.* Macmillan.

WASHTON, NATHAN S. *Teaching Science in Elementary and Middle Schools.* David McKay.

WILLIAMS, DAVID L. AND WAYNE L. HERMAN, JR. *Current Research in Elementary School Science.* Macmillan.

Index

A

D

E

F

G

H

I

J

K

L

M

N

O

P

Q

R

S

T